Physics of Thin Films

Advances in Research and Development

VOLUME 18

PLASMA SOURCES FOR THIN FILM DEPOSITION AND ETCHING

Contributors to This Volume

Richard A. Gottscho

Michael A. Lieberman

Oleg A. Popov

Suzanne L. Rohde

Christoph Steinbrüchel

Physics of Thin Films

Advances in Research and Development

PLASMA SOURCES FOR THIN FILM DEPOSITION AND ETCHING

Edited by

Maurice H. Francombe

*Department of Physics
The University of Pittsburgh
Pittsburgh, Pennsylvania*

John L. Vossen

*John Vossen Associates
Technical and Scientific Consulting
Bridgewater, New Jersey*

VOLUME 18

Academic Press
San Diego New York Boston
London Sydney Tokyo Toronto

This book is printed on acid-free paper ∞

COPYRIGHT © 1994 BY ACADEMIC PRESS, INC.

ALL RIGHTS RESERVED.
NO PART OF THIS PUBLICATION MAY BE REPRODUCED OR
TRANSMITTED IN ANY FORM OR BY ANY MEANS, ELECTRONIC
OR MECHANICAL, INCLUDING PHOTOCOPY, RECORDING, OR
ANY INFORMATION STORAGE AND RETRIEVAL SYSTEM, WITHOUT
PERMISSION IN WRITING FROM THE PUBLISHER.

ACADEMIC PRESS, INC.
A Division of Harcourt Brace & Company
525 B Street, Suite 1900
San Diego, California 92101-4495

United Kingdom Edition published by
ACADEMIC PRESS LIMITED
24-28 Oval Road, London NW1 7DX

Library of Congress Catalog Card Number: 63—16561
ISBN: 0-12-533018-9

94 95 96 97 98 99 9 8 7 6 5 4 3 2 1

PRINTED IN THE UNITED STATES OF AMERICA

Contents

Contributors ix
Preface xi

Design of High-Density Plasma Sources for Materials Processing
Michael A. Lieberman and Richard A. Gottscho

I.	Introduction	2
	A. Capacitively Coupled Radio Frequency Discharge Sources	5
	B. Limitations of Capacitively Coupled Radio Frequency Discharges	9
	C. Overview of High-Efficiency Sources	10
II.	Principles of Low-Pressure, High-Efficiency Source Design	13
	A. Unified Analysis of Source Operation	19
	B. Discharge Heating	23
III.	Electron Cyclotron Resonance (ECR) Discharges	25
	A. Source Configurations	26
	B. Electron Heating	31
	C. Resonant Wave Absorption	34
IV.	Helicon Discharges	40
	A. Helicon Configurations	41
	B. Helicon Modes	42
	C. Antenna Coupling	46
	D. Helicon Mode Absorption	50
V.	Inductive Discharges	52
	A. Inductive Source Configurations	52
	B. Power Absorption and Operating Regimes	54
	C. Source Operation and Coupling	56
	D. Low-Density Operation and Source Efficiency	58
VI.	Helical Resonator Discharges	60
VII.	Surface Wave Discharges	65

CONTENTS

- VIII. Plasma Transport ... 69
  - A. The Ion Energy Distribution Function ... 71
  - B. Methods for Measuring Ion Energy Distribution Functions ... 76
  - C. Methods for Measuring Plasma Potentials ... 80
  - D. Measurements of Energy Distributions and Potentials ... 81
  - E. Ion Energy Control ... 90
- IX. Device Damage ... 96
  - A. Atomic Displacement Damage ... 96
  - B. Contamination ... 98
  - C. Charging ... 98
  - D. Radiation ... 104
- X. Summary and Remaining Questions ... 105
- XI. Symbol Definitions ... 108
- Acknowledgments ... 112
- References ... 112

Electron Cyclotron Resonance Plasma Sources and Their Use in Plasma-Assisted Chemical Vapor Deposition of Thin Films

Oleg A. Popov

- I. Introduction ... 122
- II. ECR Fundamentals and Microwave Power Absorption ... 126
  - A. Principles of Electron Cyclotron Resonance ... 126
  - B. Microwave Power Absorption in ECR Plasmas ... 129
- III. ECR Plasma Sources: Designs and Characteristics ... 152
  - A. Basic Types of ECR Plasma Sources ... 152
  - B. Hitachi/NTT ECR Plasma Source ... 157
  - C. ECR Plasma Source Characteristics and Parameters ... 183
- IV. ECR Plasma Sources in PA CVD of Thin Films ... 208
  - A. Silicon Nitride Film ECR CVD ... 209
  - B. Silicon Oxide Films ... 216
  - C. ECR Plasma Planarization with Silicon Dioxide Films ... 221
- V. Conclusion ... 226
- Acknowledgments ... 227
- References ... 227

Unbalanced Magnetron Sputtering

Suzanne L. Rohde

- I. Introduction ... 235
- II. Motivation for Unbalanced Magnetron Sputtering ... 236
  - A. Sputtering in General ... 236
  - B. Electron Motion in a Magnetic Field ... 240

	C. Magnetron Sputtering	241
	D. Ion-Assisted Deposition	243
III.	Development of Unbalanced Magnetron Based Techniques	249
	A. Precursors to the Unbalanced Magnetron	249
	B. Unbalanced Magnetron Sputtering	255
IV.	Principles of Unbalanced Magnetron Sputtering	259
V.	Applications of UBM Sputtering	272
	A. Elemental Thin Films	272
	B. Films for Electronic and Optical Applications	274
	C. Films for Corrosion Protection	275
	D. Films for Wear and Abrasion Resistance	277
VI.	Commercial Applications of UBM Sputtering	281
VII.	Potential Future of UBM Sputtering	283
	Acknowledgments	285
	References	285

The Formation of Particles in Thin-Film Processing Plasmas

Christoph Steinbrüchel

I.	Introduction	289
II.	General Phenomena	291
III.	Particles in Deposition Plasmas	294
IV.	Particles in Sputtering Plasmas	301
V.	Particles in Reactive Ion Etching Plasmas	309
VI.	Modeling of Particles in Plasmas	312
VII.	Particle Contamination and Equipment Design	314
VIII.	Conclusions	314
	Acknowledgment	317
	References	317

Author Index . . . 319
Subject Index . . . 323

Contributors

Numbers in parenthesis indicate the pages on which the authors' contributions begin.

RICHARD A. GOTTSCHO (1), AT&T Bell Laboratories, 600 Mountain Avenue, Murray Hill, New Jersey 07974

MICHAEL A. LIEBERMAN (1), Department of Electrical Engineering and Computer Sciences, University of California, Berkeley, California 94720

OLEG A. POPOV (121), Matsushita Electric Works, R&D Laboratory, 216 West Cummings Park, Woburn, Massachusetts 01801

SUZANNE L. ROHDE (235), Department of Mechanical Engineering, University of Nebraska–Lincoln, Lincoln, Nebraska 68588

CHRISTOPH STEINBRÜCHEL (289), Department of Materials Engineering and Center for Integrated Electronics, Rensselaer Polytechnic Institute, Troy, New York 12180

Preface

The past decade has witnessed significant activity in the plasma processing and sputtering fields, associated closely with interest in high-resolution dry processing of integrated circuits, fabrication of wear-resistant coatings, and growth of compound and alloy films for new solid state applications and for magnetic recording devices. Some of the earlier developments in these fields were discussed in Volume 14 of this series, which included articles by Westwood on reactive sputtering, Siejka and Perriere on plasma oxidation, Johnson on cathodic arc plasma deposition, and by Vossen on glow-discharge cleaning of substrates. In Volume 18 of *Physics of Thin Films* we present four articles that discuss in far greater detail the design, physics, and operation of various plasma sources, and their use in the controlled growth and etching of thin films.

In the first article, on the design of high-density plasma sources for materials processing by Michael A. Lieberman and Richard A. Gottscho, the principles of low-pressure, high-efficiency source design are reviewed, and various types of discharges, e.g., electron cyclotron resonance (ECR), helicon, inductive, helical resonator, and surface wave, are discussed. Limitations of conventionally used, capacitively coupled rf discharges are described. These are, primarily, the inability to control independently the ion bombarding flux and ion acceleration energy, the associated occurrence of ion damage resulting from high voltages at the driven electrode, and poor line-width control in etched circuits. In the case of low-pressure, high-efficiency sources, power is coupled to the plasma noncapacitively, thus permitting independent control of ion flux and bombardment energy through separate variation of the wafer electrode power. Also, the low operating pressure of the plasma leads to increased anisotropy of ion transport to the device wafer surface, and hence to higher resolution in the etching of gate electrodes, contact windows, and interconnects.

The second article, by Oleg A. Popov, also addresses the topic of ECR plasma sources, but in this case for use in the plasma-assisted (or enhanced) chemical vapor deposition (CVD) of thin films. Popov's treatment includes a detailed discussion of microwave power absorption in ECR plasmas, and of the mechanisms of power dissipation. The discussion is extended to cover plasma transport, plasma chemical reactions, and the influence of design and operating parameters in commercially available systems, on deposition rate, thickness uniformity, ion bombardment, etc. Experimental data on the growth and properties of ECR plasma-CVD synthesized films of silicon nitride and silicon dioxide films, suitable for use in integrated circuits, also are presented and evaluated.

Conventional magnetron sputtering has been widely adopted for film deposition, particularly in situations where high growth rates are required and where ion bombardment at the substrate must be minimized. However, in many applications, e.g., in densification of hard, corrosion-resistant coatings, diffusion barriers or optical films, low-energy ion bombardment during growth greatly reduces voids and porosity. The third article, by Suzanne L. Rohde, on unbalanced magnetron sputtering, discusses the physics and technology of this process in relation to the need to achieve a greater range of ion-bombardment levels (while preserving high sputtering rates) through independent control of ion current density in the substrate region. Methods are reviewed for modifying (or unbalancing) magnetic field conditions, in order to control the formation of a secondary discharge in the vicinity of the substrate surface, and hence to promote ion-assisted growth of the film.

The fourth and final article, by Christoph Steinbrüchel, reviews the formation of particles in thin-film processing plasmas. The generation of particles, by mechanisms inherent to the plasma process, has been commonly observed both in plasma-assisted growth of films, in sputtering, and in reactive ion etching (RIE) of wafer surfaces. These particles, which can range in size from submicron up to several microns in diameter, are sufficiently numerous to constitute a serious device degradation phenomenon, especially in fabrication of circuit patterns with small feature dimensions. They have been observed in the plasma, using laser light scattering (LLS) techniques, as well-defined, charged clouds, at densities as high as $10^7 \, cm^{-3}$. This article proposes formation mechanisms, e.g., homogeneous nucleation within deposition plasmas, particle transfer from the target in etching plasmas, etc., and discusses theoretical aspects. Recommendations for minimizing particle generation, including plasma operating modes and chamber design, also are presented.

M. H. Francombe
J. L. Vossen

Design of High-Density Plasma Sources for Materials Processing

MICHAEL A. LIEBERMAN

*Department of Electrical Engineering and Computer Sciences,
University of California, Berkeley, California*

and

RICHARD A. GOTTSCHO

*AT&T Bell Laboratories,
Murray Hill, New Jersey*

```
 I. Introduction ............................................. 2
    A. Capacitively Coupled Radio Frequency Discharge Sources ........ 5
    B. Limitations of Capacitively Coupled Radio Frequency Discharges .... 9
    C. Overview of High-Efficiency Sources ...................... 10
II. Principles of Low-Pressure, High-Efficiency Source Design ......... 13
    A. Unified Analysis of Sources Operation ..................... 19
        1. Electron Temperature ............................. 19
        2. Ion Bombarding Energy ........................... 19
        3. Plasma Density and Ion Current Density ............... 22
    B. Discharge Heating ................................... 23
III. Electron Cyclotron Resonance (ECR) Discharges ................. 25
    A. Source Configurations ................................ 26
    B. Electron Heating .................................... 31
    C. Resonant Wave Absorption ............................ 34
IV. Helicon Discharges ..................................... 40
    A. Helicon Configurations ............................... 41
    B. Helicon Modes ..................................... 42
    C. Antenna Coupling ................................... 46
    D. Helicon Mode Absorption ............................. 50
 V. Inductive Discharges .................................... 52
    A. Inductive Source Configurations ........................ 52
    B. Power Absorption and Operating Regimes ................. 54
    C. Source Operation and Coupling ......................... 56
    D. Low-Density Operation and Source Efficiency .............. 58
VI. Helical Resonator Discharges ............................. 60
VII. Surface Wave Discharges ................................ 65
```

VIII. Plasma Transport 69
   A. The Ion Energy Distribution Function 71
      1. Ion Transport and Etching Anisotropy 73
   B. Methods for Measuring Ion Energy Distribution Functions 76
   C. Methods for Measuring Plasma Potentials 80
   D. Measurements of Energy Distributions and Potentials 81
      1. Ion Acceleration Outside the Sheath 81
      2. Transverse Ion Energy 87
   E. Ion Energy Control 90
      1. Plasma Anodization 96
IX. Device Damage 96
   A. Atomic Displacement Damage 96
   B. Contamination 98
   C. Charging 98
      1. Plasma Uniformity 99
      2. Biasing 102
   D. Radiation 104
X. Summary and Remaining Questions 105
XI. Symbol Definitions 108
   Acknowledgments 112
   References 112

I. Introduction

The advent of sub-micron electronic device fabrication has brought unprecedented demands for process optimization and control (*1,2*) which, in turn, have led to improved plasma reactors for the etching and deposition of thin films. As a result, we have witnessed the introduction of a new generation of plasma systems based on electron cyclotron resonance (ECR) heating (*3–6*). ECR plasma etching of polycrystalline Si, single crystalline Si, silicides, Al, Mo, W, SiO_2, polymers, and III–V compound semiconductors have all been reported in recent years (*7–33*). Similarly, ECR plasmas have been used to deposit amorphous Si, silicon nitride, boron carbide, and SiO_2, to name just a few materials (*34–40*). Applications of ECR plasmas beyond etching and deposition have also been reported and include ion implantation (*41–45*), surface cleaning (*46–59*), surface passivation (*60*), and oxidation (*53,61–63*). Besides ECR, many other "novel" plasma generation schemes are now being offered to satisfy manufacturers' needs in these materials processing areas. All these schemes purport to offer advantages over conventional approaches such as the capacitively coupled radio frequency discharge now used in many factories for etching and deposition of thin films during integrated circuit manufacturing.

But which scheme is best? What are the key aspects to plasma source design that affect materials processing? And why are the conventional approaches inadequate? While the answers to these questions remain elusive and are the subject of much current research, one can clearly identify commonalities and differences between the novel sources, whose most distinctive characteristic is higher efficiency than their conventional counterparts operated at low pressure. The purpose of this review is to (1) develop a unified framework from which all "high-efficiency" sources may be viewed and compared; (2) outline key elements of source design that affect processing results; and (3) highlight areas where additional research and development is needed. In so doing, we hope to assist those who use plasma for materials processing to make wise choices in constructing or purchasing sources, to guide vendors of high-efficiency sources in choosing designs that can best meet their customers' expectations, and to inspire the research community to focus on problems of technological interest.

Before such a review can be begun, several disclaimers must be made. First, the literature on applications, diagnostics, and modeling of high-efficiency sources is now so voluminous that we are not able to review or reference every paper. Rather, we have opted for highlighting key results in line with our objectives stated earlier. Second, we restrict our focus to those aspects of plasma processing that are uniquely affected by the use of high-efficiency plasmas. For example, we discuss aspects of source design that affect plasma-induced electrical damage in microelectronic circuits, but a comprehensive discussion of damage mechanisms is the subject of its own review and clearly beyond the scope of this work. Third, there are pertinent areas that while important are not yet ready for review. Foremost amongst these is the field of numerical simulation. While impressive results have been reported recently and we will draw on some of these, little has appeared in print and it is premature to review the field. Similarly, the stability of high-efficiency sources is a matter of some concern, and recent work illustrates that sudden mode changes and bistability may adversely affect materials properties, but too little has been reported and analyzed to make a thorough discussion meaningful. Finally, any review reflects the biases of the authors, and this work is no exception. Based on our interests and experience, we focus on applications of plasmas to microelectronics fabrication and, in particular, etching. Heavy emphasis is placed on simple, analytical, unifying theories and quantitative diagnostic measurements.

Why new sources? In plasma etching, the shrinking dimensions of micro-electronic devices have placed unprecedented demands on process control. Consider critical dimension (CD) control where the width of the transistor gate is specified to better than 10%. For yesterday's CD of 1 μm, this means a linewidth variation of 0.1 μm can be tolerated, but by the end of the 20th century when the CD should be only 0.25 μm, variations in CD must be less than 0.025 μm. This requires unprecedented anisotropy in the plasma etching of gate electrodes, contact windows, and metallic interconnections. To achieve such control, we need to increase the anisotropy of ion transport to the device wafer from what it is in the conventional capacitively coupled rf reactor. This means operating plasmas at lower pressures. But conventional rf sources are inefficient at low pressure, so that high powers must be used to achieve the high rates of ionization and dissociation necessary for high-throughput, low-cost manufacturing. Unfortunately, excessive power input to a capacitively coupled system leads to high ion bombarding energies that can degrade selectivity in etching and produce electrical damage that reduces device yield. Thus, new sources are needed to operate at lower pressure and higher efficiency.

In conventional rf systems, ion energy and flux are inexorably linked. But ion energy control is needed in plasma deposition to tailor film properties such as stress, composition, refractive index, crystallinity, and topography. Ion energy control is used in plasma etching to optimize selectivity and minimize atomic displacement damage while meeting linewidth and throughput specifications. Therefore, gaining superior control of ion energy and decoupling it from ion flux control is further motivation for developing new plasma sources and processing systems.

In the remainder of this section, we review briefly the properties of capacitively coupled radio frequency plasmas and elaborate further on the advantages of high-efficiency sources. In the following sections, we first discuss the fundamental principles underlying high-efficiency plasma source design and, to compare one source with another, use a simple analysis in Section II that allows estimation of electron temperature, ion bombardment energy, and plasma density in terms of the gas phase cross-sections, gas density, absorbed power, and source dimensions. In this way, we provide an approximate but common framework with which one source can be compared to another. In Sections III–VII we discuss in greater detail ECR, helicon, inductive, helical resonator, and surface wave sources, respectively. Emphasis is placed on electron heating and power absorption, since these are the primary differences between one source and another. In Section VIII, we turn to the issue of plasma transport and independent control of ion energy and flux.

Obtaining such control is largely independent of the electron heating mechanism but depends critically on source design parameters such as the magnetic field and power absorption profiles. We focus our attention in Section VIII on measurements of ion energy distributions, mostly in ECR systems since few data are available from other systems. In Sections VIII and IX, we relate ion energy and plasma uniformity, dictated by source design, to processing results such as etching anisotropy, atomic displacement damage, and charge-induced damage. In the final section, we highlight remaining issues and the areas where further investigation is needed.

Throughout this paper we strive to be consistent with dimensional analysis despite not using a consistent set of units. Generally, magnetic field is expressed in gauss, distances in meters, centimeters, or millimeters, and the electron charge in coulombs. Energies are usually given in units of volts, not electron volts, so the value of e is explicitly written. Pressures are given in Torr or milli-Torr. While this does not conform to international convention, it does conform to common usage. We apologize to the purists.

A. CAPACITIVELY COUPLED RADIO FREQUENCY DISCHARGE SOURCES

Capacitively driven rf discharges—so-called rf diodes—are the most common sources used for materials processing. An idealized source in plane parallel geometry, shown in Fig. 1a, consists of a discharge chamber containing two electrodes separated by a spacing l and driven by an rf power source. The substrates are placed on one electrode, feedstock gases are admitted to flow through the discharge, and effluent gases are removed by the vacuum pump. Coaxial discharge geometries, such as the "hexode" shown in Fig. 1b, are also in widespread use. When operated at low pressure, with the wafer mounted on the powered electrode, and used to remove substrate material, such reactors are commonly called reactive ion etchers (RIEs)—a misnomer, since the etching is generally a chemical process enhanced by energetic ion bombardment of the substrate, rather than a removal process due to reactive ions. When operated at higher pressure with the wafer mounted on the grounded electrode, such reactors are commonly referred to as plasma etchers. In terms of the physical properties of these systems, this distinction is somewhat arbitrary.

The physical operation of capacitively driven discharges is reasonably well understood. As shown in Fig. 2 for a symmetrically driven discharge

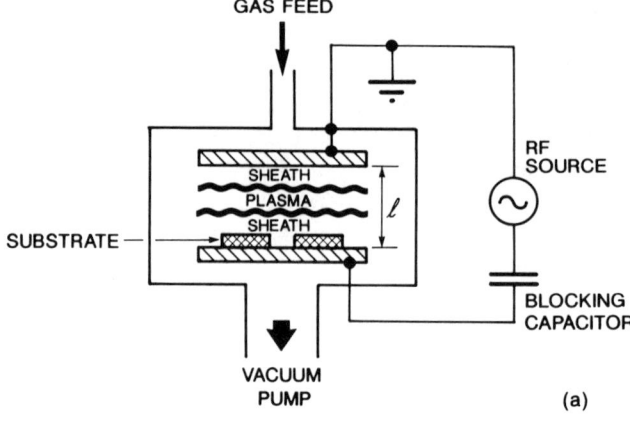

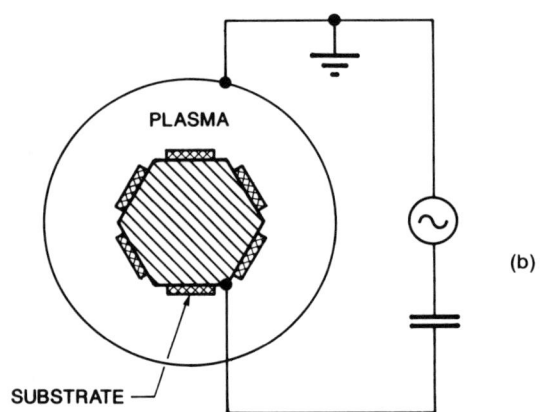

FIG. 1. Capacitive rf discharges: (a) plane parallel geometry; (b) coaxial ("hexode") geometry.

operated at frequencies between the ion and electron plasma frequencies, the mobile plasma electrons, responding to the instantaneous electric fields produced by the rf (13.6 MHz) driving voltage, oscillate back and forth within the positive space charge cloud of the ions. At 13.6 MHz, the massive ions respond only to the time-averaged electric fields. Oscillation of the electron cloud creates sheath regions near each electrode that contain net positive charge when averaged over an

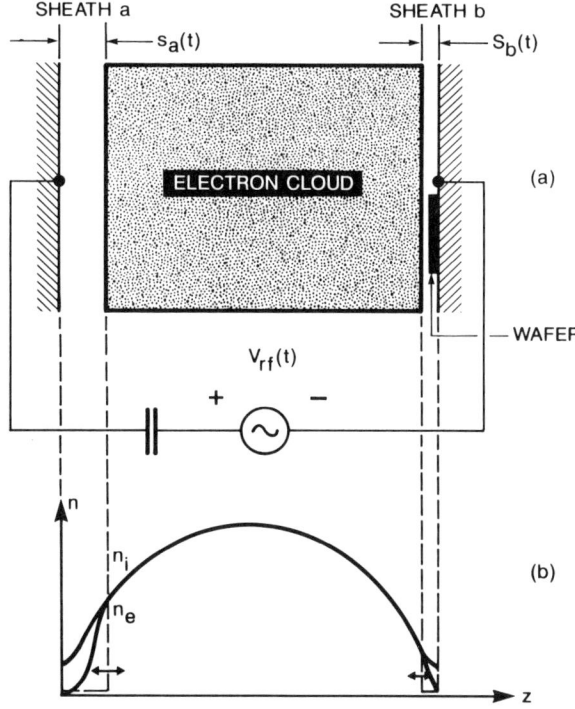

FIG. 2. Rf diode physical model.

oscillation period; i.e., the positive charge exceeds the negative charge in the system, with the excess appearing within the sheaths. This excess produces a strong time-averaged electric field within each sheath directed from the plasma to the electrode. Ions flowing out of the bulk plasma near the center of the discharge can be accelerated by the sheath fields to high energies as they flow to the substrate, leading to energetic-ion bombardment, which can enhance, inhibit, or otherwise modify surface reactions. Typical ion bombarding energies ε_i can be as high as $V_{rf}/2$ for symmetric systems (Fig. 2) and as high as V_{rf} at the powered electrode for asymmetric systems (Fig. 1b), where V_{rf}, the rf voltage amplitude (peak rf voltage) between the two electrodes, might typically vary between 100 V and 1 kV.

We note that positive ions continuously bombard the electrode over an rf cycle. In contrast, electrons are lost to the electrode only when the oscillating cloud closely approaches the electrode. During that time, the instantaneous sheath potential collapses to near-zero, allowing a sufficient number of electrons to escape to balance the ion charge delivered to the electrode. Except for such brief moments, the instantaneous potential of the discharge must always be positive with respect to any *large* electrode and wall surfaces;[1] otherwise the mobile electrons would quickly leak out. Electron confinement is ensured by the presence of positive space charge sheaths near all surfaces.

The separation of the discharge into bulk and sheath regions is an important paradigm that applies to all discharges. The bulk region is quasi- neutral, and both instantaneous and time-averaged fields are low. The bulk plasma dynamics are described by ambipolar diffusion at high pressures and by free-fall ion loss at low pressures. In the positive space charge sheaths, high fields exist, leading to dynamics that are described by various ion space charge sheath laws, including low-voltage sheaths (for high density sources) and various high-voltage sheath models (for rf diodes), such as collisionless and collisional Child laws and their modifications (*66–73*). The plasma and sheath models must be joined at their interface. The usual joining condition is to require that the mean ion velocity at the plasma-sheath edge be equal to the ion-sound (Bohm) velocity $u_B = (eT_e/M)^{1/2}$, where e and M are the charge and mass of the ion and T_e is the electron temperature in units of volts.

In the second column of Table I, typical rf diode source and plasma parameters are given. For anisotropic etching, pressures are in the range 10–100 mTorr, power densities are 0.1–1 W/cm^2, the driving frequency is typically 13.6 MHz, and multiple wafer systems are common. Plasma densities are relatively low, $\sim 10^{10}$ cm^{-3}, and mean electron energies are of order 5 V, corresponding to Maxwellian electron temperatures of order 3 V. However, non-Maxwellian electron distributions (e.g., two-temperature) are also observed, with the bulk electron temperature sometimes much less than 1 V (*74, 75*). Ion acceleration energies (sheath voltages) are high, > 200 V, and fractional ionization is low. The degree of dissociation can range widely from less than 0.1% to nearly 100% depending on gas composition and plasma conditions (*76, 77*). For deposition and isotropic etch applications, pressures tend to be higher

---

[1] Exceptions to this rule are also possible in low-frequency electronegative and dc discharges. In the former, the buildup of negative ions can reduce the plasma potential below that of large surfaces in contact with the plasma (*64*). In the latter, the plasma potential can lie between the two electrode potentials if sufficient current is drawn from the plasma (*65*).

TABLE I
Typical Parameters for High-Efficiency and Conventional rf Plasma Sources

Parameter	Units	rf Diode	High-Density Source
Pressure p	mTorr	10–1,000	0.5–50
Power P	W	50–2,000	100–5,000
Frequency f	MHz	0.05–13.6	0–2,450
Volume V	l	1–10	2–50
Cross-sectional area A	cm^2	300–2,000	300–500
Magnetic field B	kG	0	0–1
Plasma n	cm^{-3}	10^9–10^{11}	10^{10}–10^{12}
Electron temperature T_e	V	1–5	2–7
Ion acceleration energy ε_i	V	200–1,000	20–500
Fractional ionization X_{iz}	—	10^{-6}–10^{-3}	10^{-4}–10^{-1}

and frequencies sometimes lower than the commonly used standard of 13.6 MHz. For example, silicon nitride deposition used for chip encapsulation is ordinarily performed at frequencies between 50 and 500 kHz where relatively large ion bombardment energies are used to tailor film stress and stoichiometry (78).

B. Limitations of Capacitively Coupled Radio Frequency Discharges

A crucial limiting feature of rf diodes is that the ion bombarding flux $\Gamma_i = nu_B$ and the ion acceleration energy ε_i can not be varied independently. The situation is analogous to the lack of independent voltage and current control in diode vacuum tubes or semiconductor *pn* junctions. Hence, for a reasonable (but relatively low) ion flux, as well as a reasonable dissociation of the feedstock gas, sheath voltages at the driven electrode are high. For wafers placed on the driven electrode, this can result in undesirable damage, or loss of linewidth control. Furthermore, the combination of low ion flux and high ion energy leads to a relatively narrow window for many process applications. The low process rates resulting from the limited ion flux in rf diodes often mandate multiwafer or batch processing, with consequent loss of wafer-to-wafer reproducibility. Higher ion and neutral fluxes are generally required for single wafer processing in a clustered tool environment, in which a single wafer is moved by a robot through a series of process chambers. Clustered tools are used to control interface quality and are said to have the potential for significant cost savings in fabricating integrated circuits (79). Finally, low fractional ionization poses a significant problem for processes where the feedstock costs and disposal of effluents are issues.

To meet the linewidth, selectivity and damage control demands for next-generation fabrication, the mean ion bombarding energy, and its energy distribution, should be controllable independently of the ion and neutral fluxes. Some control over ion bombarding energy can be achieved by putting the wafer on the undriven electrode and independently biasing this electrode with a second rf source. Although these so-called rf triode systems are in use, processing rates are still low at low pressures and sputtering contamination is an issue.

Various magnetically enhanced rf diodes and triodes have also been developed to improve performance of the rf reactor. These include, for example, the Applied Materials AMT-5000 magnetically enhanced reactive ion etcher (MERIE) and the Microelectronics Center of North Carolina's split cathode rf magnetron. In the AMT MERIE, a dc magnetic field of 50–100 G is applied parallel to the powered electrode, on which the wafer sits. The magnetic field increases the efficiency of power transfer from the source to the plasma and also enhances plasma confinement. This results in a reduced sheath voltage and an increased plasma density when the magnetic field is applied (80, 81). However, the plasma generated is strongly nonuniform both radially and azimuthally because of $\mathbf{E} \times \mathbf{B}$ drifts, where $\mathbf{E}$ and $\mathbf{B}$ are the local electric and magnetic fields, respectively. To increase process uniformity (at least azimuthally), the magnetic field is rotated in the plane of the wafer at a frequency of 0.5 Hz. While this is an improvement, MERIE systems do not have good uniformity, which may limit their applicability to next-generation, sub-micron device fabrication. Indeed, the strongly nonuniform plasma over the wafer can give rise to a lateral dc current that can damage thin gate oxide films (see Section IX.C).

C. Overview of High-Efficiency Sources

The limitations of rf diodes and their magnetically enhanced variants have led to the development of a new generation of low-pressure, high-efficiency plasma sources. A few examples are shown schematically in Fig. 3, and typical source and plasma parameters have been given in Table 1. In addition to high density and low pressure, a common feature

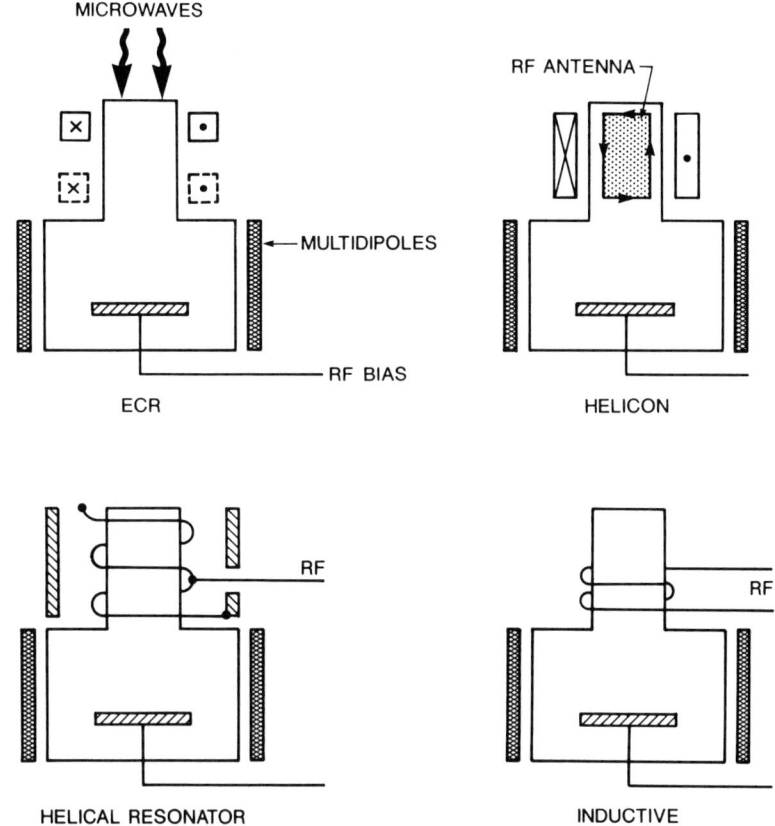

FIG. 3. Some high-density remote sources.

is that the rf or microwave power is coupled to the plasma across a dielectric window, rather than by direct connection to an electrode in the plasma, as for an rf diode. This non-capacitive power transfer is key to achieving low voltages across all plasma sheaths at electrode and wall surfaces. Dc voltages, and hence ion acceleration energies, are then typically 20–30 V at all surfaces. To control the ion energy, the electrode on which the wafer is placed can be independently driven by a capacitively coupled rf source. Hence, independent control of the ion/radical fluxes (through the source power) and the ion bombarding energy (through the wafer electrode power) is possible. This subject is discussed at greater length in Section VIII.

The common features of power transfer across dielectric windows and separate bias supply at the wafer electrode are illustrated in Fig. 3. However, sources differ significantly in the means by which power is coupled to the plasma. For the electron cyclotron resonance (ECR) source shown in Fig. 3a, one or more electromagnet coils surrounding the cylindrical source chamber generate an axially varying dc magnetic field. Microwave power is injected axially through a dielectric window into the source plasma, where it excites a right-hand circularly polarized (RHP) wave that propagates to a resonance zone, for cold electrons, at $\omega = \omega_{ce}$ where the wave is absorbed. Here $\omega = 2\pi f$ is the applied radian frequency and $\omega_{ce} = eB/m$ is the electron gyration frequency at resonance. For the typical microwave frequency $f = 2,450$ MHz used, the resonant magnetic field is $B \approx 875$ G. The plasma streams out of the source into the process chamber in which the wafer is located.

A helicon source is shown in Fig. 3b. A weak (50–200 G) dc axial magnetic field along with an rf-driven antenna placed near the dielectric cylinder that forms the source chamber allows excitation of a helicon wave within the source plasma. Resonant wave–particle interaction (Landau damping) is believed to transfer the wave energy to the plasma (*82–86*) (Section IV.D). For the helical resonator source shown in Fig. 3c, the external helix and conducting cylinder surrounding the dielectric discharge chamber form a slow wave structure, i.e., supporting an electromagnetic wave with phase velocity much less than the velocity of light. Efficient coupling of the rf power to the plasma is achieved by excitation of a resonant axial mode (Section VI). An inductive (or transformer) coupled source is shown in Fig. 3d. Here the plasma acts as a single-turn, lossy conductor that is coupled to a multiturn non-resonant rf coil across the dielectric discharge chamber; rf power is inductively coupled to the plasma by transformer action (Section V). In contrast to the ECR and helicon sources, a dc magnetic field is not required for efficient power coupling in the helical resonator or inductive sources.

Figure 3 also illustrates the use of high-density sources to feed plasma into a relatively distinct, separate process chamber in which the wafer is located. As shown in the figure, the process chamber can be surrounded by dc multidipole magnetic fields to enhance plasma confinement near the process chamber surfaces, while providing a magnetic field–free plasma environment at the wafer. Such configurations are often called "remote" sources, another misnomer since at low pressures considerable plasma and free radical production occurs within the process chamber near the wafer (see Section VIII.D). Hence, such sources are not actually remote. For reasons that are discussed further in Sections II.A.2, VIII.D, and IX.C, the source and process chambers are sometimes combined, or the wafer is placed very near to the source exit. Such configurations are useful for obtaining increased ion and radical fluxes, reducing the spread in ion energy, and improving process uniformity. But the wafer is exposed to higher levels of damaging radiation as well (Section IX).

Although the need for low pressures, high fluxes and controllable ion energies has motivated high-density source development, there are many issues that need to be resolved. A critical issue is achieving the required process uniformity over 200–300 mm wafer diameters. In contrast to the nearly one-dimensional geometry of typical rf diodes (two closely spaced parallel electrodes), high-density sources are often cylindrical systems with length-to-diameter ratios of order or exceeding unity. Plasma formation and transport in such geometries is inherently radially nonuniform. Another critical issue is efficient power transfer (coupling) across dielectric windows over a wide operating range of plasma parameters. Degradation of and deposition on the window can also lead to irreproducible source behavior and the need for frequent, costly cleaning cycles (87). Low-pressure operation leads to severe pumping requirements for high deposition or etching rates and hence to the need for large, expensive vacuum pumps. Furthermore, plasma and radical concentrations become strongly sensitive to reactor surface conditions, leading to problems of reactor aging and process irreproducibility. Finally, dc magnetic fields are required for some source concepts. These can lead to magnetic field induced process nonuniformities and damage, as seen, for example, in MERIE systems (88).

II. Principles of Low-Pressure, High-Efficiency Source Design

For the pressures of interest (see Table I), the plasma is not in thermal equilibrium, and local ionization models (89), where the ionization rate is a function of the local field and gas density only, fail. For all sources, the electrical power is coupled most efficiently to plasma electrons. In the bulk plasma, energy is transferred inefficiently from electrons to ions and neutrals by weak collisional processes; for ions, energy can also be coupled by weak ambipolar electric fields. The fraction of energy transferred by elastic collision of an electron with a heavy ion or neutral is $2m/M \sim 10^{-4}$, where m and M are the electron and heavy particle masses. Hence, the electron temperature T_e much exceeds the ion and neutral temperatures, T_i and T, respectively, in the bulk; typically, $T_e \sim 5$ V, whereas T_i and T are a few times room temperature (*90*). A more complete discussion of the ion temperature is given in Section VIII. However, dissociation and excitation processes can create a subgroup of relatively high-energy heavy particles. Also, the ambipolar electric fields accelerate positive ions toward the sheath edge, and typically, the ions in the bulk acquire a directed energy at the sheath edge of order $T_e/2$.

At these low pressures, the mean free path for ionizing electrons, with energies of 10–15 V, is typically comparable to the source dimensions. Hence, even if the electric power is deposited in a small volume within an unmagnetized source, the electron–neutral ionization rate v_{iz} is expected to be relatively uniform, since the ionization occurs on the distance scale of this mean free path. In magnetized plasmas, on the other hand, the ionization rate may be highly nonuniform as the magnetized electrons have trouble crossing field lines, so ionization along a magnetic flux tube might be uniform but significant radial nonuniformities may persist. In addition, the propagation and absorption of the exciting electromagnetic fields depend on the charge density distribution. The coupling is nonlinear and can give rise to sudden mode changes and instabilities. In some instances, the density profile can steer power into regions of higher or lower density and make the plasma more or less uniform, respectively (*91*, Section III.C).

Although the electron energy distribution function (eedf) need not be Maxwellian, recent Thomson scattering results indicate that this can be a good approximation (*92*), and at least insightful estimates of source operation can be obtained by approximating the eedf to be a Maxwellian, with T_e and the various electron collisional rates assumed to be uniform within the bulk plasma.

In high-density sources, electron–neutral collisional processes are critical not only for particle production (ionization, dissociation) but also for other collisional energy losses (excitation, elastic scattering). Ion–neutral collisions (charge transfer, elastic scattering) are also important in determining plasma transport and ion energy distribution functions (iedf) at the wafer surface. The myriad of collisional processes that can occur in heavy molecular feedstock gas mixtures can obscure the fundamental principles of high-density sources. A noble gas, such as argon, is often used as a reference for describing source operation. The relevant (second-order) rate constants K_{iz}, K_{exc}, and K_{el} for electron–neutral ionization, excitation, and elastic scattering in argon are given in Fig. 4 as a function of T_e. The corresponding rates $v(s^{-1})$ are defined

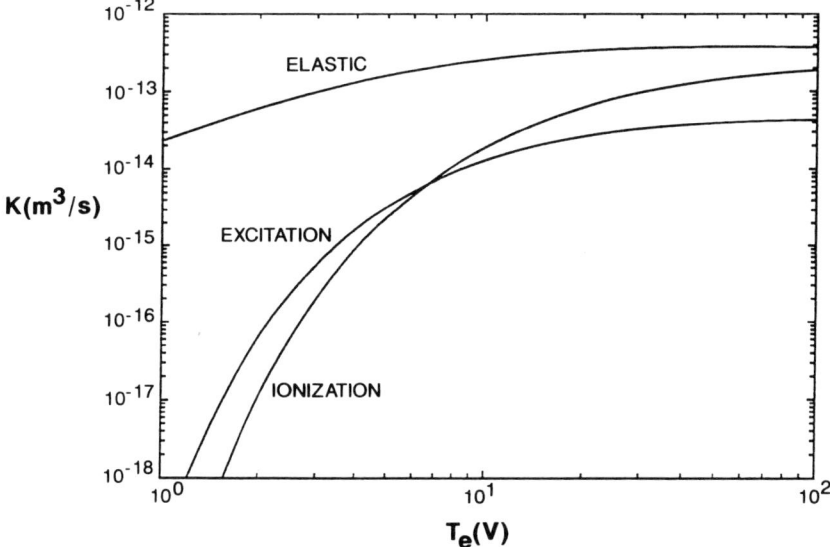

FIG. 4. Electron-collision rate coefficients in argon gas.

by

$$v = KN,$$

where $K(\text{m}^3/\text{s})$ is the rate constant and $N(\text{m}^{-3})$ is the neutral Ar concentration.

In argon, the cross-section for resonant charge transfer of Ar^+ on Ar somewhat exceeds that for elastic scattering. The combined ionic momentum transfer cross-section σ_i for these two processes is large ($\sigma_i \approx 10^{-18}\,\text{m}^2$) and relatively constant for the ion energies of interest. The corresponding ion–neutral mean free path is

$$\lambda_i = \frac{1}{N\sigma_i}.$$

For the sake of comparing the efficiency of one source with another, it is useful to define the collisional energy ε_c lost per electron–ion pair created in the system. For single-step, electron-impact ionization,

$$K_{iz}\varepsilon_c = K_{iz}\varepsilon_{iz} + K_{exc}\varepsilon_{exc} + K_{el}(2m/M)(3\,T_e/2) \tag{1}$$

where ε_{iz}, ε_{exc} and $\varepsilon_{el} \approx (3m/M)T_e$ are the energies lost by an electron as a result of ionization, excitation and elastic collisions, respectively. While the last term may appear negligible for large electron temperatures, it is important for $T_e < 2$ V, where it is dominant. The energy ε_c is lost when the electron–ion pair is subsequently lost. In a monatomic gas such as Ar, the dominant loss mechanism at low pressure is simply flow to the walls, and this loss rate in steady-state must be balanced by the rate of formation, which accounts for the factor K_{iz} on the left-hand side of (1). Note that for simplicity we have lumped all excitation channels into one effective level characterized by energy ε_{exc}. While this is crude, again it offers a simple, rapid means for comparing one source to another. In general, a detailed energy balance including many excited states and multistep ionization pathways must be considered if quantitative comparisons are to be made. The quantity ε_c, which within the framework of the assumptions just presented, is a function of T_e alone, is shown for Ar in Figure 5. For the excitation process, a composite cross-section is used from Eggarter (93), with an excitation energy of 11.97 V. For ionization, the cross-section from Peterson and Allen (94) is used. The elastic cross-section is from the data in Hayashi (95).

In addition to collisional energy losses, electrons and ions carry kinetic energy to the walls (Section VIII). For Maxwellian electrons, the mean kinetic energy lost per electron lost is $\varepsilon_e = 2T_e$. The mean kinetic energy lost per ion lost is ε_i, which is mainly due to the dc potential

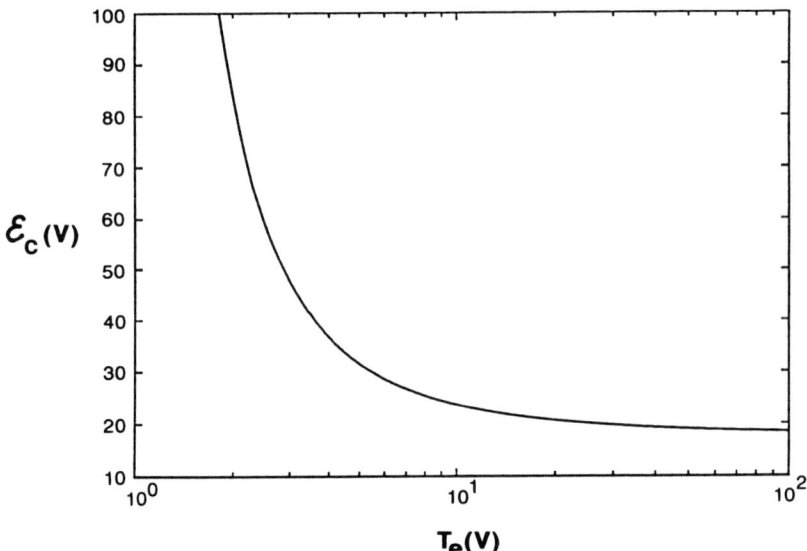

FIG. 5. Collisional energy ε_c lost per electron–ion pair created in argon gas.

across the sheath (Sections II.A.2 and VIII). Summing the three contributions yields the total energy lost per ion lost from the system:

$$\varepsilon_L = \varepsilon_c + 2T_e + \varepsilon_i. \quad (2)$$

The principle of high-density source operation, such as for the cylindrical plasma shown in Fig. 6, can be understood from the overall source power balance, written in terms of ε_L as

$$P_{abs} = en_s u_B A \varepsilon_L, \quad (3)$$

where P_{abs} is the power absorbed by the plasma, n_s is the ion density at the plasma-sheath edge, and A is the area for particle loss. The Bohm (ion loss) velocity u_B is relatively constant for a given ion mass and for the typical limited range of T_e's. Hence, n_s can be increased by reducing ε_L, reducing A, or increasing P_{abs}. All three strategies are used. First, ε_L is lowered by reducing sheath voltages from, for example, $\varepsilon_i \approx 360$ V (for rf diodes) to ~ 40 V (for high-density sources). For Ar, with $\varepsilon_c + 2T_e \approx 40$ V, this results in a five fold increase in n_s. Second, the loss area for a cylindrical unmagnetized source having radius R and length L can be effectively reduced from $2\pi R^2 + 2\pi RL$ to $2\pi R^2$ if a strong axial magnetic field is applied to inhibit radial particle loss. This reduction in A can be important for certain ECR sources. Third, P_{abs} can be increased from, say, 500 W for a typical rf diode to 2 kW or more for a high-density source, without substantially increasing the ion bombarding energy.

The relation between the density n_s at the sheath edge and the density n_0 at the plasma center is complex, because the ambipolar transport of ions and electrons spans the regime $\lambda_i \sim R, L$, depending on the pressure and the values for R and L. Assuming uniform ionization at very low pressures or small reactors, $\lambda_i \gg R, L$, the ion transport is collisionless and well described by an ion free-fall profile (96) within the bulk plasma. This profile is relatively flat near the plasma center and dips near the sheath edge, with $n_s/n_0 \approx 0.50$ for $R \gg L$ (planar geometry) and

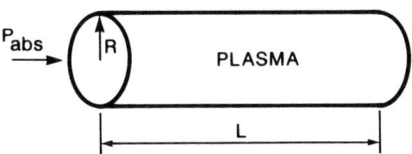

FIG. 6. Simple model of cylindrical high-density plasma source.

$n_s/n_0 \approx 0.40$ for $L \gg R$ (infinite cylinder geometry). At higher pressures or larger reactors such that $\lambda_i \ll R, L$, transport is diffusive and ambipolar. However, the usual diffusion solution for a constant diffusion coefficient (97), consisting of a J_0 Bessel function variation along r and a cosine variation along z, does not describe the profile well, because at these low (but diffusive) pressures, the magnitude of the ion transport velocity $\mathbf{u}_i$ much exceeds the ion thermal velocity u_{Ti} over most of the bulk plasma. In this regime, the ion transport is mobility-limited,

$$\mathbf{u}_i = \mu_i \mathbf{E}, \tag{4}$$

with $\mathbf{E}$ the ambipolar electric field and

$$\mu_i \approx \frac{2e\lambda_i}{\pi M |\mathbf{u}_i|} \tag{5}$$

the mobility (98). For electrons in near thermal equilibrium,

$$\mathbf{E} = -T_e \nabla n/n, \tag{6}$$

which leads to the usual Boltzmann equilibrium relating the spatially varying plasma potential Φ to the density,

$$n = n_0 e^{\Phi/T_e}. \tag{7}$$

Along with particle conservation and the assumption of uniform ionization, (4) and (7) lead to a nonlinear transport equation in the bulk plasma, which has been solved by Godyak (99) in one-dimensional planar geometry ($R \gg L$) and by Godyak and Maximov (100) for infinite cylinder geometry ($L \gg R$). The profiles are relatively flat in the center but fall steeply at the sheath edge. The profile has not been determined for a finite cylinder or for the intermediate mean free path regime $\lambda_i \sim R, L$. However, joining the collisionless and collisional results leads to the following rough estimates:

$$h_L \equiv \frac{n_{sL}}{n_0} \approx 0.86 \left(3 + \frac{L}{2\lambda_i}\right)^{-1/2} \tag{8}$$

at the axial sheath edge and

$$h_R \equiv \frac{n_{sR}}{n_0} \approx 0.80 \left(4 + \frac{R}{\lambda_i}\right)^{-1/2} \tag{9}$$

near the radial sheath edge.

A. Unified Analysis of Source Operation

Let us consider a simple high-density source model to estimate important plasma parameters and see the way these vary with power, pressure, and source geometry: the electron temperature T_e, the ion bombarding energy ε_i, the plasma density n_0, and the ion current density J_i. Referring to Fig. 6, we assume a uniform (in the bulk) cylindrical source plasma with Maxwellian electrons absorbing an electrical power P_{abs} and ionization by single-step electron impact.

1. Electron Temperature

We first determine T_e. Let ion–electron pairs be created in the bulk plasma volume by electron–neutral ionization and lost by flow to the walls. Equating the total volume ionization to the surface particle loss,

$$K_{iz} N n_0 \pi R^2 L = n_0 u_B (2\pi R^2 h_L + 2\pi R L h_R), \tag{10}$$

we solve to obtain

$$\frac{K_{iz}(T_e)}{u_B(T_e)} = \frac{1}{N d_{eff}}, \tag{11}$$

where

$$d_{eff} = \frac{1}{2} \frac{RL}{R h_L + L h_R} \tag{12}$$

is an effective plasma size, and the T_e dependence of K_{iz} and u_B is explicitly shown. Given N and d_{eff}, we can solve (11) for T_e, obtaining, for argon with K_{iz} shown in Fig. 4, the result for T_e shown in Fig. 7. We see that T_e varies over a narrow range between 2 and 5 volts for typical source pressures and sizes. We also note that the density n_0 cancels out in (10) as a result of our single-step ionization assumption. Hence, in this limit, T_e is determined by particle conservation, i.e., the ratio of the ion creation to the ion loss rate, independent of density, and therefore input power.

2. Ion Bombarding Energy

We next discuss ε_i, which is the sum of the ion energy entering the sheath and the energy gained by the ion as it traverses the sheath. The ion velocity entering the sheath is u_B, corresponding to a directed energy

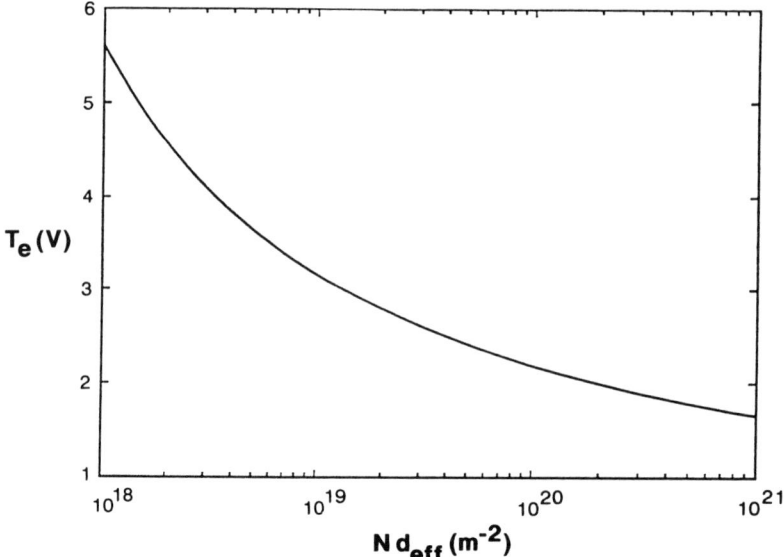

FIG. 7. Electron temperature T_e for a given neutral argon density N and effective plasma size d_{eff} in a low-pressure, high-density source.

of $T_e/2$. The sheath voltage V_s (see Fig. 8) can be estimated from particle conservation in the sheath. The wall sheath thickness s rarely exceeds a few Debye lengths λ_{De}, where $\lambda_{\text{De}} \approx 7,430\,(T_e/n_s)^{1/2}$ m, with T_e in volts and n_s in m^{-3}. Since the sheath is typically much less than a millimeter thick and is much less than a mean free path for ionization in typical

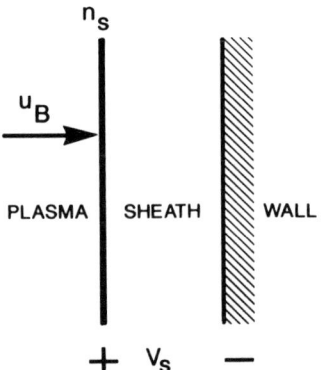

FIG. 8. The wall sheath region in a high-density source.

high-density sources, the fluxes of both ions and electrons are conserved. The ion and electron fluxes at the wall are

$$\Gamma_i = n_s u_B \tag{13}$$

and

$$\Gamma_e = \tfrac{1}{4} n_s u_e e^{-V_s/T_e}, \tag{14}$$

where

$$u_e = \left(\frac{8eT_e}{\pi m}\right)^{1/2} \tag{15}$$

is the mean electron speed. The Boltzmann factor in (14) accounts for the reduction in the electron density at the wall due to the repulsive potential within the sheath. For an insulating wall, the ion and electron fluxes must balance in the steady state. Equating (13) to (14) yields

$$V_s = \frac{T_e}{2} \ln\left(\frac{M}{2\pi m}\right), \tag{16}$$

or $V_s \approx 4.7 T_e$ for argon. Accounting for the initial ion energy, we obtain $\varepsilon_i \approx 5.2 T_e$. At a conducting wall, the fluxes need not balance, although the integrated fluxes (particle currents) must balance. However, if the fluxes are not too dissimilar, then (16) remains a good estimate because of the logarithmic dependence of V_s on the ratio of fluxes.

The ion energy ε_i can significantly exceed V_s for several reasons. In some high-density sources, the plasma flows from the source chamber into a larger-diameter process chamber (see Fig. 3). As the plasma expands into the process chamber, the plasma density drops from n_0 to, say, n_p. This leads to an additional distributed sheath potential V_d determined by the Boltzmann relation (7),

$$V_d = T_e \ln \frac{n_0}{n_p}, \tag{17}$$

which accelerates ions within the process chamber (see Section VIII.D). Ions gain the full potential V_d for sufficiently low pressures, $\lambda_i \gg L_d$, where L_d is the characteristic length over which V_d occurs, and can gain a fraction $(\sim \lambda_i/L_d) V_d$ of this potential for $\lambda_i < L_d$. As we shall see in Section VIII.D, L_d can be small compared to λ_i and characteristic reactor dimensions.

Additional ion bombarding energy can be gained near dielectric windows adjacent to the rf powered conductors that drive the source, or near separately driven rf electrodes embedded in the plasma, such as the wafer holder (Section VIII.E). In both cases, the mechanism is capacitive coupling of the rf power source to the plasma. Careful design of the coupling structure could be used to minimize or eliminate capacitive coupling from the rf powered conductors across the dielectric window to the plasma, but the design principles are not entirely clear, and some enhanced ion bombardment energies can exist in rf driven helicons, helical resonators, and inductively driven sources due to this mechanism. On the other hand, the *desired* ion bombarding energy at an rf powered wafer holder can be strongly enhanced over that obtained from (16). Letting $\tilde{V}_{pw}$ be the plasma to wafer holder rf voltage amplitude and $\bar{V}_{pw}$ be the plasma to wafer holder dc voltage, then we find that $\varepsilon_i \approx 5.2 T_e$ in the low-voltage limit $\tilde{V}_{pw} \ll V_s$, and that $\varepsilon_i \approx \bar{V}_{pw} \approx 0.8 \tilde{V}_{pw}$ in the high-voltage limit $\tilde{V}_{pw} \gg V_s$ (71). An estimate for ε_i over the entire range of driving rf voltages is given by Godyak and Sternberg (73). The additional ion energy flux $en_s u_B \varepsilon_i$ striking the wafer holder is supplied by the rf power source driving the holder in the high-voltage limit.

The thickness of the sheath in the high-voltage limit follows that of a modified ion Child law:

$$en_s u_B \approx 0.8 \varepsilon_0 \left(\frac{2e}{M}\right)^{1/2} \frac{\bar{V}_{pw}^{3/2}}{s^2}. \qquad (18)$$

If n_s, T_e and $\bar{V}_{pw} \approx 0.8 \tilde{V}_{pw}$ are known, then (18) determines s. For typical rf driven wafer holders, s is a few millimeters (101). This is still small compared to λ_i and the sheath is collisionless.

We see from the preceding discussion that estimating ion energy is not so simple, as it depends not only on electron temperature, but also on source geometry and the application of bias voltages. This subject is discussed at greater length along with a review of experimental measurements in Section VIII.

3. Plasma Density and Ion Current Density

Finally, we estimate the plasma density n_0. Accounting for possibly different values of n_s at the axial and radial sheath edges [Eqs. (8) and (9)], we solve (3) to obtain

$$n_0 = \frac{P_{abs}}{e u_B A_{eff} \varepsilon_L}, \qquad (19)$$

where the effective area is

$$A_{eff} \equiv 2\pi R(R h_L + L h_R). \qquad (20)$$

For a specified P_{abs}, and T_e determined from Fig. 7, we obtain n_0 from (19). Note that within the assumption of single-step ionization, n_0 is determined by the total power balance in the discharge and is a function of pressure through the dependence of h_L and h_R on p and through the weaker dependence of T_e on p.

As an example, let $R = 0.15$m, $L = 0.3$m, $N = 3.3 \times 10^{19}$m^{-3} ($p = 1$ mTorr at 298 K), and $P_{abs} = 800$ W. At 1 mTorr, $\lambda_i \approx 0.03$ m. Then from (8) and (9), $h_L \approx h_R \approx 0.3$, from (12) $d_{eff} \approx 0.17$ m, and from Fig. 7, $T_e \approx 4.1$ V. From Fig. 5, $\varepsilon_c \approx 42$ V. Using (2) with $\varepsilon_i \approx 5.2 T_e \approx 21$ V, we find $\varepsilon_L \approx 72$ V. The Bohm velocity is $u_B \approx 3.1 \times 10^3$ m/s, and $A_{eff} \approx 0.13$ m^2. Then (19) yields $n_0 \approx 1.8 \times 10^{17}$ m^{-3}, corresponding to a flux at the axial boundary $\Gamma_{iL} \approx 1.7 \times 10^{20}$/m^2-s or an ion current density of $J_{iL} \approx 2.7$ mA/cm^2.

If a strong dc magnetic field is applied along the cylinder axis, then particle loss to the circumferential wall is inhibited. In the limit of no radial loss, $d_{eff} = L/(2h_L) \approx 0.5$ m in (11), and we obtain $T_e \approx 3.3$ V, $\varepsilon_c \approx 46$ V, $\varepsilon_i \approx 17$ V, $\varepsilon_L \approx 70$ V, $u_B \approx 2.8 \times 10^3$ m/s, $A_{eff} = 2\pi R^2 h_L \approx 0.042$ m^2, $n_0 \approx 5.8 \times 10^{17}$ m^{-3}, and $J_{iL} \approx 7.8$ mA/cm^2. There is a significant increase in charge density and ion flux due to the magnetic field.

B. Discharge Heating

The preceding discussion provides a unified framework for qualitatively understanding rf- and microwave-driven high-density sources, at least in argon gas. However, issues such as energy transfer from power source to plasma electrons, coupling across dielectric windows, and ion flux uniformity depend on specific source concepts, geometries, and magnetic configurations. Possible electron heating mechanisms include

(a) secondary electron emission heating,
(b) stochastic (collisionless) heating,
(c) ohmic (collisional) heating, and
(d) resonant wave–particle interaction heating.

Achieving adequate electron heating is a central issue because, although the heated electrons provide the ionization required to sustain the discharge, they tend to short out the applied heating fields within the bulk plasma. Hence, electron heating in high-density sources occurs either near the plasma-sheath edge, as in (a)–(c), or by generation near the sheath edge of plasma waves that are subsequently absorbed within the bulk, as in (d).

Secondary emission heating is not believed to play a central role in low-pressure high-density sources because sheath voltages are relatively low, and hence the secondary electron energy gain is low. A possible exception is at a capacitively driven wafer holder for highly directional etch applications, such as metals, where the sheath voltages are driven to the range 50–150 V. However, the mean free path for a 100 volt electron in a 1 mTorr argon discharge is about 1.5 meters. Hence, these electrons pass only once through the system before being lost, unless the geometry and magnetic configuration are specifically designed to retain them, as in dc- and rf-driven planar magnetron discharges commonly used for sputtering thin films. These discharges are beyond the scope of this review, and the reader is referred to the relevant literature (102–105).

Stochastic electron heating has been found to be a powerful mechanism in low-pressure rf diodes. Here electrons impinging on the oscillating sheath edge suffer a change of velocity upon reflection back into the bulk plasma. As the sheath moves into the bulk, the reflected electrons gain energy; as the sheath moves away, the electrons lose energy. However, averaging over an oscillation period, there is a net energy gain. The mechanism is analogous to the energy gained by a ball when hit by a tennis racket, and the term stochastic is used to denote the probabilistic nature of the electron collision with the sheath. In a low-pressure, high-density source, this mechanism acts at capacitively coupled rf powered surfaces such as the wafer holder. For high bias voltages, $\tilde{V}_{pw} \gg V_s$, an estimate of the electron heating energy flux is (72, 106)

$$S_e \approx 0.56 \frac{m}{2e} \omega^2 \varepsilon_0 u_e \tilde{V}_{pw}, \tag{21}$$

which, when summed along with the ion energy flux $S_{ion} \approx en_s u_B \varepsilon_i$, gives the bulk (neglecting ohmic heating) of the energy flux (power density) supplied by the rf wafer-bias source. Stochastic heating is important for understanding the dynamics of rf diodes, triodes and MERIEs; there is also some evidence (107) that a similar noncollisional stochastic heating mechanism acts in low-pressure high-density inductively driven sources. This issue is considered further in a subsequent section (Section V.B).

Ohmic heating due to the in-phase components of the rf current density and rf electric field is an important mechanism for rf diodes, especially at high pressures. The time-averaged ohmic power density is

$$P_{ohm} = \tfrac{1}{2} \mathrm{Re}(\sigma_{ac}|\tilde{E}|^2), \tag{22}$$

where σ_{ac} is the ac plasma conductivity, $\tilde{E}$ the complex amplitude of the rf electric field in the bulk plasma, and Re denotes the real part. Ohmic heating is also an important energy deposition mechanism in inductive sources, where it occurs in a thin skin near the plasma-sheath edge, as will be seen (Section V.B).

While the preceding three mechanisms are central to understanding electron heating in rf diodes, the role of wave–particle interactions in electron heating in high-density sources is unique. As will be shown, a number of source concepts, particularly ECRs and helicons, rely on this as the primary heating mechanism.

III. Electron Cyclotron Resonance (ECR) Discharges

Microwave generation of plasma has been employed since the invention of high-power microwave sources in World War II. At low plasma densities, the high electric fields obtainable in a resonant microwave cavity can break down a low-pressure gas and sustain a discharge. For good field penetration in the absence of a magnetic field, $\omega_{pe} \leqslant \omega$, which sets a critical density limit $n_c \leqslant \omega^2 \varepsilon_0 m/e^2$, or, in practical units, $n_c \, (\text{m}^{-3}) \leqslant 0.012 f^2$, with f in hertz. More restrictively, for the high fields required the cavity Q must be high, further limiting the range of operation (108).

The introduction of a steady magnetic field **B**, in which there is a resonance between the applied frequency ω and the electron cyclotron frequency $\omega_{ce} = eB/m$ somewhere within the discharge, allows operation at high density and without a cavity resonance. Because of the cyclotron resonance, the gyrating electrons rotate in phase with the right-hand polarized (RHP) wave, seeing a steady electric field over many gyro-orbits. Thus, the high field of the cavity resonance, acting over a short time, is replaced by a much lower field, but acting over a much longer time. The net result is to produce sufficient energy gain of the electrons to allow ionization of the background gas. Furthermore, the injection of the microwaves along the magnetic field, with $\omega_{ce} > \omega$ at the entry into the discharge region, allows wave propagation to the absorption zone $\omega_{ce} \approx \omega$, even in a dense plasma with $\omega_{pe} > \omega$ or $n_0 > n_c$ (109).

These discharges have low ion bombarding energy, low pressure, and high fractional ionization, compared to conventional rf diodes. Consequently, ECR discharges are seeing increasing usage in the semiconductor industry for etching and deposition processes. For example, Hitachi ECR tools have been used in integrated circuit production since 1985 for six-inch polysilicon etch processes and since 1988 for eight-inch metal etch processes, with more than 500 machines in worldwide use as of the summer of 1992 (*110*).

A. Source Configurations

Figure 9a shows a typical high-profile, i.e., $L > R$, ECR system, with the microwave power injected along the magnetic field lines. The power $P_{\mu w}$ at frequency $f = \omega/2\pi$ is coupled through a vacuum end-window into a cylindrical metal source chamber, which is often lined with a dielectric to minimize metal contamination resulting from wall sputtering (Section IX.B, *87*, *101*, *111–113*). One or several magnetic field coils

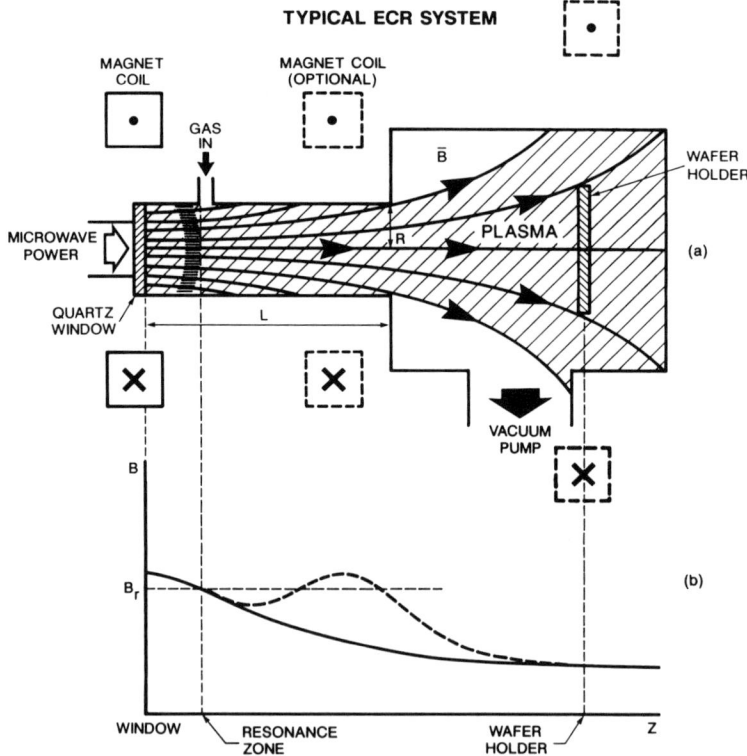

FIG. 9. A typical high-profile ECR system: (a) geometric configuration; (b) axial magnetic field variation, showing one or more resonance zones.

are used to generate a nonuniform, axial magnetic field $B(z)$ within the chamber. The magnetic field strength is chosen to achieve the ECR condition, $\omega_{ce}(z_{res}) \approx \omega$, where z_{res} is the axial resonance position. When a low-pressure gas is introduced, the gas breaks down and a discharge forms inside the chamber. The plasma streams or diffuses along the expanding magnetic field lines into a process chamber toward a wafer holder. Energetic ions and free radicals generated within the entire discharge region (source and process chambers, Section VIII) impinge on the wafer. A magnetic field coil at the wafer holder is often used to modify the uniformity of the etch or deposition process.

Typical parameters for ECR discharges used for semiconductor materials processing are shown in the last column of Table I. The electron cyclotron frequency $f_{ce}(\text{MHz}) \approx 2.8B$, with B in gauss. For $f_{ce} = f = 2,450 \text{ MHz}$, we obtain a resonant magnetic field $B_{res} \approx 875\text{G}$. A typical source diameter is 15 cm.

In some cases, there are multiple resonance positions, as shown in Fig. 9b. A uniform profile is never used for a high-profile system because of the difficulty of maintaining exact resonance and the possibility of overheating the electrons. The monotonically decreasing profile ($dB/dz < 0$) shown as the solid line in Fig. 9b, with one resonant zone near the window, is often used. The mirror profile shown as the dashed line in Fig. 9b has one resonant zone near the window and two additional zones under the second magnet. This profile can yield higher ionization efficiencies, because of enhanced confinement of hot (superthermal) electrons that are magnetically trapped between the two mirror (high-field) positions. However, the high profile (long length) of the source chamber leads to enhanced radial diffusion at high pressures and consequently may reduce plasma densities at the wafer holder.

A typical microwave power system is shown in Fig. 10. A dc power supply drives a magnetron source[2] coupled to the discharge by means of a TE_{10} waveguide transmission system. This consists of a circulator, to divert reflected power to a water-cooled, matched load; a directional coupler, to monitor the transmitted and reflected power; a multiscrew tuner, to match the source to the load through the dielectric window, achieving a condition of low reflected power; and, often, a mode converter, to convert the TE_{10} linear polarized, rectangular waveguide mode to an appropriate mode in the cylindrical source chamber. The

---

[2] Samukawa (114) has recently shown that power fluctuations with magnetron sources can result in unstable operation and broadening of the ion energy distribution functions. Operation with a klystron microwave source helped to avoid these problems.

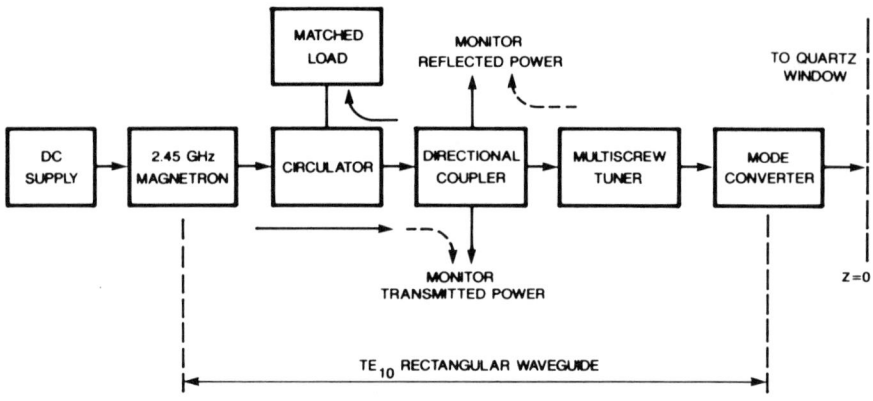

FIG. 10. Typical ECR microwave system.

simplest mode converter (Fig. 11a) is from TE_{10} rectangular to TE_{11} circular mode. At 2,450 MHz, the minimum source chamber diameter for TE_{11} mode propagation (in vacuum) is 7.18 cm (115). However, the electric field profile and corresponding power flux is peaked on axis and is not azimuthally symmetric for this mode, leading to possible non-axisymmetric processing profiles on the wafer. A common converter to an axisymmetric mode configuration (Fig. 11b) is from TE_{10} rectangular to TM_{01} circular mode, having a minimum diameter for mode propagation of 9.38 cm at 2,450 MHz. The profile is ringlike, with a vanishing on-axis power flux. The electric field for both modes is linearly polarized, consisting of equal admixtures of RHP and LHP waves. The basic power absorption mechanism is the absorption of the RHP wave on a "magnetic beach," where the wave propagates from higher to lower magnetic field to the resonance $\omega_{ce}(B) \approx \omega$. The fate of the LHP wave is unclear, but it is probably inefficiently converted to a RHP wave as the power reflects from surfaces or a critical density layer in the source. A more efficient scheme is to use a microwave polarizer and convert from TE_{10} rectangular to a TE_{11} circular mode structure that rotates in the right-hand sense at frequency ω (116). This yields a time-averaged azimuthally symmetric power profile peaked on axis and having an on-axis electric field that is right-hand polarized. Hence, most of the power can be delivered to the plasma in the form of the RHP wave alone.

ECR process tools come in a variety of "flavors." A basic distinction is in coupling the microwave power to the resonance zone. The three

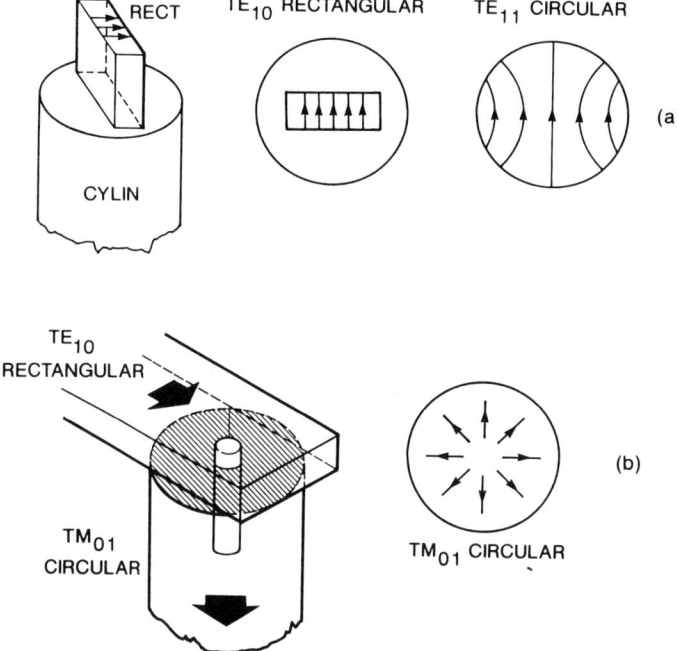

FIG. 11. Microwave field patterns for ECR excitation.

categories are (1) traveling wave propagation mainly along **B** (wavevector **k** ∥ **B**), (2) propagation mainly across **B** (**k** ⊥ **B**), and (3) standing wave excitation (mainly cavity coupled). While these distinctions are useful, most ECR sources rely on the "magnetic beach" absorption of the RHP wave. Additionally, the sources are not neatly broken into these categories; e.g., wave propagation is at an angle to **B**, and absorption can involve standing waves (*117*).

Various ECR configurations are shown in Fig. 12. A high-profile (far from the wafer) source with microwave injection along **B** is shown in Fig. 12a. The resonance (heating) zone can be ring- or disk-shaped (the latter is shown) and may be as much as 50 cm from the wafer. Expansion of the plasma from the resonance zone to the wafer reduces the ion flux and increases the ion impact energy at the wafer. Hence, high-profile sources have given way to low-profile (close to the wafer) sources shown in Fig. 12b, where the resonance zone may be only 10–20 cm from the wafer. Uniformity is controlled at least in part by shaping the axial magnetic field. Uniformity can be further improved and density in-

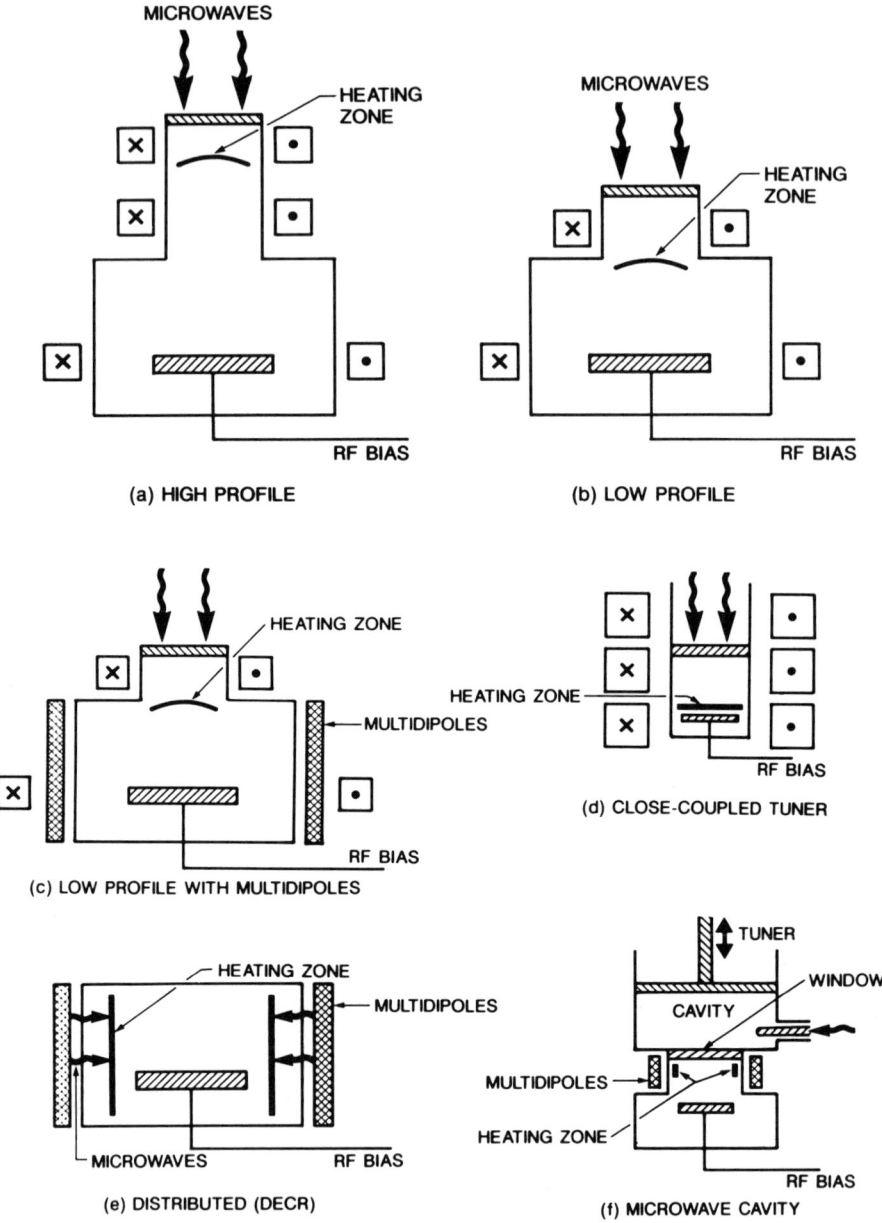

FIG. 12. Common ECR configurations: (a) high-profile; (b) low-profile; (c) low-profile with multidipoles; (d) close-coupled; (e) distributed (DECR); (f) microwave cavity excited.

creased by adding 6–12 linear multidipole permanent magnets around the circumference of the process chamber (*118*), as shown in Fig. 12c. As a variation, a strong (rare earth) permanent magnet can also replace the source coil (*119*). Another approach to achieving adequate uniformity and density is to combine the source and process chambers and place the resonance zone close to the wafer, leading to the close-coupled configuration shown in Fig. 12d (*31*). Uniformity requirements can be met by using a relatively flat, radially uniform resonance zone (*120*).

The multipole, distributed ECR system shown in Fig. 12e is powered by microwave injection perpendicular to the strong, permanent magnet, multipole magnetic fields. Typically, four or more microwave applicators are arranged around the circumference to achieve adequate uniformity (*121*). Each applicator creates an approximately linear resonance zone near the process chamber wall as shown.

A microwave cavity source is shown in Fig. 12f. The coaxial feed is tuned using a sliding short on top and a stub tuner from the side (*3,5*). In earlier, lower-density versions, a grid was used below the plasma generation region, providing microwave containment while allowing the plasma to diffuse out. The linear resonance zones, similar to those in the DECR (Fig. 12e), are generated by a set of 8–12 strong permanent magnets arranged around the circumference of the source chamber as shown (*122*).

B. Electron Heating

The basic principle of ECR heating is illustrated in Fig. 13. A linearly polarized microwave field launched into the source chamber can be decomposed into the sum of two counter-rotating circularly polarized waves. Assuming a sinusoidal steady state with the incident wave polarized along $\hat{x}$,

$$\mathbf{E}(\mathbf{r}, t) = \text{Re}[\hat{x} E_x(\mathbf{r}) e^{j\omega t}], \tag{23}$$

we have

$$\hat{x} E_x = (\hat{x} - j\hat{y}) E_{\text{rhp}} + (\hat{x} + j\hat{y}) E_{\text{lhp}}, \tag{24}$$

where $\hat{x}$ and $\hat{y}$ are unit vectors along x and y and where E_{rhp} and E_{lhp} are the amplitudes of the RHP and LHP waves, with $E_{\text{rhp}} = E_{\text{lhp}} = E_x/2$. The electric field vector of the RHP wave rotates in the right-hand direction (counterclockwise around $\mathbf{B}_0$) at frequency ω, while an electron in a uniform magnetic field $\mathbf{B}_0$ also gyrates counterclockwise at fre-

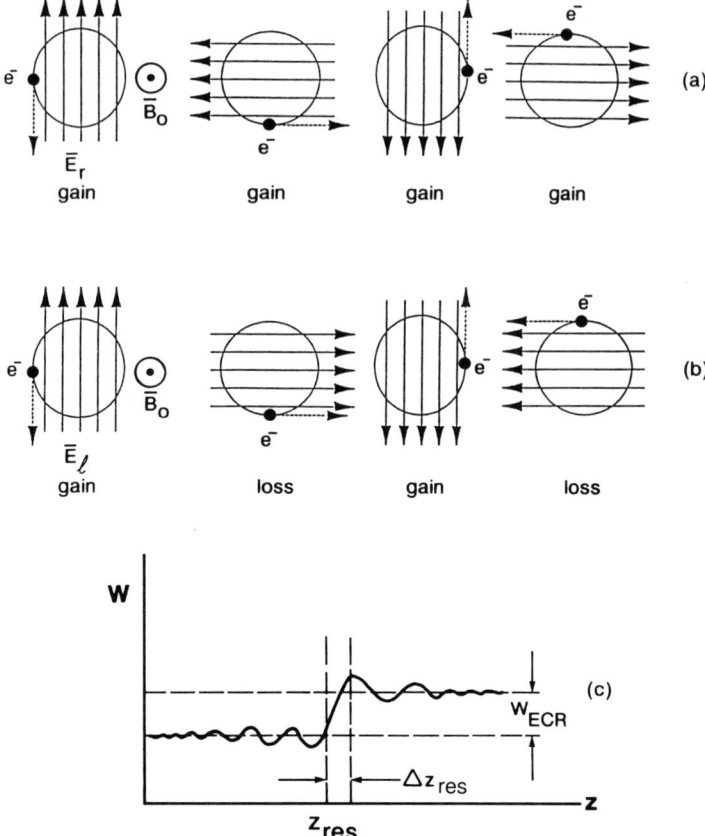

FIG. 13. Basic principle of ECR heating: (a) continuous energy gain for right-hand polarization; (b) oscillating energy gain for left-hand polarization; (c) energy gain in one pass through a resonance zone.

quency ω_{ce}. Consequently, as shown in Fig. 13a, for $\omega_{ce} = \omega$, the force $-eE$ accelerates the electron along its circular orbit, resulting in a continuous transverse energy gain. In contrast, as shown in Fig. 13b, the LHP wave field produces an oscillating force whose time average is zero, resulting in no energy gain.

To determine the overall heating power, the nonuniformity in the magnetic field profile $B(z)$ must be considered. For $\omega_{ce} \neq \omega$, an electron does not continuously gain energy, but rather its energy oscillates at the difference frequency $\omega_{ce} - \omega$. As an electron moving along z passes through resonance, its energy oscillates as shown in Fig. 13c, leading to an average transverse energy W_{ECR} gained in one pass. For low power absorption, where the electric field at the resonance zone is known, the heating can be estimated as follows. We expand the magnetic field near resonance as

$$\omega_{ce}(z') = \omega(1 + \alpha z'), \tag{25}$$

where $z' = z - z_{res}$ is the distance from exact resonance, $\alpha = \partial \omega_{ce}/\partial z'$ is proportional to the gradient in $B(z)$ near the resonant zone, and we approximate $z'(t) \approx u_{res}t$, where u_{res} is the parallel speed at resonance. The energy gain can be written in the form $W_{ECR} = \frac{1}{2}m(\Delta u)^2$, where from Newton's second law $\Delta u = (e/m)Et_{res}$, and t_{res} is the effective time in resonance. To estimate t_{res} we note that an electron passing through the zone coherently gains energy for a time t_{res} such that

$$|\omega - \omega_{ce}(u_{res}t_{res})|t_{res} \approx 2\pi. \tag{26}$$

Inserting (25) into (26), using the definition for z' and solving for t_{res}, we obtain

$$t_{res} = |2\pi/\omega\alpha u_{res}|^{1/2}. \tag{27}$$

The effective resonance width (see Fig. 13c) is

$$\Delta z_{res} = u_{res}t_{res}, \tag{28}$$

which, for typical ECR parameters, is ~ 0.5 cm. The energy gain per pass is thus

$$W_{ECR} = \frac{\pi e^2 E_{rhp}^2}{m\omega |\alpha u_{res}|}. \tag{29}$$

The absorbed power per unit area, or energy flux, is found by integrating (29) over the flux of electrons incident on the zone, yielding

$$S_{ECR} = \frac{\pi n e^2 E_{rhp}^2}{m\omega |\alpha|}. \tag{30}$$

A more careful derivation of this result, including the effect of nonconstant u_{res} during passage through resonance, is given by Jaeger et al. (123). We see that S_{ECR} is proportional to the density, the scale length α^{-1} of the magnetic field variation, and the square of the RHP electric field amplitude at the resonance, and is independent of the axial electron velocity.

C. Resonant Wave Absorption

A serious limitation on the result (30) is that it assumes the electric field within the resonance zone is constant and known from the input power. That this cannot be true in the case of strong absorption is clear, since the absorbed power cannot exceed the incident power. The resolution of this difficulty lies in the attenuation of the wave in the resonance zone, so that the resonant value of E_{rhp} is in fact much smaller than the value of the incident E_{rhp}.

The propagation and absorption of microwave power in ECR sources is an active area of research and is not fully understood. For excitation at an end window (Figs. 12a–d), the waves in a cylindrical magnetized plasma are neither exactly RHP nor propagating exactly along $\mathbf{B}_0$. The waves are not simple plane waves, and the mode structure in a magnetized plasma of finite dimension must be considered. Nevertheless, the essence of the wave coupling, transformation and absorption at the resonance zone can be understood by considering the one-dimensional problem of an RHP plane wave propagating strictly along $\mathbf{B}_0$. For right-hand polarization,

$$\mathbf{E}_{rhp} = \mathrm{Re}[(\hat{x} - j\hat{y})E_{rhp}(z)e^{j\omega t}], \tag{31}$$

where E_{rhp} is the spatially varying electric field amplitude. The wave equation for plane waves propagating along $\mathbf{B}_0$ parallel to z can be written (124, 125)

$$\frac{d^2 E_{rhp}}{dz^2} + k_0^2 \kappa_r E_{rhp} = 0, \tag{32}$$

where far from resonance such that $\omega - \omega_{ce} \gg \nu_c$, the electron–neutral collision frequency,

$$\kappa_r = 1 - \frac{\omega_{pe}^2(z)}{\omega(\omega - \omega_{ce}(z))} \tag{33}$$

is the relative dielectric constant, and $k_0 = \omega/c$, with c the velocity of light. κ_r varies with z because of the dependence of ω_{pe}^2 on the density $n(z)$ and of ω_{ce} on the magnetic field $B(z)$. If the variation of κ_r with z is weak, then a Wentzel–Kramers–Brillouin (WKB) wave expansion can be made (126),

$$E_{rhp} = E_{rhp0}(z) \exp\left[-j \int^z k_r(z')\, dz'\right], \tag{34}$$

where

$$k_r(z) = k_0 \kappa_r^{1/2}(z) \tag{35}$$

is the spatially varying propagation constant, with $\lambda_r = 2\pi/k_r$ the local wavelength. The WKB wave propagates without reflection or absorption for $\kappa_r > 0$, since k_r is real, and the wave is evanescent for $\kappa_r < 0$, since k_r is imaginary.

To illustrate the propagation and absorption of a wave traveling into a decreasing magnetic field from a region where $\omega_{ce} > \omega$, we let $\omega_{pe}^2 = $ constant and ω_{ce} vary linearly with z as given in (25). Then κ_r is plotted versus $\omega_{ce}(z)/\omega$ in Fig. 14a for low density ($\omega_{pe} < \omega$) and in Fig. 14b for high density ($\omega_{pe} > \omega$). The wave travels from right (upstream of the resonance, $\omega_{ce} > \omega$) to left ($\omega_{ce} < \omega$) on this figure. The wave is

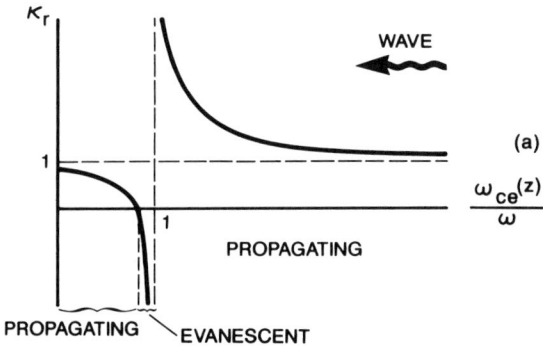

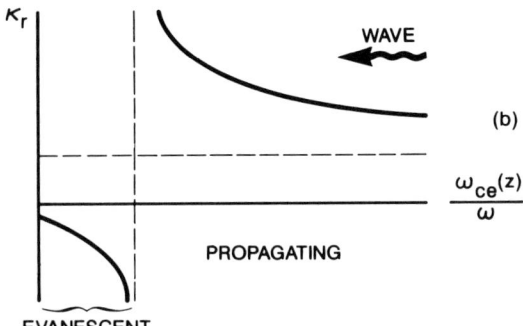

FIG. 14. κ_r versus ω_{ce}/ω for (a) low density, $\omega_{pe} < \omega$, and (b) high density, $\omega_{pe} > \omega$.

evanescent downstream of the resonance in the region

$$1 - \frac{\omega_{pe}^2}{\omega^2} < \frac{\omega_{ce}}{\omega} < 1 \tag{36}$$

and is propagating otherwise. For $\omega_{pe} \ll \omega$, the region of evanescence is thin (in z), and the wave can tunnel through this region to propagate again further downstream. As ω_{pe} increases toward ω, less power can tunnel through. For $\omega_{pe} > \omega$, the wave is always evanescent downstream of the resonance.

For WKB wave propagation, the time-averaged power per unit area carried by the wave is

$$S_r = \tfrac{1}{2} Z_0^{-1} \kappa_r^{1/2} E_{rhp0}^2, \tag{37}$$

where $Z_0 = (\mu_0/\varepsilon_0)^{1/2} \approx 377$ ohms is the impedance of free space. The WKB solution is valid only when the wavelength variation is small:

$$|d\lambda_r/dz| \ll 2\pi, \tag{38}$$

which is clearly invalid near resonance where $|d\lambda_r/dz| \to \infty$, and some or all of the wave power is absorbed there, depending on whether or not significant tunneling occurs.

For constant density and linear magnetic field variation, Budden (*125*) solved (32) to determine the transmitted, reflected, and absorbed power for a wave incident on the resonance zone from the high field side, obtaining

$$P_{abs}/P_{inc} = 1 - e^{-\pi\eta}, \tag{39}$$

$$P_{trans}/P_{inc} = e^{-\pi\eta}, \tag{40}$$

$$P_{refl}/P_{inc} = 0, \tag{41}$$

where

$$\eta = \frac{\omega_{pe}^2}{\omega c |\alpha|}. \tag{42}$$

Hence, the wave power is either absorbed at the resonance or tunnels through to the other side, with no power reflected. Taking a typical case for which $\alpha = 0.1 \, \text{cm}^{-1}$ and $k_0 = 0.5 \, \text{cm}^{-1}$, we find that $\eta > 1$ corresponds to $\omega_{pe}^2/\omega^2 > 0.2$. Thus, at 2,450 MHz we expect most of the incident power will be absorbed for a density $n_0 > 1.5 \times 10^{10} \, \text{cm}^{-3}$. Since from (19) and (20) the bulk density scales as $n_0 \sim P_{abs}$ at low pressures and as $n_0 \sim p^{1/2} P_{abs}$ at high pressures, we obtain the region of good power absorption sketched in Fig. 15 (*112*). For parameters well

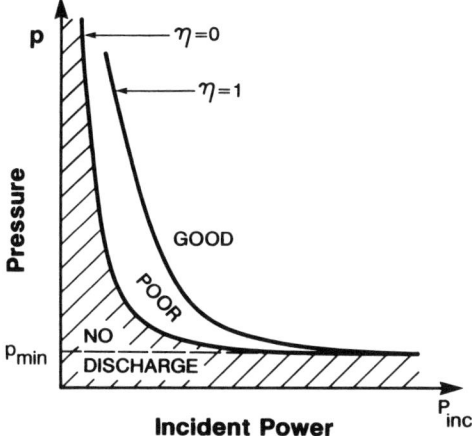

FIG. 15. Parameters for good ECR source operation: pressure vs. power.

within this region, the incident microwave power is efficiently absorbed over the entire cross-section of the resonance zone. For operation outside this region, considerable microwave power can impinge on the wafer.

The minimum P_{inc} for an ECR discharge to be sustained can similarly be found. Expanding (39) in the limit of $n_0 \to 0$ yields $P_{abs} = \pi \eta P_{inc}$. Substituting this into (19), we obtain the minimum value of P_{inc} to sustain the discharge. At a given pressure, this minimum is found to be a factor of two below the $\eta = 1$ condition for good power absorption, as illustrated in Fig. 15. We should also note that the discharge cannot be sustained if the pressure drops below some minimum value p_{min}, because the particle balance equation (11) has no solution for T_e. This limit is also illustrated in Fig. 15.

The size, shape, and location of the resonant zone is set by the magnet coil configuration and the magnet currents. The zone shape and location can also be modified by the Doppler effect for electrons incident on the zone. The actual resonance position is determined by the Doppler-shifted frequency (91)

$$\omega + k_r u_{res} = \omega_{ce}(z_{res}).$$

At high densities, from (35), k_r can be large near the zone, leading to a large Doppler shift. For example, for $k_r = 6.3 \, \text{cm}^{-1}$ ($\lambda_r = 1 \, \text{cm}$), a typical value at the edge of the resonance zone, and $u_{res} = 10^8 \, \text{cm/s}$ (3 volt electron), we obtain $k_r u_{res}/\omega \approx 0.094$. Hence, the resonant magnetic field is 910 gauss for this electron and not 875 gauss. For $\alpha = 0.1\,\text{cm}^{-1}$, this leads to a shift in the zone location of 0.4 cm. By using a coaxial electrostatic probe to sample the microwave field in an ECR and beating that signal against a reference signal from the incident microwaves (Fig. 16), Stevens et al. (91) have recently measured the microwave field amplitude as a function of position in an ECR source and verified that the resonant zone is Doppler shifted, in their case to $\sim 975\,\text{G}$, as shown in Fig. 17.

Axial and radial density and magnetic field variations can lead to wave refraction effects that alter the power flux profile as the wave propagates to the resonance zone. A density profile that is peaked on axis leads to a dielectric constant κ_r that is peaked on axis. This in turn can lead to a self-focusing effect that can increase the sharpness of the microwave power profile as the wave propagates to the zone, adversely affecting uniformity. The mechanism is analogous to the use of a graded dielectric constant optical fiber to guide an optical wave. However, the ECR refraction problem is much more complicated because the density profile is not known *a priori* and the magnetized plasma medium cannot be represented as an isotropic dielectric. A simplified picture of the refraction is obtained in the geometrical optics limit by examining the trajectories of optical rays as they propagate. The ray dynamics are derivable from the dispersion equation and have a Hamiltonian form

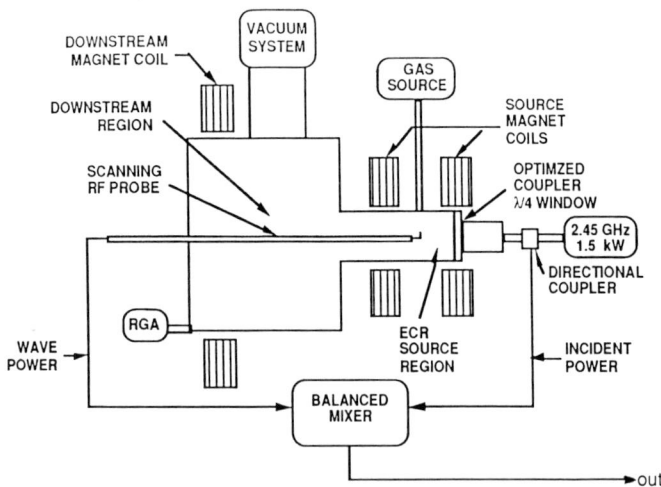

FIG. 16. Schematic illustration of circuit used to measure wave electric field amplitude, kindly provided by Stevens et al. (91).

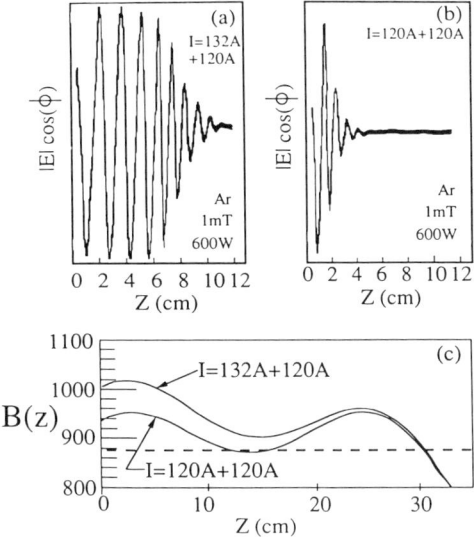

FIG. 17. Electric field amplitude of RHP wave for two different magnetic field configurations. Note that the wave is damped at the Doppler shifted resonant field of ~975 gauss. Adapted from Ref. 91 with permission.

(127, 128), with (k_z, z) and $(k_\perp, r)$ canonically conjugate variable pairs. For high densities and magnetic fields ($\omega_{pe}, \omega_{ce} \gg \omega$) and propagation at an angle to the magnetic field, the dispersion equation reduces to that of whistler waves (Ref. 126, p. 55):

$$kk_z/k_0^2 = \omega_{pe}^2/\omega\omega_{ce}, \qquad (43)$$

where $k = (k_\perp^2 + k_z^2)^{1/2}$ is the wavevector magnitude and $k_\perp$ and k_z are the radial and axial components. Choosing $\omega_{pe}^2/\omega_{ce}$ to have radial variation only, independent of z, Hamilton's equations show that k_z is conserved along the path of a ray (129). If $\omega_{pe}^2/\omega_{ce}$ is a decreasing function of r, then (43) shows that $k_\perp$ decreases with increasing r, implying that the ray bends toward the axis, a focusing action. On the other hand, for some parameter choices, e.g., $\omega_{pe} \sim \omega \sim \omega_{ce}$, a refraction of the wave away from the axis has been found by numerical integration of the ray equations, leading, for this particular case, to an increased uniformity of the power flux profile (91).

For some source concepts (e.g., DECR in Fig. 12e), the microwave power is injected perpendicular to the magnetic field, and not parallel to the field. In this case, the feed structure excites the so-called extraordinary (X) wave, which in the WKB limit has a resonance at the upper hybrid frequency $\omega_h = (\omega_{pe}^2 + \omega_{ce}^2)^{1/2}$, where the wave power is absorbed (128, 109). Since ω_h depends on both ω_{pe} and ω_{ce}, we see that the shape and location of the resonance zone depends on the density as well as the magnetic configuration. Furthermore, the X-wave is evanescent for frequencies such that $\omega_h < \omega < \omega_R$, where

$$\omega_R = \tfrac{1}{2}[\omega_{ce} + (\omega_{ce}^2 + 4\omega_{pe}^2)^{1/2}].$$

For a fixed driving frequency $\omega > \omega_{ce}$, there can be an evanescent layer that the X-wave must tunnel through on its journey from the feed structure to the zone. For $\omega_{pe} \gg \omega$, the tunneling is negligible and the wave cannot propagate to the zone. This can limit the density obtainable in these sources to order $2 \times 10^{12}\, cm^{-3}$ at 2,450 MHz, although the limitation is not severe for typical processing applications. Microwave cavity sources (Fig. 12f) can suffer from similar limitations. On the other hand, densities as high as $3 \times 10^{13}\, cm^{-3}$ have been generated using RHP wave injection along **B** (130).

Although there are a number of commercially available ECR sources of considerable sophistication, they are generally not well characterized or understood. While the claim is sometimes made that ECR technology is complicated and expensive, we note that plasma generation is a relatively small part of the plasma processing system used in manufacturing. Other source concepts may provide a less expensive, more reliable technology for future high-density processing applications, and we turn to some of these now.

IV. Helicon Discharges

Helicon generation of plasmas was first employed by Boswell (131), following a 10-year history of helicon propagation studies, first in solid state and then in gaseous plasmas (132–134). Boswell and his group at the Australian National University have done the most extensive experimental studies, and Chen (86, 135) has given the most complete theory of helicon propagation and absorption. Recent experiments have shed further light on mode excitation and absorption (84, 85, 135, 136). Etching of silicon in helicon SF_6 discharges was first performed by Boswell and Henry in 1985 (137), and the first helicon reactor specifically designed for materials processing was operated by Perry and Boswell in 1989 (138).

Helicons are propagating wave modes in a finite diameter, axially magnetized plasma column. The electric and magnetic fields of the modes have radial, axial, and usually, azimuthal variation, and they propagate in a low- frequency, low magnetic field, high-density regime characterized by

$$\omega_L \ll \omega \ll \omega_{ce}, \qquad (44)$$

$$\omega_{pe}^2 \gg \omega \omega_{ce}, \qquad (45)$$

where

$$\frac{1}{\omega_L^2} = \frac{1}{\omega_{pi}^2} + \frac{1}{\omega_{ce}\omega_{ci}}$$

defines the lower hybrid frequency ω_L, with ω_{pi} and ω_{ci} the ion plasma frequency and ion gyrofrequency, respectively. The driving frequency is typically 1–50 MHz, with 13.56 MHz used for processing discharges. The magnetic fields vary from 50 to 100 G for processing discharges, while fields up to 1,000 G have been employed for some fundamental plasma studies. Charge densities range from 10^{11} to 10^{14} cm^{-3}, with 10^{11}–10^{12} cm^{-3} typical for processing.

A. Helicon Configurations

Helicons are excited by an rf-driven antenna that couples to the transverse mode structure across an insulating chamber wall. The mode then propagates along the column, and the mode energy is absorbed by plasma electrons due to collisional (resistive) or collisionless (Landau) damping. All helicon applications to materials processing to date have utilized a process chamber downstream from the source. A typical helicon system is shown in Fig. 3.

The plasma potential in helicon discharges is typically low, of order 15–20 volts, as for ECRs (139). However, the magnetic field is much lower than the 875 G required for ECRs, and the helicon power is supplied by rf rather than microwave sources. The smaller magnetic field, in particular, may provide lower cost of ownership for the helicon when compared to the ECR source. However, as we will see, the resonant coupling of the helicon mode to the antenna can lead to non-smooth variation of density with source parameters, known as "mode jumps," restricting the operating regime for a given source design.

Helicons are relatively undeveloped sources for materials processing.

No sources were being used on production lines in 1992. However, several equipment vendors produce complete source systems for research and advanced development. Since close-coupled geometries have not been developed for helicons, as they have for ECRs (see Fig. 12d), the transport and diffusion of the source plasma into the process chamber is a significant issue (Section VIII). The process chamber can have multidipole confinement magnets to increase uniformity or can have a wafer-level magnet coil (e.g., as in Fig. 9) to keep the source plasma more tightly focused, thus increasing the etch rate with some reduction in uniformity (*139*).

The rf power system driving the helicon antenna can be of conventional design (as for RIE reactors). A 500–2,000 W, 50 ohm, 13.56 MHz supply can be used to drive the antenna through a set of meters to measure incident and reflected power, followed by a matching network to minimize the reflected power seen by the supply. The matching network can be an L design with two variable capacitors and the antenna itself as the inductive element. The antenna can also be driven through a balanced transformer so that the antenna coil is isolated from ground. This reduces the maximum antenna–plasma voltage by a factor of two, thus also reducing the undesired capacitive current coupled to the plasma by a factor of two.

B. Helicon Modes

Before we can consider helicon source design, which is mostly a matter of choosing antenna dimensions in addition to the radius and length of the source and the magnetic field profile, we must understand the helicon mode structure to which the antenna couples. Helicon modes are a superposition of low-frequency whistler waves propagating at a common (fixed) angle to $\mathbf{B}_0$. Hence, although helicons have a complex transverse mode structure, they have the same dispersion equation as whistlers, which is repeated here from Section III.C:

$$kk_z/k_0^2 = \omega_{pe}^2/\omega\omega_{ce}, \tag{46}$$

where

$$k = (k_\perp^2 + k_z^2)^{1/2} \tag{47}$$

is the wavevector magnitude, $k_\perp$ and k_z are the radial and axial components, and $k_0 = \omega/c$. The helicon modes are mixtures of electromagnetic ($\nabla \cdot \mathbf{E} \approx 0$) and quasistatic ($\nabla \times \mathbf{E} \approx 0$) fields having the form

$$\tilde{\mathbf{E}}, \tilde{\mathbf{H}} \sim \exp j(\omega t - k_z z - m\theta),$$

where here the integer m specifies the azimuthal mode. For an insulating (or conducting) wall at $r = R$ and assuming a uniform plasma density, the boundary condition on the total radial current density amplitude $\tilde{J}_r = 0$ (or $\tilde{E}_\theta = 0$) leads to

$$mkJ_m(k_\perp R) + k_z J'_m(k_\perp R) = 0, \tag{48}$$

where the prime denotes a derivative of the Bessel function, J_m, with respect to its argument. For a given frequency ω, density n_0, and magnetic field B_0, (46)–(48) can be solved to obtain $k_\perp, k_z$, and k.

Helicon sources based on excitation of the $m = 0$ mode and the $m = 1$ mode have been developed. Since the $m = 0$ mode is axisymmetric and the $m = 1$ mode has a helical variation, both modes generate time-averaged, axisymmetric field intensities. The transverse electric field patterns and the way these propagate along z are shown in Fig. 18a for the $m = 0$ mode and in Fig. 18b for the $m = 1$ mode (86, 135). Undamped helicon modes have $\tilde{E}_z = 0$, i.e., the quasistatic and electromagnetic

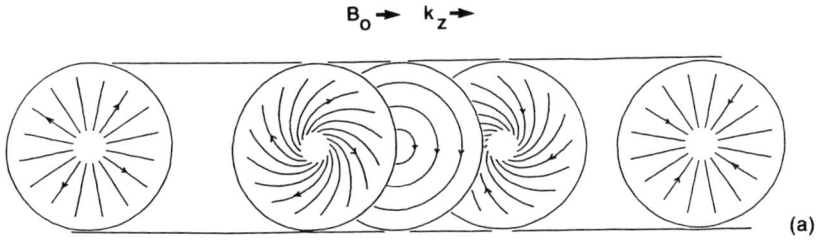

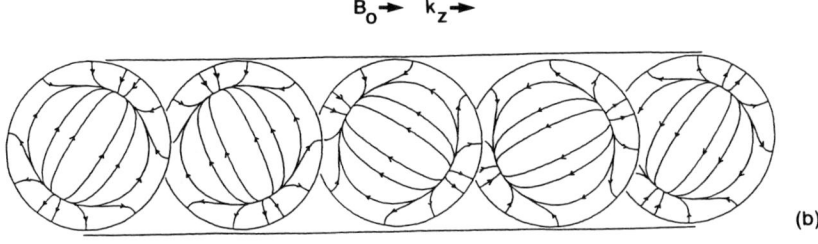

FIG. 18. Transverse electric fields of helicon modes at five different axial positions: (a) $m = 0$; (b) $m = 1$. Reprinted from Ref. 86 with permission.

components of $\tilde{E}_z$ exactly cancel. The antenna couples to the transverse electric or magnetic fields to excite the modes.

Equation (48) can be solved for $k_\perp R$ as a function of k_z/k. There are an infinite number of solutions corresponding to different radial field variations, and in any real system a mixture of modes is very likely excited. For simplicity, let us consider the first radial mode, shown in Fig. 18. For $m = 0$, we find

$$k_\perp R = 3.83 \qquad (m = 0)$$

for any k_z/k. For $m = 1$, we solve numerically to obtain the graph shown in Fig. 19, with the limiting values

$$k_\perp R = 3.83 \qquad (m = 1, k_\perp \gg k_z),$$
$$k_\perp R = 2.41 \qquad (m = 1, k_z \gg k_\perp).$$

To design an antenna for efficient power coupling, we must solve (46)–(48) and determine $k_\perp$ and k_z. This can be done analytically for two limiting regimes:

(a) Low-density with $k_\perp \gg k_z$, and
(b) high-density with $k_z \gg k_\perp$.

Rewriting (46) in more physical terms:

$$kk_z = \frac{e\mu_0 n_0 \omega}{B_0}. \qquad (49)$$

Let us estimate the condition $n = n_0^*$ for $k_z = k_\perp$ for the $m = 1$ mode. We have $k_z = k_\perp \approx 2.5/R$ from Fig. 19, and $k = \sqrt{2}k_z$. Choosing typical processing source parameters of $R = 5$ cm, $f = 13.56$ MHz, and $B_0 = 200$ G, we obtain $n_0^* \approx 4.0 \times 10^{12}$ cm^{-3}. Hence, for this source with $n_0 \ll n_0^*$, we have $k_\perp \gg k_z$ and, from (47), $k \approx k_\perp$. For this case, (49) yields the axial wavelength of the helicon mode for low-density operation:

$$\lambda_z = \frac{2\pi}{k_z} = \frac{3.83}{R} \frac{B_0}{e\mu_0 n_0 f}. \qquad (50)$$

This regime is of limited interest for materials processing because setting the antenna length $\ell \sim \lambda_z$ (see Section IV.C) requires $R \ll \ell < L$. Hence, the source would be long and thin and uniformity over a large area would be compromised.

For $n_0 \gg n_0^*$, we have $k_z \gg k_\perp$ and $k \approx k_z$. In this high-density regime, we find

$$\lambda_z = \left(\frac{2\pi B_0}{e\mu_0 n_0 f}\right)^{1/2}. \qquad (51)$$

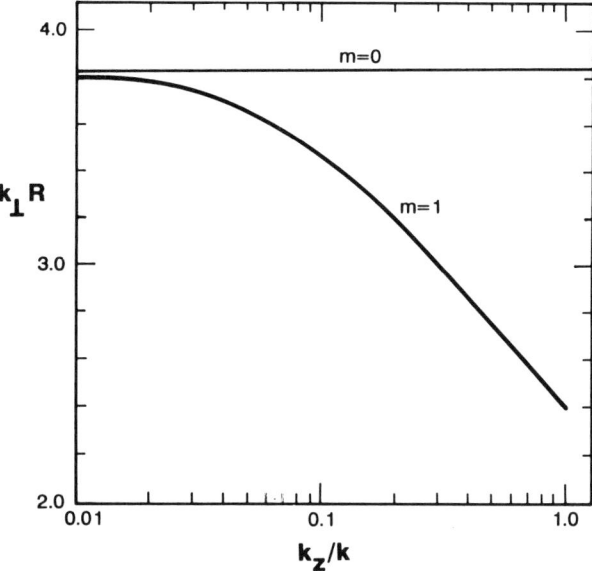

FIG. 19. $k_\perp R$ versus k_z/k for helicon modes.

This regime is of marginal interest because it requires $\ell \ll R$, a short fat antenna, which leads to inefficient coupling of power from the antenna to the plasma because, for a given current, only a small axial voltage is induced, leading to a small axial charge separation to drive the helicon mode. The regime of most interest for materials processing sources is $n_0 \sim n_0^*$, for which $k_z \sim k_\perp$; hence, we have $R \sim \ell \sim L$, yielding a low-profile source configuration. This regime is not easy to analyze. For $m = 1$, the solution must be found numerically. One usually chooses $k_\perp$ somewhat larger than k_z; hence we use (50) for simple estimates of source operation (86). Komori et al. (84) have measured the helicon wave magnetic field using a magnetic pickup coil (140, 141) and circuit similar to that used by Stevens et al. (91) for measuring the microwave electric field (Fig. 16). The dependence of λ_z on B_0/n_0 shown in Fig. 20 roughly follows (50).

Recall from Section II.A.3 that the bulk density n_0 is determined by the absorbed power P_{abs} and the pressure p as specified in (19) and (20). Once $B_0, f,$ and R (for low density) are chosen, then (50) or (51) determine λ_z. Ideally, the antenna must be designed to excite modes having that partcular λ_z. At first sight, this seems to limit source operation to one particular density unless B_0 or f can be conveniently varied. Fortunately, antennas excite a range of λ_z's, thus allowing source operation over a range of n_0's.

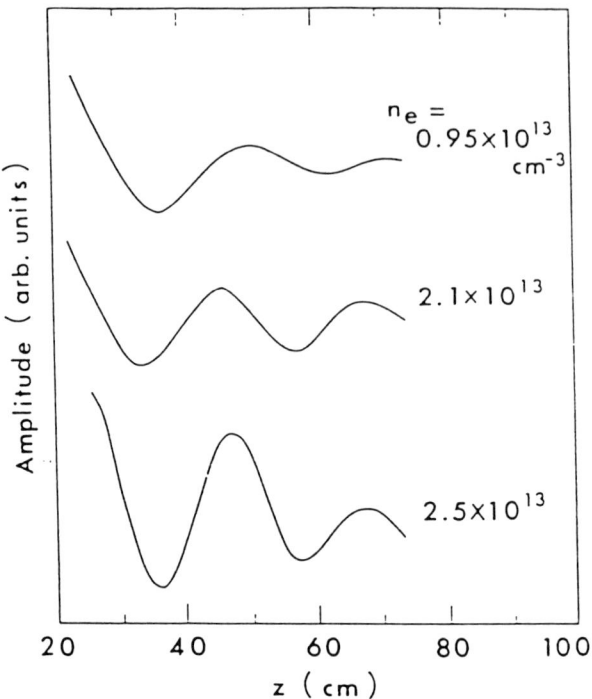

FIG. 20. Magnetic field amplitude for helicon waves at three different values of B_0/n. Reprinted from Ref. 84 with permission.

C. Antenna Coupling

A typical antenna used to excite the $m = 1$ mode is shown in Fig. 21. Other antennas are described by Chen (135). Looking at the $x-y$ transverse coordinates shown in the figure, we see that this antenna generates a $\tilde{B}_x$ field over an axial antenna length ℓ, which can couple to the transverse magnetic field of the helicon mode. The antenna also induces a current within the plasma column just beneath each horizontal wire, in a direction opposite to the currents shown. This current produces charge of opposite signs at the two ends of the antenna, which in turn generates a transverse quasistatic field, $\tilde{E}_y$, which can couple to the transverse quasi-static fields of the helicon mode (see Fig. 18b). The conditions for which each form of coupling dominates are not well understood.

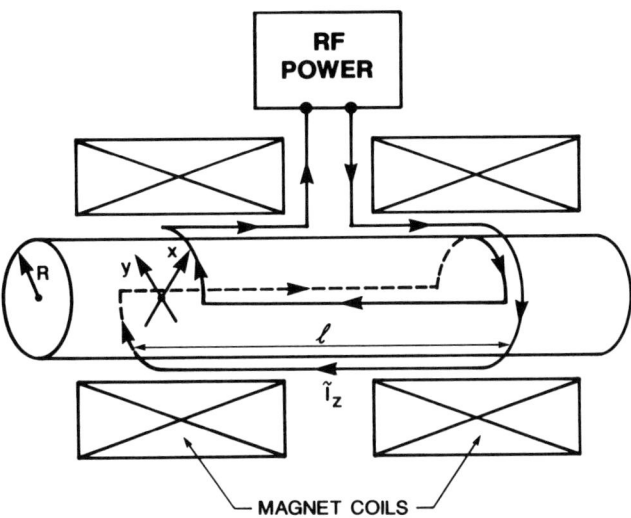

FIG. 21. Illustrating the antenna for $m = 1$ helicon mode excitation (after Boswell et al., 1982).

To illustrate the wavelength matching condition for helicon excitation, we consider an ideal antenna field for quasi-static coupling,

$$\tilde{E}_y(z) \sim \tilde{E}_{y0}[\delta(z + \ell/2) - \delta(z - \ell/2)], \tag{52}$$

where δ is the Dirac delta function. This ideal field is sharply peaked near the two antenna ends, as shown schematically in Fig. 22a. Taking the Fourier transform,

$$E_y(k_z) = \int_{-\infty}^{\infty} dz\, \tilde{E}_y(z) \exp(-jk_z z),$$

and squaring this to obtain the spatial power spectrum of the antenna, we obtain

$$E_y^2(k_z) = 4\tilde{E}_{y0}^2 \sin^2 \frac{k_z \ell}{2}, \tag{53}$$

which is plotted in Fig. 22b. We see that the antenna couples well to the helicon mode for $k_z \approx \pi/\ell$, $3\pi/\ell$, etc., corresponding to $\lambda_z \approx 2\ell, 2\ell/3$, etc. The coupling is poor for $k_z \approx 0, 2\pi/\ell, 4\pi/\ell$, etc., corresponding to $\lambda_z \to \infty$, $\lambda_z \approx \ell$, $\lambda_z \approx \ell/2$, etc.

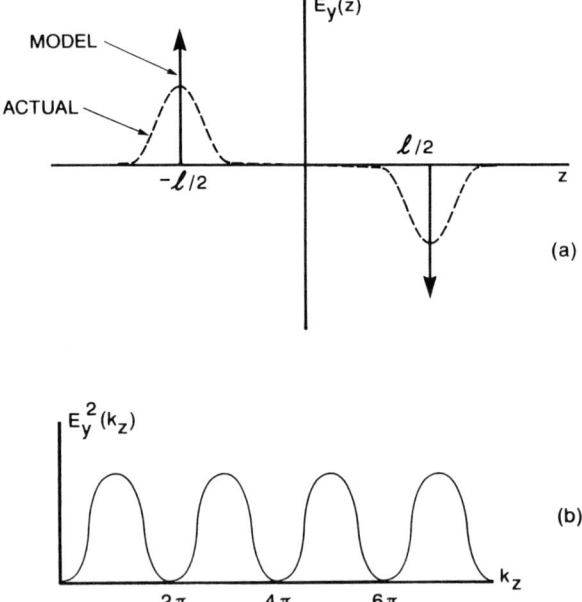

FIG. 22. Illustrating the quasistatic antenna coupling field $\tilde{E}_y$: (a) ideal and actual field; (b) spatial power spectrum of the field.

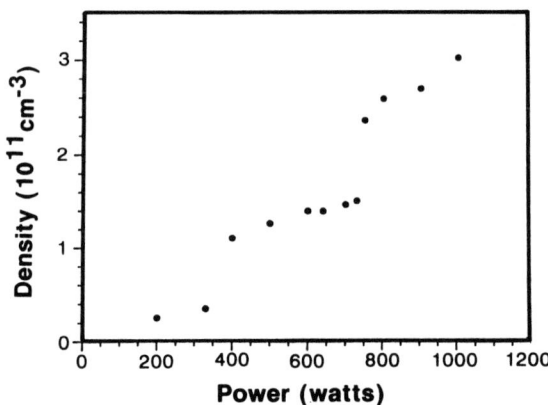

FIG. 23. Measured density as a function of input power for $B_0 = 80\,\text{G}$ at 5 mTorr in argon. Reprinted from Ref. 139 with permission.

Figure 23 (*139*) shows the effect of the antenna coupling on the density n_0 as the power P_{inc} supplied to the antenna is increased, based on a 36 GHz microwave interferometer measurement of n_0. For $P_{inc} < 350$ W, n_0 determined from power balance (19) is low, leading to $k_z \ll \pi/\ell$ and, from (53), poor coupling to the helicon mode. The discharge in this regime is probably capacitively driven, with a relatively high antenna voltage (~ 2 kV) and plasma potential (> 30 V). The transition to helicon mode operation with $k_z \approx \pi/\ell \approx 0.4$ $k_\perp$ for $P_{inc} \approx 400$ W and $n_0 \approx 1.4 \times 10^{11}$ cm^{-3} is clearly seen. A second transition is seen to $k_z \approx 3\pi/\ell \approx k_\perp$ with $n_0 \approx 2.7 \times 10^{11}$ cm^{-3}. Standing helicon wave effects may also play a role in this transition (*139*), as described in Section IV.D. Figure 24 (*142*) shows the roughly linear scaling of n_0 with B_0 predicted from (50) or (51), for a different source than that of Fig. 23. Again we see the density steps imposed by the antenna coupling condition. Depending on the specific experimental configuration—for example, the distance between the antenna and the outer surface of the source dielectric cylinder—the density steps are not always as evident as shown in these data (*143*). They may also be produced by large relaxation oscillations as the discharge "hunts" between helicon and inductive excitation modes (*83*). The antenna can also be designed to couple efficiently to a wide range of k_z's, reducing the importance of mode jumps in the density range of interest.

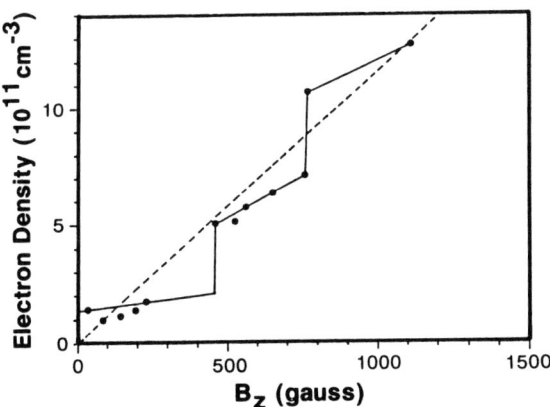

FIG. 24. Measured density as a function of magnetic field at a fixed input power. The dashed line represents the resonance condition imposed by the antenna. Reprinted from Ref. 139 with permission.

Similar effects can be expected for $m = 0$ mode helicons. This mode can be excited by an antenna consisting of two circular coils of radius R, separated by a length ℓ, carrying oppositely directed currents.

D. Helicon Mode Absorption

The helicon mode energy is believed to be transferred to the plasma electrons as the mode propagates along the column by collisional or collisionless (Landau) damping. The former mechanism transfers the energy to the thermal (bulk) electron population, while the latter mechanism can act to preferentially heat a nonthermal electron population to energies greatly exceeding the bulk electron temperature. There is considerable evidence (86, 135, 142) that collisional absorption is too weak to account for energy deposition at low pressures ($< 10\,\text{mTorr}$ argon), although this mechanism may dominate at higher pressures. Landau damping is a process by which a wave transfers energy to electrons having velocities near the phase velocity $u_{\text{ph}} = \omega/k_z$ of the wave. Chen (86) has estimated the effective collision frequency ν_{LD} for Landau damping of the helicon mode as

$$\nu_{\text{LD}} \approx 2\sqrt{\pi}\omega\zeta^3 \exp(-\zeta^2), \qquad \zeta \gg 1, \tag{54}$$

$$\nu_{\text{LD}}(\text{max}) \approx 1.45\omega, \tag{55}$$

where $\zeta = \omega/(k_z u_{\text{Te}})$, with $u_{\text{Te}} = (2eT_e/m)^{1/2}$ the electron thermal velocity. From (50) or (51) we see that for $\zeta \gg 1$, ν_{LD} increases with increasing electron density at constant magnetic field. However, in typical helicon sources where ζ may be less than or of order unity, ν_{LD} can decrease with increasing n. The total effective collision frequency can be written as

$$\nu_T = \nu_{\text{eT}} + \nu_{\text{LD}},$$

where ν_{eT} is the sum of the electron–neutral and electron–ion collision rates. The axial decay length α_z^{-1} for helicon mode damping is

$$\alpha_z^{-1} \approx \frac{\omega_{\text{ce}}}{k_\perp \nu_T} \tag{56}$$

for low density ($k_\perp \gg k_z$), and

$$\alpha_z^{-1} \approx \frac{2\omega_{\text{ce}}}{k_z \nu_T} \tag{57}$$

for high density ($k_z \gg k_\perp$). For efficient power transfer to the plasma electrons, we require that $\alpha_z^{-1} \leqslant L$, where L is the helicon chamber source length. However, if this condition is not satisfied, then power may still be efficiently absorbed by means of helicon standing waves along the source (length L), or source and process chambers (total length L_T), leading to additional resonant absorption effects when $\lambda_z \sim 2L$ or $\lambda_z \sim 2L_T$ (*139, 142, 144*).

By choosing the antenna length ℓ such that $k_z \approx \pi/\ell$, it is possible to heat electrons, by Landau damping, whose energies are near that corresponding to the wave phase velocity

$$e\varepsilon = \tfrac{1}{2}m(\omega/k_z)^2. \tag{58}$$

If ε is chosen near the peak of the ionization cross-section (~ 50 volts in argon), then the collisional energy ε_c lost per electron–ion pair created can be reduced to a low value, of order the ionization energy ε_{iz}. It follows from (19) that this can lead to a significant increase in density for the same absorbed power. However, the effective collision frequency v_{LD} falls precipitously for $\omega/k_z \gg u_{Te}$, leading to a low spatial decay rate and requiring $L \gg R$, of limited interest for materials processing sources having $L \sim R$. Also, there is some experimental evidence (*145*) that, if the antenna is designed to excite superthermal electrons ($\zeta \gg 1$), then k_z adjust itself as the wave propagates away from the antenna so as to excite thermal electrons ($\zeta \approx 1$) downstream of the antenna. Hence, it may not be easy to achieve excitation of superthermal electrons. Evidence of Landau damping has been reported by Komori et al. (*84*) and Lowenhardt et al. (*85*), but other absorption mechanisms, such as nonlinear excitation of plasma instabilities, may also play a role in helicon mode energy transfer (*142*).

In principle, the radial power deposition is different for the $m = 0$ and $m = 1$ modes. The $m = 0$ mode deposits its energy preferentially on the axis, while the $m = 1$ mode has a maximum power deposition at $r = 0.48R$ (*86, 135, 146*).

As an example of helicon design, let $R = 5$ cm, $L = 20$ cm, $B_0 = 200$ G, $N = 3.3 \times 10^{13}$ cm^{-3} (1 mTorr), $\omega = 85 \times 10^6$ s^{-1} (13.6 MHz), and $P_{abs} = 3{,}000$ watts. At 1 mTorr, $\lambda_i \approx 3$ cm. Then from (8) and (9), $h_L \approx h_R \approx 0.33$, and from (12), $d_{eff} \approx 6.1$ cm. For argon we then obtain from Fig. 7 that $T_e \approx 5.2$ V, and from Fig. 5, that $\varepsilon_c \approx 39$ V. Using (2), we find $\varepsilon_L \approx 76$ V. The Bohm velocity is $u_B \approx 3.5 \times 10^5$ cm/s, and from (20), $A_{eff} \approx 260$ cm^2. Then from (19), we obtain $n_0 \approx 2.7 \times 10^{12}$ cm^{-3}. We see that $n_0 < n_0^*$ at $B_0 = 200$ G. From (50), we find $\lambda_z = 20.8$ cm, and hence we choose an antenna length $\ell = \lambda_z/2 = 10.4$ cm to optimize power coupling. We note that $\omega/k_z = 2.8 \times 10^8$ m/s, compared with the electron thermal velocity $u_{T_e} = 1.4 \times 10^8$ cm/s. Hence, $\zeta \approx 2$, not far from the peak of the Landau damping rate for thermal electrons.

V. Inductive Discharges

Inductive discharges are nearly as old as the invention of electric power, with the first report of an "electrodeless ring discharge" by Hittorf in 1884 (*147*). He wrapped a coil around an evacuated tube and observed a discharge when the coil was excited with a Leyden jar. A subsequent 50-year controversy developed (*148, 149*) as to whether these discharges were capacitively driven by plasma coupling to the low- and high-voltage ends of the cylindrical coil, as in an rf diode, or were driven by the induced electric field inside the coil. This issue was resolved with the recognition that the discharge was capacitively driven at low plasma densities, with a transition to an inductive mode of operation at high densities (*150*). Succeeding developments, which focused on pressures exceeding 20 mTorr in a cylindrical coil geometry, are described in a review article by Eckert (*151*). The high pressure regime was intensively developed in the 1970s with the invention of the open air induction torch and its use for spectroscopy. In the late 1980s, the planar coil configuration was developed (*152, 153*), renewing interest in the use of high density inductive discharges for materials processing at low pressures (< 50 mTorr). It is this regime that is the primary focus here.

A. Inductive Source Configurations

The two exciting coil configurations, cylindrical and planar, are shown in Fig. 25 for a low-profile source. The planar coil is a flat helix wound from near the axis to near the outer radius of the source chamber ("electric stovetop" coil shape). Planar and cylindrical coils can also be united to give "cylindrical cap" or "hemispherical" coil shapes. Multidipoles can be used around the process chamber circumference to increase radial plasma uniformity, as shown. The planar coil can also be moved close to the wafer surface, resulting in a close-coupled or planar source geometry ($L \leq R$) having good uniformity properties even in the absence of multidipole confinement (*153*). In the close-coupled configuration, the coil can be wound nonuniformly or driven with radially varying currents to control the radial plasma uniformity.

Similar to helicon antennas, inductive coils can be driven by a

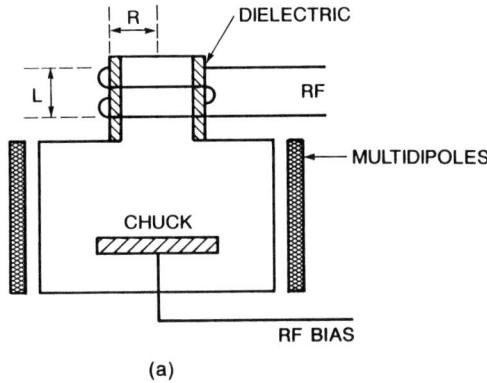

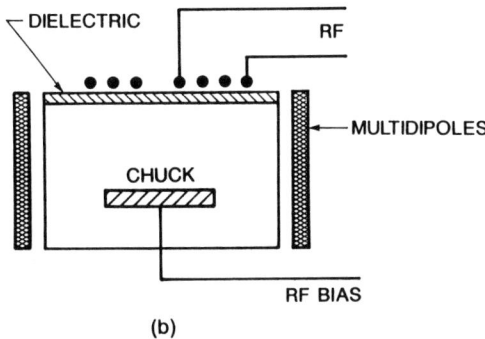

FIG. 25. Schematic of inductively driven sources in (a) cylindrical and (b) planar geometries.

13.56 MHz, 50 ohm rf supply through an L matching network. The coil can be driven push–pull using a balanced transformer, which places a virtual ground in the middle of the coil and reduces the maximum coil-to-plasma voltage by a factor of two. This reduces the undesired capacitively coupled rf current flowing from coil to plasma by a factor of two. An electrostatic shield placed between the coil and the plasma can further reduce the capacitive coupling if desired, while allowing the inductive field to couple unhindered to the plasma.

Plasma in an inductive source is created by application of rf power to a non-resonant, inductive coil, resulting in the breakdown of the process gas within or near the coil by the induced rf electric field. The plasma potential in these discharges is typically less than 30–40 volts (*153*). Multidipole confinement can even further reduce these potentials. The plasma created in the source region streams toward a wafer holder that can be independently biased by application of rf power using a separate generator. Inductive sources have potential advantages over other high-density sources, including simplicity of concept, no requirement for dc magnetic fields (as required for ECRs and helicons), rf rather than microwave source power, and nonresonant source operation, which can enable efcient power absorption over a wide range of source parameters.

In contrast to ECRs and helicons, which can be configured to achieve densities $n_0 \geq 10^{13}\,\text{cm}^{-3}$, we will see (Section V.D) that inductive sources may have natural density limits, $n_0 \leq n_0 \leq 10^{13}\,\text{cm}^{-3}$, for efficient power transfer to the plasma. However, the density regime $10^{11} \leq n_0 \leq 10^{13}\,\text{cm}^{-3}$, for efficient inductive source operation, as much as a factor of 100 times higher than for rf diodes, is of considerable interest for low-pressure processing.

Inductive sources for materials processing applications are in their infancy. The first commercially available system, the so-called TCP (transformer coupled plasma), was announced in the summer of 1992 (*153*). Other vendors are beta-testing similar products for release in 1992 or 1993.

B. Power Absorption and Operating Regimes

In the inductively coupled plasma, power is transferred from the electric fields to the plasma electrons by collisional (ohmic) dissipation. At low pressures ($< 10\,\text{mTorr}$), a collisionless heating process may also act (*107*), in which bulk plasma electrons "collide" with the oscillating, inductive fields and may be accelerated and thermalized much like the stochastic heating near capacitive rf sheaths we discussed in Section II.B. Here, we concentrate on the ohmic heating process, for which the interaction time of the electrons with the inductive fields is long compared to an rf period.

The spatial decay constant α_z for an electromagnetic wave incident on a uniform density plasma along $\hat{z}$ is then

$$\alpha_z = \text{Im}\left(\frac{\omega}{c}\kappa_p^{1/2}\right), \tag{59}$$

where

$$\kappa_p = 1 - \frac{\omega_{pe}^2}{\omega(\omega - j\nu_c)} \approx -\frac{\omega_{pe}^2}{\omega^2(1 - j\nu_c/\omega)} \tag{60}$$

is the relative plasma (Lorentz) dielectric constant, valid in the absence of a magnetic field (compare to κ_r in (33)), ω_{pe} is the plasma frequency near the plasma boundary, and v_c is the electron–neutral collision frequency. For $v_c \ll \omega$, we obtain

$$\alpha_z = \frac{\omega_{pe}}{c} = \frac{1}{\delta_p}, \qquad (61)$$

where δ_p is the collisionless skin depth:

$$\delta_p = \left(\frac{m}{e^2 \mu_0 n_s}\right)^{1/2}. \qquad (62)$$

For $v_c \gg \omega$, we obtain

$$\alpha_z = \frac{1}{\sqrt{2}} \frac{\omega_{pe}}{c} \left(\frac{\omega}{v_c}\right)^{1/2} = \frac{1}{\delta_c}, \qquad (63)$$

where δ_c is the collisional skin depth:

$$\delta_c = \left(\frac{2}{\omega \mu_0 \sigma}\right)^{1/2}, \qquad (64)$$

where

$$\sigma = \frac{e^2 n_s}{m v_c} \qquad (65)$$

is the dc conductivity of the plasma. We therefore distinguish two pressure regimes:

(a) low pressure, $v_c \ll \omega$, skin depth δ_p;
(b) high pressure, $v_c \gg \omega$, skin depth δ_c.

For each pressure regime, we also distinguish two density regimes:

(a) high density, $\delta \ll R, L$;
(b) low density, $\delta \gg R, L$.

At 13.56 MHz in argon, we find $v_c = \omega$ for $p^* \approx 25\,\text{mTorr}$. Let us consider the low-pressure regime (a) with $p \ll p^*$. For a cylindrical coil with $R \approx 10\,\text{cm}$, or for a planar coil with $L \approx 10\,\text{cm}$, we find from (62) that $\delta_p = R, L$ for $n_0^* \approx 3 \times 10^9\,\text{cm}^{-3}$. Hence, for a typical low-pressure processing discharge with $n_0 \geqslant 3 \times 10^{10}\,\text{cm}^{-3}$, we have $v_c \ll \omega$ and $\delta_p \ll R, L$ as the regime of operation. We briefly discuss the $v_c \gg \omega$ and $\delta_p \gg R, L$ regime in Section V.D when we consider the minimum current and power necessary to generate an inductively coupled plasma.

C. Source Operation and Coupling

Although most systems are operated with planar coils (see Fig. 25b), finite geometry effects make these configurations difficult to analyze. To illustrate the general principles of inductive source operation, we concentrate on the cylindrical source (Fig. 25a) in the long, thin geometry $L \gg R$. We take the source coil to have η turns at radius $b > R$. For ohmic heating in the plasma skin,

$$P_{abs} = \frac{1}{2}\frac{J^2}{\sigma} 2\pi R L \delta_p, \tag{66}$$

where J is the rf-induced current density in the skin near $r = R$ (opposite in direction to the applied current in the coil). Letting $I_p = J L \delta_p$ be the total induced rf current and defining the plasma resistance through $P_{abs} = \frac{1}{2}I_p^2 R_p$, we obtain

$$R_p = \frac{2\pi R}{\sigma L \delta_p}. \tag{67}$$

The plasma inductance L_p is found using $\Phi = L_p I_p$, where Φ is the total magnetic flux linked by the skin current. Using $\Phi = \mu_0 \pi R^2 H_z$, where $H_z = J \delta_p$ is the magnetic field produced by the skin current, we obtain

$$L_p = \frac{\mu_0 \pi R^2}{L}. \tag{68}$$

Letting the coil have η turns at a radius $b \geqslant R$, where $b - R$ is the "thickness" of the dielectric interface separating coil and plasma, then we can model the source as the TCP shown in Fig. 26. Evaluating the inductance matrix for this transformer, defined through (154, p. 27)

$$V_{rf} = j\omega L_{11} I_{rf} + j\omega L_{12} I_p, \tag{69}$$

$$V_p = j\omega L_{21} I_{rf} + j\omega L_{22} I_p, \tag{70}$$

we obtain

$$L_{11} = \frac{\mu_0 \pi b^2 \eta^2}{L}, \tag{71}$$

$$L_{12} = L_{21} = \frac{\mu_0 \pi R^2 \eta}{L}, \tag{72}$$

$$L_{22} = L_p = \frac{\mu_0 \pi R^2}{L}. \tag{73}$$

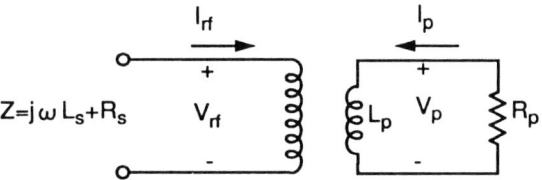

FIG. 26. Equivalent transformer coupled circuit model of an inductive discharge.

Using $V_p = -I_p R_p$ (see Fig. 26) in (70) and inserting into (69), we can solve for the impedance seen at the coil terminals:

$$Z_s = \frac{V_{rf}}{I_{rf}} = j\omega L_{11} + \frac{\omega^2 L_{12}^2}{R_p + j\omega L_p}. \tag{74}$$

For $\delta_p \ll R$, it can easily be seen that $R_p \ll \omega L_p$. Hence, expanding the denominator in (74), we obtain

$$L_s = \frac{\mu_0 \pi R^2 \eta^2}{L}\left(\frac{b^2}{R^2} - 1\right), \tag{75}$$

$$R_s = \eta^2 \frac{2\pi R}{\sigma L \delta_p}, \tag{76}$$

where $Z_s = R_s + j\omega L_s$. The power balance

$$P_{abs} = \tfrac{1}{2} I_{rf}^2 R_s \tag{77}$$

then yields the required rf source current, and the rf voltage is determined from

$$V_{rf} = I_{rf} Z_s. \tag{78}$$

As an example, let $R = 10$ cm, $b = 15$ cm, $L = 20$ cm, $\eta = 3$ turns, $N = 1.7 \times 10^{14}$ cm^{-3} (5 mTorr, 298 K), $\omega = 85 \times 10^6$ s^{-1} (13.6 MHz), and $P_{abs} = 600$ watts. At 5 mTorr, $\lambda_i \approx 0.6$ cm. Then from (8) and (9), $h_L \approx h_R \approx 0.19$, and from (12), $d_{eff} \approx 17.9$ cm. For argon we then obtain from Fig. 7 that $T_e \approx 2.6$ V, and from Fig. 5, that $\varepsilon_c \approx 58$ V. Using (2), we find $\varepsilon_L \approx 77$ V. The Bohm velocity is $u_B \approx 2.5 \times 10^5$ cm/s, and from (20), $A_{eff} \approx 350$ cm^2. Then from (19), we obtain $n_0 \approx 5.6 \times 10^{11}$ cm^{-3}. Estimating v_c for argon from Fig. 4, we find $v_c \approx 1.4 \times 10^7$ s^{-1}, and from (65) with $n_s = h_L n_0$, $\sigma \approx 214$ mho/m. Using (62), we obtain $\delta_p \approx 17$ mm. Evaluating (75) and (76), we find $R_s \approx 7.8$ ohms and $L_s \approx 2.2\,\mu$H, such that $\omega L_s \approx 190$ ohms. Equations (77) and (78) then yield $I_{rf} \approx 12.4$ A and

$V_{rf} \approx 2360$ V. The high inductive voltage required for this three-turn source can be supplied from a 50 ohm rf power source through a capacitive matching network.

D. Low-Density Operation and Source Efficiency

Since the dc conductivity $\sigma \propto n_0$, and $\delta_p \propto n_0^{-1/2}$, it is apparent from (75) and (76) that at fixed driving current I_{rf}, we have the scaling

$$P_{abs} \propto n_0^{-1/2}. \tag{79}$$

However, at low densities, such that $\delta_p \gg R$, the conductivity is low and the fields fully penetrate the plasma. In this case, applying Faraday's law to determine the induced electric field E_θ within the coil, we obtain

$$E_\theta(r) = \tfrac{1}{2}j\omega r \mu_0 \eta I_{rf}/L,$$

and, writing $J = j\omega\varepsilon_0\kappa_p E$ for $v_c \ll \omega$, we have $J \propto n_0 r I_{rf}$. Evaluating the power absorbed for this case,

$$P_{abs} = \frac{1}{2}\int_0^R \frac{J^2(r)}{\sigma} 2\pi r L\, dr,$$

yields

$$P_{abs} \propto n_0 \tag{80}$$

in this low-density regime. Comparing (79) and (80), we see that for fixed I_{rf}, P_{abs} versus n_0 must have a maximum near $\delta_p \sim R$, as sketched in Fig. 27 for several different values of I_{rf}. Now consider the power balance requirement (19), which is plotted as the straight line in the figure. The intersection of the line with the curves defines the equilibrium point for discharge operation. We see that inductive source operation is impossible if the source current I_{rf} is below some minimum value I_{min}. In this regime, any discharge must be capacitive. The required I_{min} is similar to the required minimum value of P_{inc} to sustain an ECR discharge, as shown in Fig. 15.

Let us note that the driving coil (primary of the transformer shown in Fig. 26) has some resistance R_{coil}. Hence, even if the discharge is extinguished ($n_0 = 0$), there is a minimum power $P_{inc} = \tfrac{1}{2}I_{min}^2 R_{coil}$ supplied by the source before the inductive discharge can form.

Because $P_{abs} \propto n_0^{-1/2}$ at high densities, we see from Fig. 27 that the power transfer efficiency P_{abs}/P_{inc} falls continually as n_0 is increased, hence limiting source operation at very high densities because of power

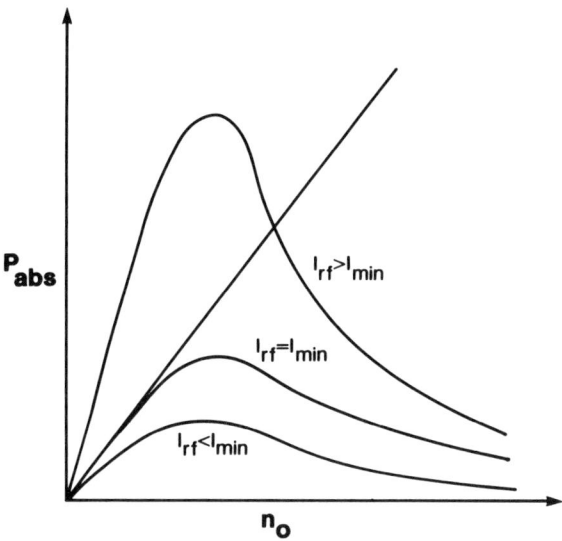

FIG. 27. Power versus density from the inductive source characteristics (curves) and from the plasma power balance (straight line). The curves are drawn for different values of the driving current I_{rf}.

supply limitations. As pointed out by Piejak et al. (107), the poor power transfer to the plasma at very low and at very high densities is analogous to the well-known property of an ordinary transformer with an open and a shorted secondary winding. In both cases no power is dissipated in the load (here the plasma), but in both cases there is power dissipated in the primary winding (here the coil) because of its inherent resistance. Piejak et al. (107) have given a complete analysis of an inductive discharge in the high-pressure regime in terms of measurable source voltages and currents based on this analogy.

We see that $\delta_p \sim R$ is the preferred operating regime for maximum power efficiency in the low-pressure regime. In fact, at high pressure, Thomson (148) obtained the similar condition $\delta_c \approx 0.57R$ for maximum power efficiency in a uniform density source.

Other issues of inductive source operation include finite geometry effects $(L \sim R)$, planar coil source operation, collisionless heating at low pressures, capacitive operating mode and startup, and self-resonant coil effects due to stray coil capacitances. Some of these issues are addressed in the literature (107, 151, 155, 156).

VI. Helical Resonator Discharges

While helical resonators have long been used as electronic circuit elements (*157*), they have only recently been used for efficient plasma generation at pressures as low as 10^{-5} Torr. High pressure discharges (~ 1 Torr) were first applied by Steinberg and Steinberg (*158*) and used for downstream stripping. The concept was further refined and applied to low pressures discharges (0.1–1 mTorr) by Flamm et al. (*159*) and was used by Flamm (*160*) and by Cook et al. (*18, 161*) for polysilicon gate etching and for downstream deposition of silicon dioxide and silicon nitride films.

Helical resonator plasma operate at radio frequencies (3–30 MHz) with simple hardware, do not require a dc magnetic field (as for ECRs and helicons) and exhibit high Q (600–1,500 typically without the plasma present) and high characteristic impedance (Z_0). These resonators are slow wave structures, supporting an electromagnetic wave propagating along the z axis with phase velocity $u_{\text{ph}} \ll c$. As shown in Fig. 28, the source consists of a coil surrounded by a grounded coaxial cylinder. The composite structure becomes resonant when an integral number of quarter waves of the rf field fit between the two ends. When this condition is satisfied, the intense electromagnetic fields within the helix can sustain a plasma with negligible matching loss at low gas pressure.

As with inductive discharges, which they resemble, helical resonator discharges can be operated in two regimes:

(a) capacitive coupling, low power, low density and high plasma potential;
(b) inductive coupling, high power, high density and low plasma potential.

In the capacitive regime, the discharge is driven by rf current flowing through the plasma from the high-voltage to the low-voltage end of the coil, and energy transfer to the plasma electrons is through the $\tilde{E}_r$ and $\tilde{E}_z$ fields. This regime is similar to that of an rf diode.

The inductive regime is of major interest for high-density plasma assisted materials processing. To force operation in this regime, an electrostatic shield can be added between the helix and the plasma column to reduce the capacitive coupling to a negligible value. The shield is typically a metal cylinder slotted along z that allows the inductive field $\tilde{E}_\theta$ to penetrate into the plasma, while shorting out the capacitive $\tilde{E}_r$ and $\tilde{E}_z$ fields.

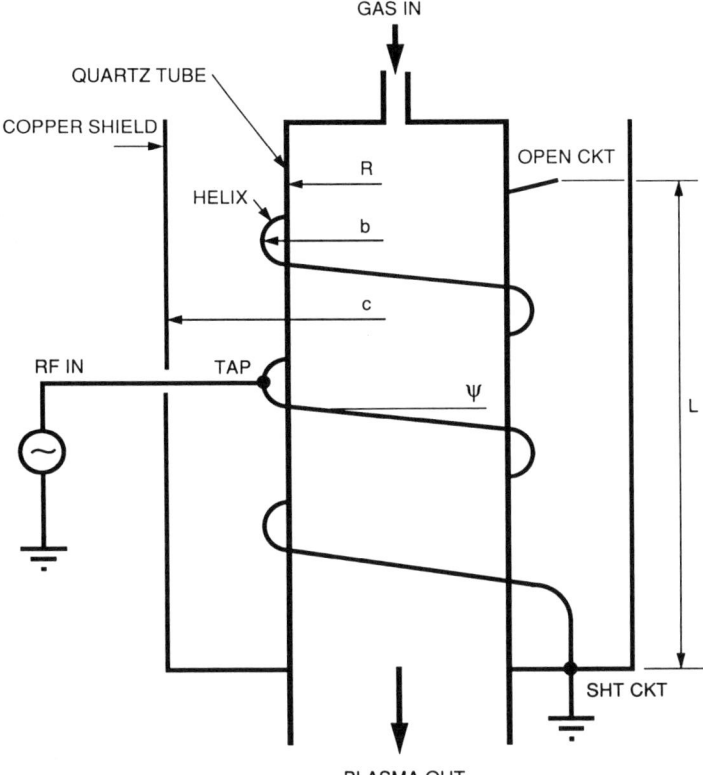

FIG. 28. Schematic of a helical resonator plasma source.

There have been few fundamental studies of helical resonator discharges (*162,163*). There is one commercial manufacturer (*164*) of sources for research and advanced development applications. Sources as large as 25 cm in diameter are available, driven by up to 5 kW of rf power at 13.56 MHz and producing plasmas with densities (in Ar) exceeding $2 \times 10^{12}\,\text{cm}^{-3}$.

The basic design parameters for a helical resonator discharge consist of pressure, rf power, source length, plasma radius, helix radius, outer cylinder radius, winding pitch angle, and excitation frequency. This is a complicated system that is not well understood. A first step is to determine the helical slow-wave modes and their interaction with the plasma. This has not been done for an electrostatically shielded discharge, so we illustrate the approach for an unshielded plasma column. The dispersion equation k_z versus ω, and the relationship among the field quantities, can be found in the approximation of a uniform, collisionless ($\omega \gg \nu_c$) plasma having relative dielectric constant $\kappa_p = 1 - \omega_{pe}^2/\omega^2$ (see (60)) by using a "developed sheath helix" model, in which the rf current in the helical wires is replaced by a continuous current sheet ("sheath") and the cylindrical (r, θ, z) geometry is unfolded into a rectangular (x, y, z) geometry ("developed"). This is a standard analytical technique for treating helical systems (165) that retains most essential physics. The details of the calculation are given in Lieberman et al. (162). In the absence of the plasma, it can be shown that there is a single mode that tends to propagate along the helical wire, as expected for this two-conductor transmission line system.[3] For typical source parameters, $R = 3$ cm, $b = 5$ cm, $c = 10$ cm, $L = 30$ cm, and $\psi = 0.1$ radians, where the parameters are defined in Fig. 28, we determine the propagation at low plasma density $n_0 = 10^9$ cm^{-3}, at high density $n_0 = 10^{11}$ cm^{-3}, and without plasma $n_0 = 0$. Figure 29 gives k_z versus $f = \omega/2\pi$ with n_0 as a parameter. For comparison, the upper line shows a wave following the geometrical helix pitch,

$$k_{zh} = \frac{\omega}{c \tan \psi}, \qquad (81)$$

and the lower line shows a light wave $k_{z0} = \omega/c$. Without a plasma, there is only one mode of propagation, with k_z somewhat smaller than k_{zh}; i.e., the wave velocity ω/k_z is somewhat larger than $c \tan \psi$. As n_0 increases, the wave speeds up, and as $n_0 \to \infty$, $\omega/k_z \to c$. For this "coax" mode, at large n_0, the plasma is at a high voltage with respect to the outer cylinder.

A second "helix" mode appears when n_0 is such that $\omega_{pe} > \omega$, a condition that is always met for typical discharge operation. Hence, both modes coexist during typical operation. The wave velocity for the second mode is always smaller than the helix velocity $c \tan \psi$. The mode appears as a resonance $k_z \to \infty$ at n_0 such that $\omega_{pe} = \omega$, and the wave speeds up as n_0 increases. For the helix mode at large n_0, the plasma and outer cylinder are at nearly the same voltage, and the helix is at a high voltage with respect to them both. We expect to see two modes in the high-density limit because the plasma acts like a conducting cylinder;

---

[3]This is analogous to wave propagation in a coaxial line, with the helical coil being the inner conductor and the grounded cylinder being the outer conductor.

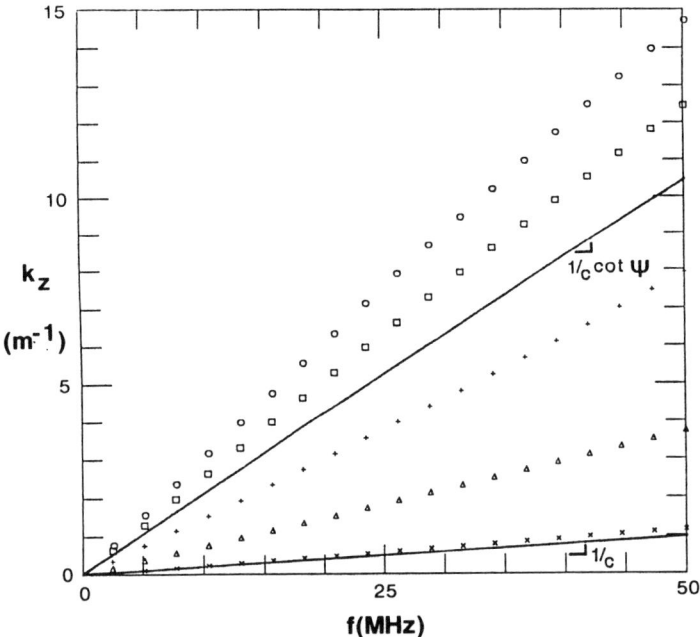

FIG. 29. Axial wavenumber k_z versus frequency f for the coax and helix modes, with density n_0 as a parameter. (○) helix mode, $n_0 = 10^9$ cm^{-3}; (□) helix mode, $n_0 = 10^{11}$ cm^{-3}; (+) coax mode, $n_0 = 0$; (△) coax mode, $n_0 = 10^9$ cm^{-3}; (×) coax mode, $n_0 = 10^{11}$ cm^{-3}.

hence, we have a three-conductor transmission line system in this limit.

At high densities, the axial wavenumbers for the two modes are very different. For example, at $n_0 = 10^{11}$ cm^{-3}, $k_z(\text{coax}) \approx 0.5$ m^{-1} and $k_z(\text{helix}) \approx 5.5$ m^{-1}. Since the source length L is chosen to be roughly a quarter wavelength at the helix geometrical pitch, $k_{zh}L \approx \pi/2$, the coax mode is not resonantly excited [$k_z(\text{coax}) \ll k_{zh}$]. However, this mode does play a role in source operation at startup. During typical source operation, only the helix mode is resonant, and it dominates the source operation.

As an example, for $L = 30$ cm and $k_zL = \pi/2$, we obtain $k_z \approx 5.2$ m^{-1}. Then from (81), we estimate the resonant frequency $f \approx 25$ MHz at high densities. Other methods for estimating the frequency are also available (166). In particular, end effects can change the resonant frequency.

Once the resonant frequency for quarter-wavelength operation is determined, then the fields within all regions inside the helical resonator can be found. The characteristic impedance of the helical transmission line can then be found:

$$Z_0 = \frac{1}{3}\tilde{\zeta}\left(\frac{\mu_0}{\varepsilon_0}\right)^{1/2}\frac{\eta b}{L}, \qquad (82)$$

where η is the number of helical turns, and $\tilde{\zeta}$ is a geometrical factor of order unity. Typically, $Z_0 \sim 1{,}000$ ohms. From the fields in the plasma, the absorbed power can be found. This has been done for a quasi-static field approximation in the capacitively coupled regime where both ohmic and stochastic heating (see Section II.B) contribute to the power absorbed by the electrons (162). The calculation has not been performed for the inductive regime, where only $\tilde{E}_\theta$ contributes to the absorbed power. However, the analysis should be similar to that used for conventional inductive sources (see Section V).

Power can be simply coupled from an external generator to the resonator, and the condition for a match (critical coupling) can be estimated approximately from a perturbation analysis. Consider the rf generator and its transmission line to have characteristic impedances Z_S, with one side of the transmission line connected to the helix at the tap position z_T and the other side connected to the outer shield, as shown schematically in Fig. 28. Since the helix characteristic impedance Z_0 given in (82) is typically large compared to Z_S, we expect a match to occur with the tap made near the shorted end of the helical resonator, where the voltage is small and the current is large.

From perturbation theory, the conductance seen at the position of the tap is

$$G_T = \frac{2P_{abs}}{|V_T|^2}, \qquad (83)$$

where P_{abs} is the total rf power absorbed and

$$V_T = V_m \sin k_z z \qquad (84)$$

is the helix voltage at the tap. For a match we require $G_T = Z_S^{-1}$. Substituting (84) in (83) and expanding for $k_z z_T \ll 1$, we obtain

$$V_m^2 k_z^2 z_T^2 = 2P_{abs} Z_S. \qquad (85)$$

For our example with $k_z = \pi/(2L) \approx 0.052\,\text{cm}^{-1}$, and with $P_{abs} \approx 51$ W, $V_m \approx 1{,}610$ V and $Z_S = 50$ ohms, we obtain $z_T \approx 0.85$ cm, corresponding to a tap between one and two turns.

VII. Surface Wave Discharges

Electromagnetic surface waves that propagate along a cylindrical plasma column can be efficiently absorbed by the plasma, hence sustaining a discharge. Surface waves, which are propagating modes having strong fields only near the plasma column surface, were first described by Smullin and Chorney (*167*) and Trivelpiece and Gould (*168*). The first surface-wave sustained discharge was operated by Tuma in 1970 (*169*). Moisan and his group at the Université de Montréal have extensively analyzed the concept and developed efficient wave launching systems over a wide frequency range (1 MHz–10 GHz). Surface wave sources have been reviewed by Moisan and Zakrzewski (*170*). Although there are some applications to materials processing (*171–173*), the absorption length α_z^{-1} for the surface modes tends to be long, such that $L \gg R$ for these discharges. Hence, they are not suitable as low-profile ($L \sim R$) sources for wide-area materials processing, and they have found wider application for ion sources, lasers and spectroscopy. Surface wave discharges having diameters as large as 15 cm have been operated, although diameters of 3–10 cm are more commonly used. The simplest sources operate without an imposed axial magnetic field. The surface wave can be made to propagate in the traveling wave mode or suffer reflections at the ends of the column, thereby increasing plasma density. In the former case, which is experimentally simpler, the sources must be driven at microwave frequencies in the range of 1–10 GHz to obtain the high densities of interest here.

The simplest description of surface wave propagation is that corresponding to a nonmagnetized plasma column of radius R confined by a thick dielectric tube (radius $b \gg R$) having relative dielectric constant κ_d. The aximuthally symmetric $m = 0$ mode has $\tilde{H}_z = 0$ and

$$\tilde{E}_z = \tilde{E}_{z0} \frac{I_0(k_{\perp p} r)}{I_0(k_{\perp p} R)} \exp j(\omega t - k_z z), \qquad r < R, \qquad (86)$$

$$= \tilde{E}_{z0} \frac{K_0(k_{\perp d} r)}{K_0(k_{\perp d} R)} \exp j(\omega t - k_z z), \qquad r > R, \qquad (87)$$

where

$$k_{\perp p}^2 = k_z^2 - k_0^2 \kappa_p, \qquad (88)$$

$$k_{\perp d}^2 = k_z^2 - k_0^2 \kappa_d, \qquad (89)$$

and where κ_p, given by (60), is the plasma relative dielectric constant. I_0

and K_0 are the modified Bessel functions of the first and second kind, $k_{\perp p}$ and $k_{\perp d}$ are the transverse wavenumbers inside and outside the plasma, respectively, and k_z is the complex axial propagation constant. We note from the form of the Bessel functions that the fields decay away from the surface of the plasma in both directions.

The transverse fields are obtained from $\tilde{E}_z$ using Maxwell's equations. In particular, we find

$$\tilde{H}_\theta = -\frac{j\omega\varepsilon_0\kappa}{k_\perp^2}\frac{\partial \tilde{E}_z}{\partial r}$$

in the two regions. The continuity of the tangential magnetic field $\tilde{H}_\theta$ then yields the dispersion equation

$$\frac{\kappa_p}{k_{\perp p}R}\frac{I_0'(k_{\perp p}R)}{I_0(k_{\perp p}R)} = \frac{\kappa_d}{k_{\perp d}R}\frac{K_0'(k_{\perp d}R)}{K_0(k_{\perp d}R)}. \tag{90}$$

From (86) and (87), it is clear that $k_\perp R \gg 1$ for the surface mode to decay rapidly. Using the asymptotic expansions of the Bessel functions $I_0'/I_0 = 1$ and $K_0'/K_0 = -1$, we obtain

$$\kappa_p k_{\perp d} = -\kappa_d k_{\perp p}. \tag{91}$$

Substituting (88) and (89) into (91) and solving for k_z yields

$$k_z = k_0\left(\frac{\kappa_p \kappa_d}{\kappa_p + \kappa_d}\right)^{1/2}. \tag{92}$$

For the case of no loss, $\nu_c = 0$, we obtain from (92) that

$$k_z = \kappa_d^{1/2}\frac{\omega}{c}\left(\frac{\omega_{pe}^2 - \omega^2}{\omega_{pe}^2 - (1 + \kappa_d)\omega^2}\right)^{1/2}.$$

Figure 30 shows k_z versus ω for the lossless case. Note that in surface wave plasmas, ω_{pe} varies with z while ω is constant. We see that k_z is real for $\omega \leq \omega_{res}$, where $\omega_{res} = \omega_{pe}/(1 + \kappa_d)^{1/2}$ gives the resonance $k_z \to \infty$ of the surface wave. For $\omega \ll \omega_{res}$, we see that $k_z \approx \omega/c$. However, in this low-frequency limit, the ordering $k_\perp R \gg 1$ is not valid, and the complete dispersion equation (91) must be solved numerically. The result is similar to that shown in Fig. 30. Under traveling wave conditions, the minimum value of ω_{pe} (obtained at the end of the column) is close to ω_{res}. Hence, for high-density sources, the frequencies of interest are above 1 GHz—i.e., microwave frequencies.

Fixing ω for the source, we introduce the resonance value of the

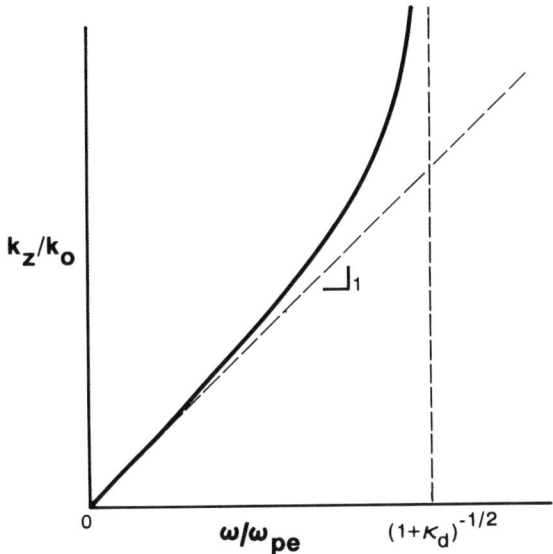

FIG. 30. Surface wave dispersion k_z versus ω for the $m = 0$ mode.

density $n_{\text{res}} = \varepsilon_0 m \omega^2 (1 + \kappa_d)/e^2$. Then the surface wave propagates for densities $n_0 \geq n_{\text{res}}$. The source operation for the usual case of a long, thin source, $L \gg R$, follows from the general principles described in Section II.A. In particular the local power balance along z determines the density n_0 for a given absorbed power P'_{abs} per unit length along the column, as in the derivation leading to (19). Letting P_w be the power carried by the wave along the column at the position z, at which the density is n_0, then

$$P'_{\text{abs}}(n_0) = 2\alpha_z(n_0) P_w, \qquad (93)$$

where α_z is the axial attenuation constant of the wave fields at the density n_0. Equating P'_{abs} to the power P'_{lost} lost per unit length,

$$P'_{\text{lost}}(n_0) = e n_0 u_B A'_{\text{eff}} \varepsilon_L, \qquad (94)$$

where $A'_{\text{eff}} = 2\pi R h_R$ is the effective (radial) loss area per unit length, we obtain $n_0(z)$ for a given wave power $P_w(z)$.

The mode attenuates as it propagates along z because of a nonzero v_c. Letting $v_c \ll \omega$ in (60), substituting this in (92), and taking the imaginary part, we obtain the attenuation constant $\alpha_z(n_0) = -\operatorname{Im} k_z$ at a fixed ω. The expression is complicated and we give only the scaling for n_0 greater than, but not too near, resonance:

$$\alpha_z \propto \frac{n_0 v_c}{(n_0 - n_{res})^{3/2}}. \tag{95}$$

At resonance, Eq. (95) is not valid and there is a finite α_z, while for $n_0 < n_{res}$, the wave does not propagate and the absorbed power falls sharply. For this variation of α_z, P'_{abs} is plotted versus n_0 for several different values of P_w in Fig. 31. The linear variation of P'_{lost} given by (94) is also plotted on the figure. The intersection of P'_{abs} with P'_{lost} determines the equilibrium density along the column. It can be seen that there is a minimum value $P_{wmin}(z)$ below which a discharge at that z cannot be sustained. A discharge forms near the position of surface wave excitation $z = 0$ for $P_{wmax} > P_{wmin}$. As the wave propagates, P_w attenuates along z because of wave absorption. A discharge cannot be sustained when P_w falls below P_{wmin} at $z = z_{max}$. Hence, the discharge exists as a finite length plasma column over $0 < z < z_{max}$. Typical plasma column variations of n_0 and P_w are shown in Fig. 32.

We note in Fig. 31 that there are generally two intersections of $P'_{abs}(n_0)$ with $P'_{lost}(n_0)$. The lower-density intersection is an unstable equilibrium because a fluctuation that decreases n_0 leads to $P'_{abs} < P'_{lost}$,

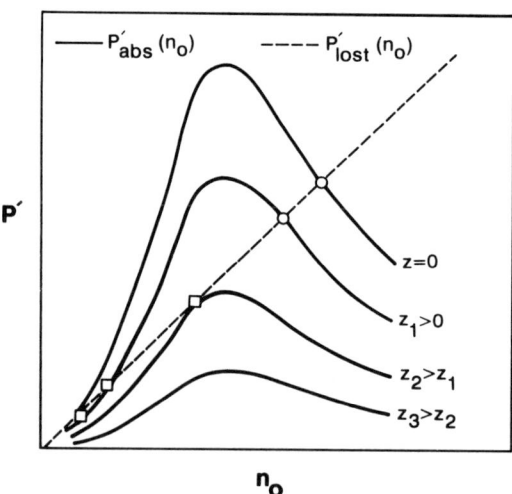

FIG. 31. Illustrating the determination of the equilibrium density in a surface wave discharge. The high-density intersection of the P'_{abs} and P'_{lost} gives the equilibrium density. Reprinted from Ref. 170 with permission.

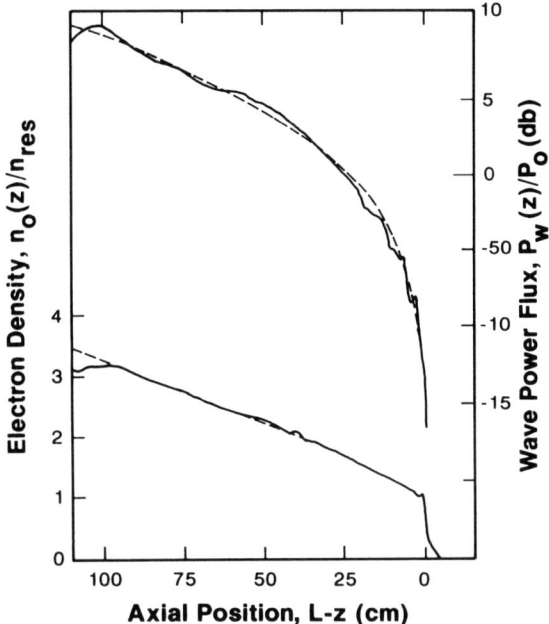

FIG. 32. Density n_0 and wave power P_w versus z for a typical surface wave source. Reprinted from Ref. 170 with permission.

thus further decreasing n_0. The higher-density intersection is stable by similar reasoning.

High-power wave launchers and matching networks for surface wave discharges have been developed (*170*). The addition of an axial magnetic field allows further choice of propagating modes (*174*), but has not been applied to materials processing.

VIII. Plasma Transport

In the preceding sections, we have outlined simple, unifying, analytical theories with which plasma source design can be understood and the generation rates and densities of ions and electrons can be quickly estimated. This analysis could be extended by using the electron temperature from (11) and suitable cross-sections to estimate generation and loss rates, and thereby densities, for reactive neutral species. We now turn to plasma transport: effects on materials processes, measurements, and strategies for control.

Consider etching. We can understand the influence of plasma transport on etching rates by reviewing a simple model proposed by Mayer and Barker (*175*) to explain the so-called ion–neutral synergistic effect (*176*,177). The etching rate may be expressed as the ion flux times the average volume removed per ion. Alternatively, one may express the etching rate as the neutral flux times the reaction probability times the volume removed per reaction event. Such rate expressions, however, cannot be directly applied because the yield per ion and the reaction probability of neutrals both depend on the neutral and ion fluxes. This linkage can be broken by expressing rates in terms of the ion- and neutral-flux–dependent surface coverage of reactants. Assuming that the yield per ion is proportional to the ion energy times the surface coverage of the chemically assisting neutral species, the etching rate is given by

$$\text{ER} = v_i \Theta \varepsilon_i \Gamma_i, \qquad (96)$$

where v_i is the volume removed per unit bombardment energy (cm^3/eV) for a saturated surface, and Θ is the surface coverage. For this simple model, we assume that v_i is independent of ε_i. To complete the model, we assume Langmuir adsorption kinetics where the reactive sticking probability of neutrals is proportional to the number of bare sites on the surface. Thus, the etching rate is also given by

$$\text{ER} = v_n SC_0 (1 - \Theta) \Gamma_n, \qquad (97)$$

where v_n is the volume removed per etching neutral (cm^3), SC_0 is the reactive sticking probability on a bare surface (unitless), and Γ_n is the neutral flux (cm^{-2}s^{-1}) to the surface. Equating the preceding rate expressions, we obtain an expression for the surface coverage as a function of the ion energy flux to neutral flux ratio:

$$\Theta = \frac{1}{1 + v_i \varepsilon_i \Gamma_i / (v_n SC_0 \Gamma_n)} \qquad (98)$$

Substituting this into (96), we obtain our final expression for the etching rate as a function of the ion and ion and neutral fluxes:

$$\text{ER} = \frac{v_i \varepsilon_i \Gamma_i}{1 + v_i E_i \Gamma_i / (v_n SC_0 \Gamma_n)}. \qquad (99)$$

Ion–neutral synergy is clearly evident in (99): If the neutral flux is negligible, $v_n SC_0 \Gamma_n = 0$, the etch rate becomes vanishingly small (we have neglected sputtering); similarly, when the ion energy flux is negligible, $\varepsilon_i \Gamma_i = 0$, the etch rate again vanishes (we have neglected thermally activated neutral etching). Thus, the total etch rate with both neutrals and ions is greater than the etching rates with either species alone (176,177). The validity of (99) has been verified for many material/ion/assisting-species combinations, including Si/He$^+$/Cl$_2$ (178), Si/Ne$^+$/Cl$_2$ (178), Si/Cl$_2^+$/Cl$_2$ (179), Si/Ar$^+$/Cl$_2$ (175,178,180), SiO$_2$/CF$_x^+$/CF$_4$ (179,181), and organic polymer/O$_2^+$/O$_2$ (182). The measured dependence of the surface coverage on the ion to neutral flux ratio has also confirmed the predictions of (98) for the Si/Ar$^+$/Cl$_2$ system (175,180).

There are two interesting limiting cases of (99). When $v_n SC_0 \Gamma_n \gg v_i \varepsilon_i \Gamma_i$, the surface is saturated with neutrals and the etching rate ER = $v_i \varepsilon_i \Gamma_i$ depends only on the ion energy flux and is independent of the neutral flux as assumed, for example, by Shaqfeh and Jurgensen (183). In this case, ion generation and transport in the plasma will govern etching rates and etching uniformity. In the opposite case where $v_n SC_0 \Gamma_n \ll v_i \varepsilon_i \Gamma_i$, the surface is starved for neutrals and the etching rate ER = $v_n SC_0 \Gamma_n$, is proportional only to the flux of assisting neutral species. Which regime is dominant? That is dictated in large part by source design and operation.

The ion–neutral synergism is useful for understanding rate enhancement when both ions and neutrals impinge on surfaces. It is also easy to understand, then, the origin of anisotropy in plasma etching: Because ion transport is anisotropic, etch rate enhancement via the ion–neutral synergism is also anisotropic. However, other factors must also be considered. For example, the expression for the etching rate just given ignores the possibility that there is a threshold energy below which the reactive ion sputter yield v_i is negligible (184,185). Such an effect is clearly evident in the Cl$_2$ helicon etching of SiO$_2$ as a function of bias shown in Fig. 33 and can play a role in determining selectivity of etching one material with respect to another. Similarly, damage created by atomic displacement when an energetic ion impinges on a surface may be minimized or even eliminated by tuning ion energies below the damage threshold. In general, the threshold energies for etching and damage are poorly known, but one of the guiding principles in and major motivations behind new plasma source development is that ion energies must be "fine" tuned to take advantage of threshold effects.

A. THE ION ENERGY DISTRIBUTION FUNCTION

In Section II.A.2 we estimated the mean ion energy and mean ion velocity at the sheath edge as a function of electron temperature, source

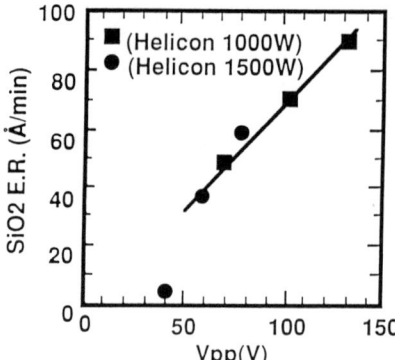

FIG. 33. Etch rate of SiO_2 as a function of peak-to-peak bias voltage in helicon discharge through chlorine. A threshold is clearly observed near 40 V (184). Figure kindly provided by and reprinted with permission from Plasma Materials and Technology.

design, and rf biasing. However, these simple scaling relationships do not provide estimates of the ion energy distribution function, iedf, or spread in ion energy. This is by no means a trivial matter. What good is it, for example, to fine tune ε_i to just above threshold for etching poly-Si and just below threshold for etching SiO_2 when the breadth of the iedf is five times broader than its mean?

In equilibrium, the iedf is characterized simply by the ion temperature, which, for convenience, we generalize to $T_i \equiv \frac{1}{3} M \langle (u_i - \langle u_i \rangle)^2 \rangle / e$ (in volts) for use in describing the iedf in nonequilibrium systems. In the magnetized, low-pressure, high-efficiency plasmas considered here, the iedf is frequently anisotropic so that different "temperatures" $T_{i\perp}$ and T_{iz} are needed to characterize the iedf perpendicular and parallel, respectively, to the magnetic field axis. Even in unmagnetized plasmas, one expects an anisotropic iedf as ions are accelerated along a specific direction. The ion velocity distribution function, ivdf, may also depart significantly from a Maxwellian form, in which case the temperatures just defined are inadequate to describe the distribution function: Higher order moments are also required.

Source design and operation determine the iedf. Ions gain directed energy by acceleration in an electric field, created, for example, by an expanding magnetic field, by sheaths at bounding surfaces, by double layers and striations separating different plasma regions, by pre-sheaths, and by applied bias voltages (Section II.A.2). Ions lose energy by collisions, primarily with neutral atoms and molecules. Two collisional processes are dominant: charge exchange and elastic scattering. Charge exchange is effective in transforming hot, fast ions into hot, fast neutrals and cold, slow neutrals into cold, slow ions. Elastic scattering is effective in transforming directed energy gained by the ions falling through a potential into random energy with a large component perpendicular to the strong electric field. Both processes tend to broaden the iedf and the ion angular distribution function, iadf, thereby possibly compromising selectivity, anisotropy, rate, and damage control.

Besides collisional energy transfer, the iedf can also be broadened (so that precision energy control is lost), as a result of ionization occuring over a distributed sheath created, for example, by plasma expansion or by fluctuations in the plasma potential caused by power source fluctuations or inherent plasma instabilities (*114*).

Another limit to ion energy control in plasma processing has been little considered until recently (*143*): the creation and buildup of energy in excited states. All atomic ions have metastable excited electronic states that typically can pool energies of 10 eV or more above the ground state. Recent experimental work indicates that at least 25% of the ions in a low-pressure argon helicon plasma are in excited states with more than 16 eV energy above the ground state ion (*143*). Molecular ions can also effectively pool energy in vibrational and rotational modes as well as in excited electronic states. The pooling of energy in excited states should be most important in low-pressure, high-density plasmas where deactivation by collision with neutrals is relatively rare compared to frequent production by collision with electrons. How important 10 eV or more of excess, internal energy can be in etching and deposition applications must depend on the specific materials involved, but is basically an open question for which little or no data exist. Besides metastable states, in high-density plasmas one must also anticipate the production and buildup of large concentrations of doubly ionized ions. These will impact device surfaces with at least two times the kinetic energy of their singly ionized counterparts. To make matters even more complicated, metastable and doubly charged ions may have significantly different iadfs, thereby making CD as well as selectivity and damage more difficult to control.

1. *Ion Transport and Etching Anisotropy*

It is instructive to consider the consequences of a finite iadf on etch anisotropy in the regime where the ion energy flux is rate limiting, i.e., $v_n SC_0 \Gamma_n \gg v_i \varepsilon_i \Gamma_i$. Ions impinging on the surface at oblique angles may accelerate etching of the sidewall and compromise pattern transfer fidelity. A useful metric for describing the iadf is the normalized, energy-flux–weighted, cumulative angular distribution function (66, 183):

$$C(\theta) \equiv \frac{\int_0^\theta \int_0^\infty d\theta \sin\theta\, d\varepsilon_i \cos\theta \gamma(\theta, \varepsilon_i)\varepsilon_i(\theta)}{\int_0^{\pi/2} \int_0^\infty d\theta \sin\theta\, d\varepsilon_i \cos\theta \gamma(\theta, \varepsilon_i)\varepsilon_i(\theta)}, \qquad (100)$$

where θ is the polar angle from the surface normal and $\gamma(\theta, \varepsilon_i)\cos(\theta)$ is the differential ion flux normal to the surface. Physically, $C(\theta)$ is proportional to the etching rate of a surface perpendicular to the accelerating electric field and shadowed by a cone of half angle θ when the etching is ion energy-flux–limited (183). Thus, measurement or calculation of $C(\theta)$ is necessary for accurate simulation and prediction of etched profiles. $C(\theta)$, in turn, is determined by the transport of ions to and then through the sheath to the wafer surface.

Since the sheaths are so thin in high-efficiency plasmas (Section II.A.2), transport of ions through the sheath is effectively collisionless: $s/\lambda_i \ll 1$. This is one of the characteristics of high-density plasmas that distinguishes them from the conventional rf diode and is a direct consequence of the high charge density and lower rf bias voltages at lower pressures. Because the sheath is collisionless, the perpendicular ion velocity at the sheath edge is preserved as the ions are accelerated to the wafer surface. Thus, $T_{i\perp}$ largely determines the anisotropy of the ion transport and the angular dependence of $C(\theta)$. Consider a simple example, where the spread in u_{iz}, ion speed normal to a surface in the xy plane, is negligible compared to the velocity gained by acceleration across the sheath, $u_{is} \equiv (2eV_s/M)^{1/2}$, and the distribution of perpendicular velocities, $f(u_{i\perp})$, is Gaussian with temperature $T_{i\perp}$. In this case,

$$C(\theta) = \frac{\int_0^\theta d\theta \tan\theta \exp(-\beta\tan^2\theta)/\cos^3\theta}{\int_0^{\pi/2} d\theta \tan\theta \exp(-\beta\tan^2\theta)/\cos^3\theta}, \qquad (101)$$

where $\beta \equiv V_s/T_{i\perp}$. This function and its derivative are plotted in Fig. 34 for $\beta = 10$ and 100 along with the corresponding function for an isotropic angular distribution, i.e., $T_{iz} = T_{i\perp}$ and $u_{is} = 0$. Clearly, the effects of transverse ion energy can be significant, and the design of plasma sources must take into account the mechanisms by which ions gain energy transverse to the surface normal. Before discussing these design aspects, we digress to consider how measurements of the iedf, ivdf, and iadf are made.

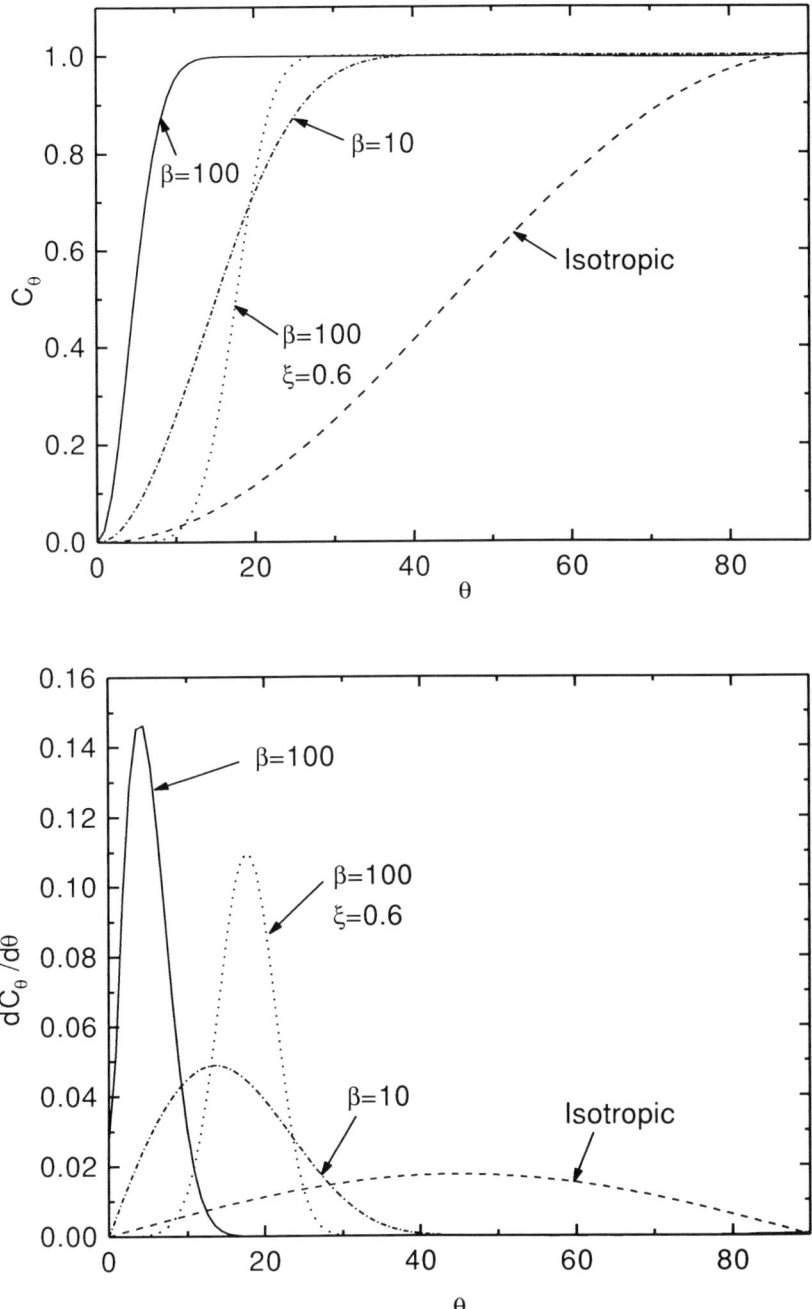

FIG. 34. Normalized, energy-flux-weighted, (a) cumulative and (b) differential angular distribution functions versus incident, polar angle.

B. METHODS FOR MEASURING ION ENERGY DISTRIBUTION FUNCTIONS

Electrostatic energy analysis and Doppler-shift spectroscopy are the two methods used most often for measuring ion energy and velocity distributions, respectively. Electrostatic energy analyzers consist of a pinhole to sample the ions, a grid at the same potential as the sampling electrode to provide a field-free drift, grids to repel electrons and accelerate ions, grids to retard ions below a cutoff energy, and a detector (Fig. 35). If desired, ions can be mass filtered first and then energy analyzed. The distribution function is obtained by differentiating the measured ion current as a function of retarding voltage. Advantages of this technique stem from its universality and simplicity. Recently, Liu *et al.* (*186*) have used an electrostatic energy analyzer with a sectored current collector to determine not only the iedf but also the iadf at each energy, albeit with fairly low angular resolution, in an rf diode reactor.

In high charge density plasmas, these measurements are more challenging than usual if quantitatively meaningful data are to be obtained. For example, the pinhole acts as an energy-dependent lens and restricts the solid angle from which ions are accepted, thus discriminating against detection of low-energy, obliquely incident ions. The usual solution to this problem is to use pinholes that are small compared to the sheath thickness, but because the Debye and sheath lengths are so short in high-density sources (see Section II.A.2), the pinhole must be $\ll 50\,\mu$m to minimize artifacts. Unless the sampling is done through the wafer

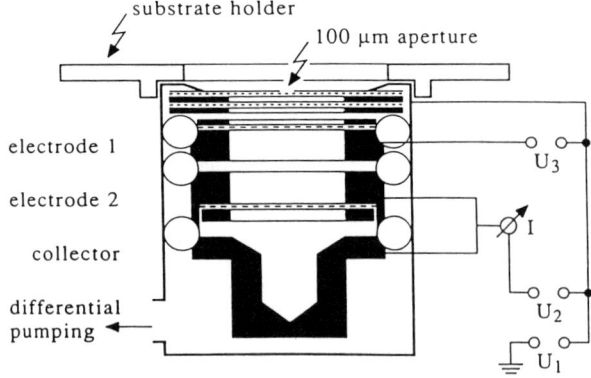

FIG. 35. Schematic illustration of gridded energy analyzer used for measurements of ion energy distribution functions. Reprinted from Ref. 199 with permission.

platen, the technique is also inherently intrusive: The sampling probes are relatively large, drawing significant current, and inducing a sheath and pre-sheath that can perturb the plasma over large distances. In a magnetized plasma, the ion sampling probe will subtend only a subset of flux tubes, resulting in perturbations that are radially isolated but that propagate large distances along the flux tubes. In strong magnetic fields, ions are easily deflected in the analyzer and extra care must be taken to ensure that angle and energy distribution functions are not distorted. Finally, ion sampling in the plasma using probes lacks good spatial resolution. Despite these deficiencies and caveats, electrostatic energy analysis is well suited for sampling through the wafer platen, which yields the most vital information from a processing perspective: the iedf and iadf at the wafer.

Since the sheath is collisionless, however, it can be useful to measure the ivdf throughout the plasma. This can be done with great precision by using spectroscopic methods that rely on the shift in absorption or emission frequency (Doppler effect) that occurs when the ion is moving with respect to the reference frame:

$$\tilde{\sigma} = \tilde{\sigma}_0(1 - u_z/c), \quad (102)$$

where $\tilde{\sigma}$ is the Doppler-shifted absorption or emission frequency; $\tilde{\sigma}_0$ is the corresponding line-center (rest velocity) frequency; u_z is the ion velocity along the light beam propagation direction, taken here to be along the $\hat{z}$ axis; and c is the speed of light. In both absorption and emission experiments, the measured line profile as a function of frequency is given by an integral over the perpendicular velocity components, u_x and u_y and the spatial volume sampled:

$$I(v)\,dv \propto \int\int\int\int\int n_0(x,y,z) f(u_x,u_y,u_z,x,y,z)\,du_x du_y dx\,dy\,dz\,du_z, \quad (103)$$

where the limits of integration correspond to the volume sampled. Both the ion density n_0 and the ivdf f are assumed to depend on position, i.e., the plasma is not uniform. Because of the integrations in (103), Doppler-shifted absorption and emission data must be measured along many axes and then suitably transformed to obtain a truly one-dimensional ivdf at any given position (*187*). In the simplest case of cylindrical symmetry, for example, an Abel inversion of line profiles obtained from line-of-sight along a cord must be used to obtain the radial variation of the radial component of the ivdf. Without inversion, the data are of little quantitative value in distinguishing, for example, between random and directed energy.

The multiple integral in (103) is easily simplified experimentally by detecting absorption via the appearance of fluorescence, i.e., laser-induced fluorescence (LIF). In this case, the fluorescence induced by absorption of laser light is detected perpendicular to the laser propagation direction and imaged onto a spatial filter that discriminates against all but a small portion of the $\hat{z}$ axis (Fig. 36). Typical spatial resolution along $\hat{z}$ or $\hat{x}$—dictated by laser beam diameter, the magnification provided by the fluorescence collecting lenses, and the dimensions of the slits onto which the fluorescent image is projected—is 0.5–2 mm, sufficiently small for probing all but the sheath regions of low-pressure high-density discharges that the integrals over x, y, and z in (103) are eliminated. Thus, the LIF method provides a precision measurement of the ivdf, albeit still averaged over the velocity components perpendicular to the laser propagation direction,[4] at a specific point in the plasma, and

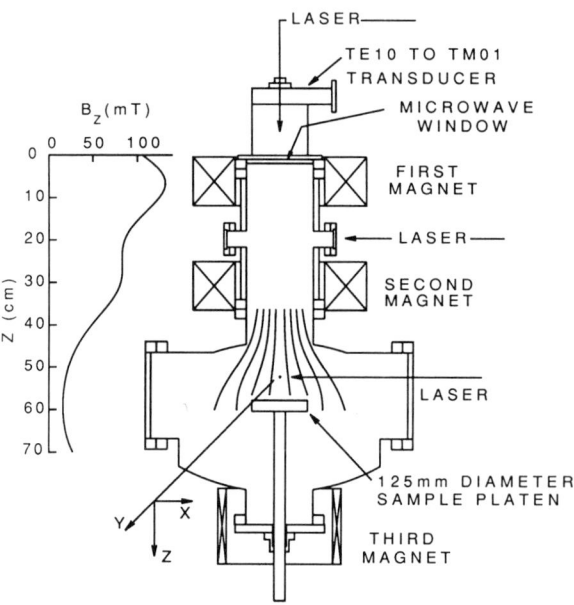

FIG. 36. Schematic illustration of apparatus used to measure spatially resolved ion velocity distribution functions by laser-induced fluorescence (101).

---

[4]To obtain ion trajectories, one must resort to optical tomography as described by Koslover and McWilliams (187). This approach, while elegant, requires optical access not readily available in processing reactors.

it thereby provides information on distributed sheaths, distributed ionization, plasma uniformity, etc.

Because ground state ions typically undergo rapid charge exchange with ground state neutrals, information can be lost concerning ion formation and transport when ground state ions are probed using LIF. The component of ion velocity measured will be dictated by the energy gained from the field since the last collision:

$$u_i \approx (2eE\lambda_i/M)^{1/2}. \tag{104}$$

Thus, such measurements tend to sample the local field. For example, den Hartog et al. (*188*) showed that above 1.0 mTorr, the drift of N_2^+ downstream from an ECR source is given by

$$u_i = \mu(E)E, \tag{105}$$

where $\mu(E)$ is the field-dependent mobility (5), dictated in large part by the charge exchange cross-section. In fact, using (105), den Hartog et al. deduced the electric field from the measured ion Doppler shift.

By contrast, consider the collisionless limit where the Doppler shift is now a measure of the energy gained, $e\Delta V$, by the ions between the point of formation and the point of observation. If ions are formed over a large enough region so that the plasma potential varies significantly, a distribution of Doppler shifts will be measured. If a net shift is observed, it means that the ions have fallen through a potential, which in turn implies, through the Boltzmann relation (7), a plasma density gradient. Of course, the collisionless limit may be realized by probing ground state ions at low pressures ($\ll 1$ mTorr) or by probing excited ionic states with smaller collisional cross-sections. Such an approach has been taken by Sadeghi et al. (*101, 189, 190*), who probed metastable ionic states of Ar and Cl. Because charge exchange is a two-electron process ($3p^4 3d \Leftrightarrow 3p^6$ in Ar) for the metastable state and a one-electron process ($3p^5 \Leftrightarrow 3p^6$ in Ar) for the ground state, it is reasonable to expect the charge exchange cross-section to be smaller for the metastable state. Furthermore, metastable states, and excited states in general, can often be quenched easily at higher pressures, for example by the nearly resonant Penning ionization process in Ar:

$$Ar_m^+ + Ar \rightarrow Ar^+ + Ar^+ + e^-. \tag{106}$$

Thus, such destructive collisions help to preserve the collisionless ivdf by selectively removing ions that have undergone collision.

C. Methods for Measuring Plasma Potentials

The variation in plasma potential caused by expansion, pre-sheath formation, distributed sheath formation, etc., can be measured either by spatially resolving ivdf's (*101*) or by using electrostatic (Langmuir, emissive, double,...) probes. It is beyond the scope of this work to discuss electrostatic probe methodology and the reader is referred to the reviews by Chen (*191*) and Hershkowitz (*192*) and the book by Swift and Schwar (*193*). Of particular note is the recent review by Godyak *et al.* (*194*), who describe the great care that must be taken when using probes to measure plasma properties in the presence of rf excitation or bias. In addition, several caveats are offered concerning the use of Langmuir probes in measuring plasma parameters in high-density, magnetized plasmas. As Chen points out (*191*), making meaningful probe measurements in magnetized plasmas is extremely difficult. An anomalously small electron to ion saturation current ratio is a clear indication that the probe characteristic is distorted by the magnetic field. When in the electron saturation current regime, the probe will tend to deplete electrons along a flux tube but not outside this flux tube, thereby giving an anomalously small electron saturation current and perturbing the discharge. It is often possible to see visually the distortion of the plasma created by insertion of a probe; in etching and deposition plasmas, it is also possible to image the probe onto the thin film being processed.

These distortions are important when interpreting the probe current–voltage characteristic in terms of electron "temperature" or the electron energy distribution function (eedf) or the plasma potential. For example, the plasma potential is routinely estimated by linearly extrapolating, on a semilog plot, the current from the electron current saturation regime into the electron retarding current regime (*191*), and similarly extrapolating linearly the current from the retarding current regime into the saturation current regime. Where the two lines intersect is commonly taken to be the plasma potential. Clearly, if the electron current is artificially suppressed for the reasons discussed earlier, the plasma potential might be in error. With these caveats in mind, we still use literature estimates of plasma potential and electron temperature determined in this way. From comparisons with recent numerical simulations (*195*) we find that the qualitative trends are useful in gaining insight into high-density plasma generation and transport.

D. Measurements of Energy Distributions and Potentials

1. Ion Acceleration Outside the Sheath

Most measurements of high-efficiency plasma iedfs have been done for diverging field ECR systems. The work of Matsuoka and Ono (*196, 197*) is typical (Fig. 37). Microwaves are launched from a cavity into a high magnetic field region so that the RHP wave propagates and then is absorbed, heating electrons in the process (Section III). Because the magnetic field continues to decrease and, equivalently, expand, the plasma expands, the plasma density decreases and an ambipolar field is created that accelerates ions along the magnetic field gradient (Section II.A.2).

At some point downstream, ions are sampled through a 50 μm pinhole and energy analyzed using two grids and a collector (Fig. 35). Although the relatively large orifice diameter and the use of arbitrary units for spatial distance makes this work of dubious quantitative value, the trends are still notable and are borne out in many other experiments (*101, 185, 188, 189, 198–201*).

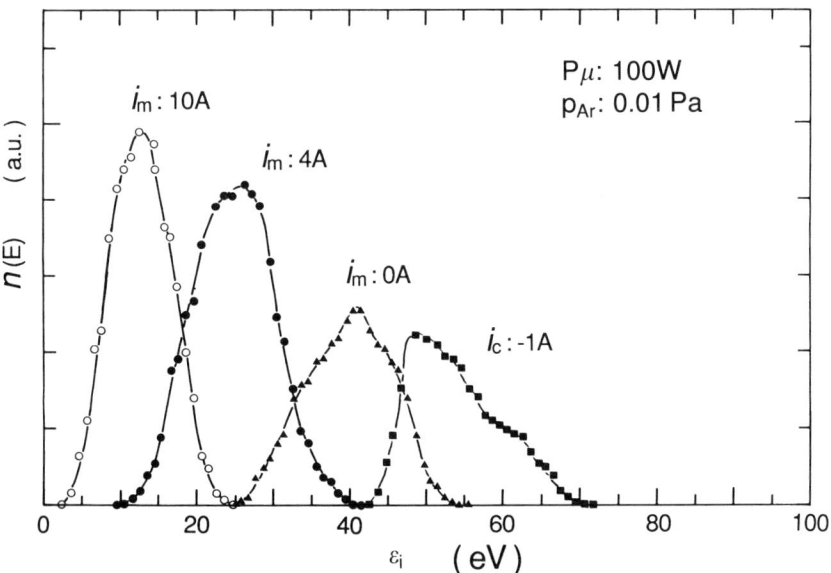

FIG. 37. Change in the bombarding ion energy distribution as the wafer-level coil current i_m is varied. Reprinted from Refs. 196 and 197 with permission.

Matsuoka and Ono focused primarily on the effects of magnetic field configuration and pressure. By varying the current in an electromagnet located near the sampling plane, they modified the divergence of the magnetic field: from a mirror to a cusp. Figure 37 shows their iedf's for different subcoil currents. As the field is collimated, T_{iz} decreases and ε_i shifts to lower values. At the same time, the ion current density increases, the plasma potential (deduced from Langmuir probe current–voltage characteristics) decreases, and the plasma potential gradient or electric field decreases. These effects are all consistent with reduced plasma expansion. The decrease in T_{iz} results from the smaller electric field so that distributed ionization no longer broadens the iedf. By contrast, the largest ε_i and T_{iz} are obtained when the subcoil magnet is used to produce a cusp before the sampling orifice. Note, however, that the iedf is clearly not Gaussian under these conditions and T_{iz} alone is not sufficient to describe the distribution function. Under these conditions, the plasma expansion is largest as the magnetic field decreases to zero and then reverses on the other side of the cusp.[5]

Regardless of the magnetic field configuration, both ε_i and T_i decrease as the pressure is increased and charge exchange cools the iedf (185, 196–199, 202, 203). Using Doppler-shifted LIF, Woods et al. (188, 200, 201) have examined the ground state N_2^+ ivdf as a function of pressure downstream from a diverging field ECR source similar to Matsuoka and Ono's and also find (Figs. 38 and 39) that ε_i and T_i decrease with increasing pressure. The LIF data further show a bimodal distribution function[6] resulting from charge exchange (204): The slow component has been created by charge exchange of fast ions with slow neutrals.

Acceleration of ions caused by plasma expansion can clearly be seen in the metastable LIF measurements of Sadeghi et al. (101, 189) (Fig. 40) as well in electrostatic analyzer data (198, 199, 203, 205). The ivdf measured where the source expands into the downstream region shows two components that are attributed to ions created in the source, the fast component, and ions created at the junction between source and reactor

---

[5]Note that ions and electrons do not follow field lines through a cusp since the field decreases to zero (189).

[6]Because of kinematic compression when the ions fall through the sheath to the pinhole, the slow and fast velocity components tend to merge after the sheath is traversed. In addition, because of the finite size of the pinhole, one must be concerned about the energy or velocity dependence to Matsuoka and Ono's detection efficiency. In any case, the iedf's reported by Matsuoka and Ono, as well as others (101, 189, 201, 203, 204), are clearly asymmetric, suggesting more than one velocity or energy component consistent with the LIF measurements.

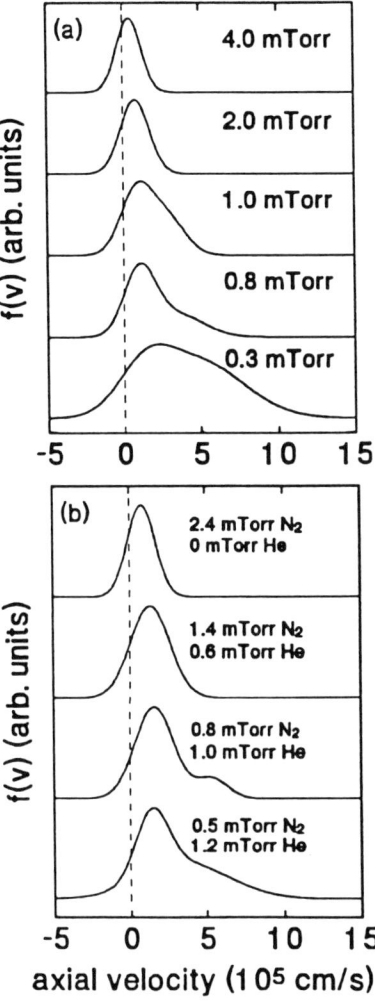

FIG. 38. Ion velocity distribution functions at several pressures downstream from (a) N_2 and (b) N_2/He ECR plasmas. Reprinted from Ref. 201 with permission.

where the plasma expansion commences. The fast component has already gained an energy of approximately 13 eV parallel to the magnetic field axis as a result of the large potential difference between the source and downstream regions caused by the plasma expansion. This effect has been simulated recently by Porteous et al. (195) using a so-called hybrid

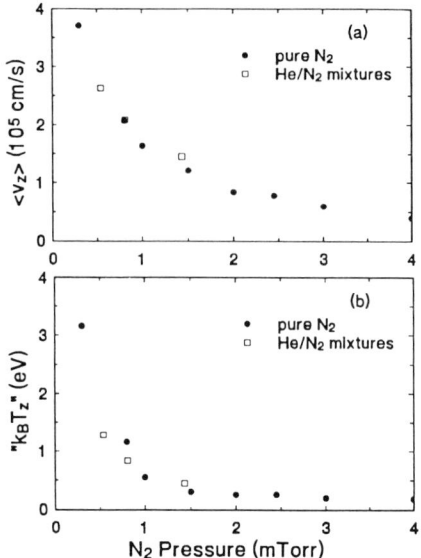

FIG. 39. (a) Average ion velocity along **B**, v_z, measured by Doppler-shifted laser-induced fluorescence; (b) corresponding average energy, "$k_B T_z$"; and (c) perpendicular ion temperature $T_{i\perp}$; all as a function of pressure downstream from N_2 and N_2/He ECR sources. Reprinted from Ref. 201 with permission.

approach where electrons are treated as a fluid and ions as particles. Their result for the plasma potential is reproduced here in Fig. 41. When a collimating magnet is used downstream as in the Matsuoka and Ono experiments, this potential drop and the corresponding energy of the fast component both decrease. While the origin of this "distributed sheath" or double layer lies with the plasma expansion, the magnitude is also affected by a neutral density gradient from the source to the reactor (*101,112,113,195*) that arises from the depletion of neutrals due to the high ionization rate in the source, as well as ion acceleration and neutral heating. It is this gradient coupled with the decreasing electron density caused by the plasma expansion that causes the secondary source of ionization at the expansion point: here, the neutral density rises rapidly while the electron density is decreasing, thus leading to a local maximum in the ionization rate, shown schematically in Fig. 42. The ground state ion formation rate should also be enhanced at this point, and these results suggest that many of the ions impacting device wafers downstream from the source will have been created downstream and not in the plasma source!

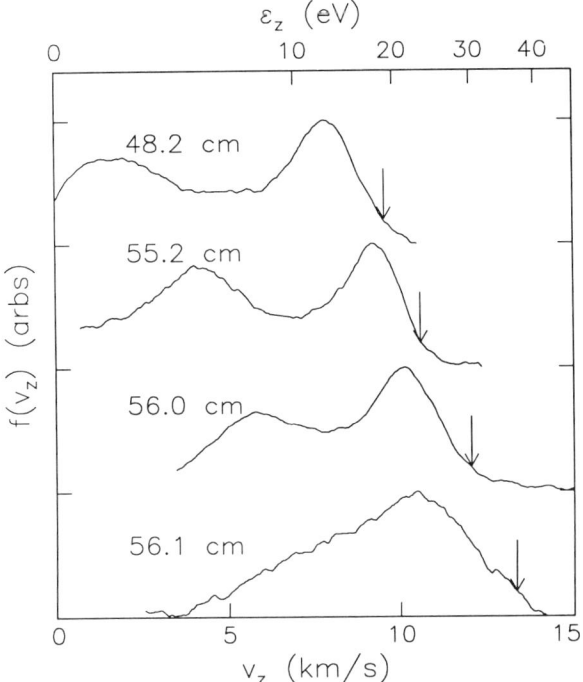

FIG. 40. Parallel ion velocity distribution functions at different positions from a wafer platen situated downstream from an Ar/He ECR source (*101*).

The effect of the wafer platen pre-sheath can also be seen in the metastable LIF data of Sadeghi *et al.* (*101, 189*) (Fig. 40). The velocities of both slow and fast components increase systematically as the distance from the source increases and the platen is approached, and, assuming collisionless transport for the metastable ions at this low pressure (0.5 mTorr), a pre-sheath electric field of ~ 0.5 V/cm is determined, in excellent agreement with recent numerical simulations (*195*).

What are the consequences of ion acceleration outside the sheath? The primary result is to broaden the ivdf normal to the surface, but if the sheath potential is large, this will have a negligible effect on etch anisotropy. This is easily seen by approximating $C(\theta)$ using a Gaussian distribution with temperature T_{iz} for the parallel component. Although the parallel distribution is clearly not Gaussian and therefore T_{iz} is not sufficient to describe the distribution, this simple analysis is useful for illustrating the effects of a finite width of the parallel ivdf. We find that

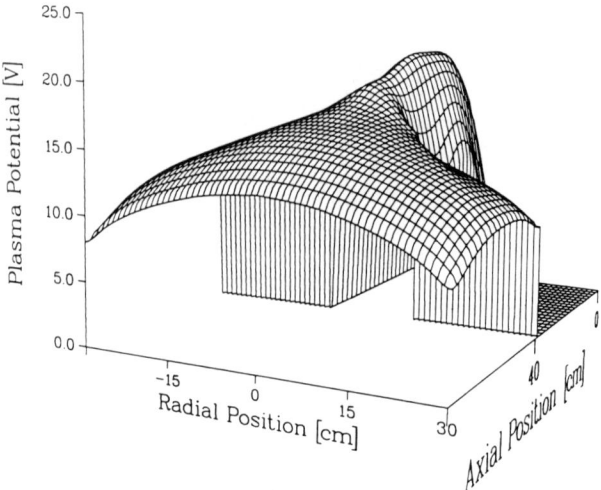

FIG. 41. Simulated plasma potential as a function of axial and radial position for an ECR source. Note the distributed sheath or potential step from the narrow source region, between axial position 0 and 40 cm, and the downstream region beyond 40 cm. Reprinted from Ref. 195 with permission.

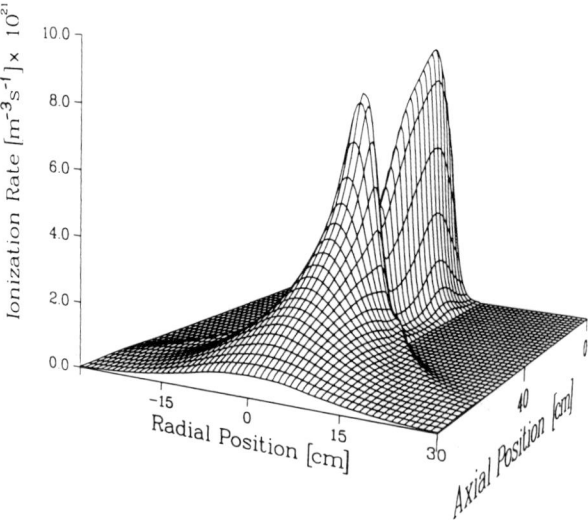

FIG. 42. Simulated ionization rate as a function of radial and axial position for the same system as in Fig. 41. Note the local maximum in the ionization rate where the plasma expands, which results from a neutral density gradient. Reprinted with permission from Ref. 195.

broadening in the parallel ivdf, for $T_{iz}/T_{i\perp} \sim 10$, has a *negligible* effect on the anisotropy. However, the broadening of the parallel ivdf can degrade etching selectivity and increase atom displacement damage (Section IX) when processes are designed to operate near threshold. Therefore, sources should be designed to minimize both T_{iz} and $T_{i\perp}$. This is most easily done by using the close-coupled configurations (Fig. 12) that eliminate acceleration caused by plasma expansion. By making the plasma as uniform as possible, broadening of the ivdf resulting from distributed ionization can be minimized. However, the pre-sheath field cannot be eliminated and to the extent that ionization occurs in this region, some residual broadening of the ivdf is unavoidable.

2. Transverse Ion Energy

Source designs that employ plasma expansion create further problems by broadening the perpendicular ivdf. This is clearly seen in the spatially resolved ivdf's measured using the LIF technique (Fig. 43): Both ground state and metastable state ions show systematic Doppler shifts with increasing distance radially from the center axis of the source (*101,189,201*). These shifts correspond to acceleration along a radial electric field that in turn is created because of the plasma density gradient in the radial direction. Plasma expansion aggravates the magnitude of this density gradient and the corresponding potential gradient. Note that ions have little trouble crossing magnetic field lines since they are only weakly magnetized.

By collimating the downstream field, the shift in the ivdf is clearly reduced (Fig. 43), although, curiously, the broadening of the ivdf can increase as a result of local ionization. By adding the collimating field and increasing the confinement downstream, the ionization rate downstream is also increased. But ions created downstream have not been accelerated to the same extent radially as those ions that stream out of the source. Thus, collimating the field and increasing the confinement of plasma produces a slow, somewhat isotropic velocity component downstream.

Large radial density gradients and the resultant radial acceleration of ions can severely affect etching anisotropy, particularly on the outer edges of a wafer where the radial velocity component is largest (Fig. 43). This again is illustrated by calculating $C(\theta)$ using a shifted Gaussian distribution for the $u_{i\perp}$ velocity component (Fig. 34): Besides β, we need another scaling parameter, $\xi = 2u_{\perp}^{(0)}/u_{is}$, or the ratio of the directed radial velocity to the velocity gained by acceleration through the sheath. In

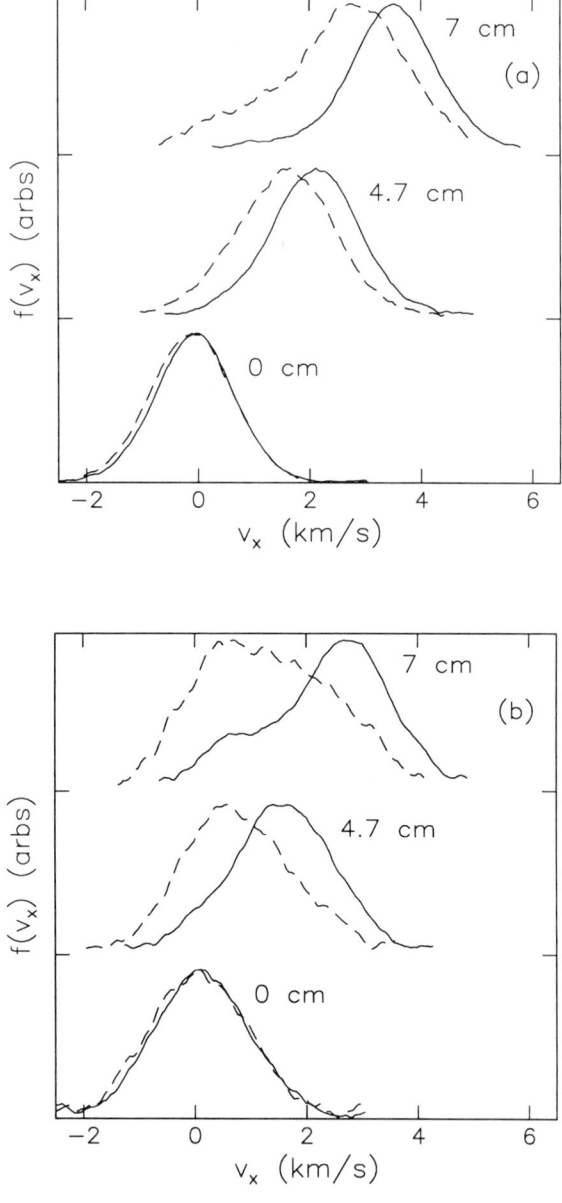

FIG. 43. Perpendicular ion velocity distribution functions at different radial positions (101). When a collimating magnetic field is applied (dashed curves), the mean ion velocity is reduced but the spread in ion energies increases. (a) 0.38 mTorr; (b) 0.82 mTorr.

Fig. 34, we have chosen $\beta = 100$ as before and $\xi = 0.6$. For a sheath voltage of 50 V, this corresponds to $T_{i\perp} = 0.5$ V and a radially directed energy of 4.5 V or a directed velocity of 4.6 km s^{-1}, which is comparable to that observed at only 7 cm from the axis in Fig. 43. Note that $C(\theta)$ exhibits a threshold for this case as the directed velocity exceeds the ion thermal speed and no ions can impact the surface at normal (0°) incidence (Fig. 34). However, for etched profile simulation, this situation is considerably more complex as the velocity distribution for finite radius is no longer axisymmetric: The profile can be etched at a net angle with respect to the surface normal, and $C(\theta)$ is no longer directly correlated with the etch rate induced by ions within a cone angle θ. To our knowledge, this problem has not been treated in simulations of etched profiles.

Besides the directed radial velocity component, in magnetized plasmas there is also rotational motion, $u_{i\varphi}$, caused by $\mathbf{E} \times \mathbf{B}$ and diamagnetic drifts (*109*). The diamagnetic drift arises from the diffusive flux along the radial density gradient. Downstream, where the magnetic field is usually weak, this rotational motion is a small fraction of the "thermal" spread in perpendicular velocity (*101, 189*). However, in the source of an ECR where the magnetic fields are on the order of 1,000 G, the rotational velocity can be many times the "thermal" velocity (*195*).

The last source of transverse energy is the "random" or "thermal" component. The distributions are Gaussian about their mean except for large radial positions (Fig. 43). However, the origin of this broadening is unclear. Given an electron temperature of 5 eV, an ion temperature of 0.25 eV, a plasma density of 10^{12} cm^{-3}, and an ion–electron energy transfer cross-section of 10^{-14} cm^{-2} and assuming ambipolar diffusive loss for the ions, only $\sim 10^{-3}$ eV of energy can be gained by the ions via elastic collisions with electrons before the ions are lost to the walls (Section II.A.1, *206*); this is insufficient to account for the ion temperatures measured (Figs. 39 and 43). For lower charge densities and a smaller cross-section, the energy transfer from electrons to ions is even less. Most likely, the "random" ion energy observed results from elastic collisions between ions and neutrals that convert the directed ion motion, discussed earlier, into random perpendicular motion. Because ionization takes place over dimensions that are large compared to the uniformity in the plasma potential, a distribution of ion energies in each direction also results as ions flow from one region to another; again, making the plasma more uniform will largely eliminate this source of broadening.

As shown in Fig. 39 for ground state N_2^+, $T_{i\perp}$ decreases monotonically as the pressure is increased, most likely as a result of charge exchange collisions. This means that linewidth control in anisotropic etching should be optimized at an intermediate pressure: at low pressures, $T_{i\perp}$ is larger, but if the pressure is increased too much, the sheath becomes collisional and linewidth control is sacrificed once more (66).

E. Ion Energy Control

As we have said, a primary motivation for replacing conventional rf diode systems with high-efficiency sources is the need for independent control of ion energy and flux so that rate, selectivity, and film properties can be optimized. By placing the wafer on a platen to which a bias voltage is applied while generating the plasma with a high-density source, this problem is nominally circumvented.

For most applications, rf biasing has been employed, although rf is no longer essential, as it is with the conventional diode, since the plasma is maintained using a separate source. In conventional RIE, rf is used to maintain the plasma in the presence of insulating thin films on one of the electrodes. Because the measurements are easier to make, however, resort is made to dc biasing when the iedfs from high-efficiency plasmas have been sampled. To our knowledge, Holber and Foster (185) and Sadeghi et al. (101) are the only ones to have examined the iedf and ivdf, respectively, with an applied rf bias. In the experiments of Holber and Foster, the electrostatic energy analyzer is biased along with the sampling electrode to measure the iedf's through the rf modulated sheath. The results are reproduced here in Fig. 44, where distributions are shown for two different frequencies: 0.5 and 20 MHz. In the high-frequency case, the iedf is relatively narrow with a width of ~ 5 eV and an average energy of ~ 35 eV, corresponding to the streaming energy of 15 eV, resulting from expansion of a divergent-field ECR (see Section VIII.D.1), and a dc bias across the rf sheath of 20 V.[7] At the lower frequency, the ion transit time across the sheath is long compared to the rf period and the ion energy is modulated as the ion traverses the sheath. This leads to substantial broadening, ~ 37 eV, with a width determined by the peak-to-peak rf voltage. The maximum ion energy in this case corre-

---

[7]The experiments of Sadeghi et al. (101) provide further qualitative evidence for ion energy tuning by application of a 13.56 MHz bias, but the LIF method lacks sufficient spatial resolution to resolve the sheath and the ivdfs measured appear anomalously broadened as a result of the spatial averaging.

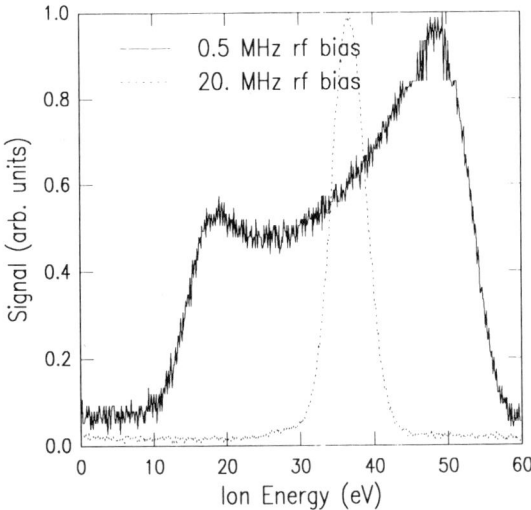

FIG. 44. Ion energy distributions through an rf sheath biased at 0.5 and 20 MHz. Reprinted with permission from Ref. 185.

sponds to the peak rf voltage plus the streaming energy. Clearly, lower frequency bias should be avoided if precision ion energy control is desired to optimize selectivity and minimize atomic displacement damage. However, we shall see shortly (Section IX) that, in fact, the commercial ECR system sold by Hitachi as well as other prototype systems have been operated at frequencies as low as 400 kHz to minimize charge-up damage.

Although the foregoing results indicate that rf biasing is useful for tuning ion energies, as desired, other experiments indicate that the extent to which ion energy can be precisely controlled depends on system design. Consider the recent results of Reinke *et al.* (*199*), who used a dc bias to tune the ion energy, sampled through a pinhole with an electrostatic energy analyzer (Fig. 35). The mean ion energy is only tuned with the dc bias when the biased electrode is "sufficiently" small (Fig. 45). In Reinke *et al.*'s experiment, "sufficiently" small is an electrode whose diameter is 6 cm or less. In other experiments, Shirai and Gonda (*207*) used a 3×3 cm dc biased plate, and Iizuka and Sato (*208*) used a 25 cm diameter electrode and found that the plasma and floating potentials, as measured using Langmuir probes, did not depend on the bias voltage for negative bias voltages, indicating that the biased electrode sheath in these cases were effectively modulated along with ion

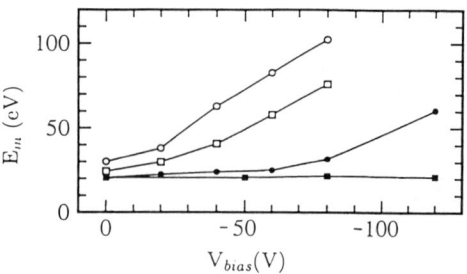

FIG. 45. Variation of maximum ion energy through dc biased sheath as a function of electrode area. Smaller electrodes exhibit larger variation of ion energy with bias voltage. Reprinted with permission from Ref. 199.

energy. In Shirai and Gonda's experiment, the electrode was smaller than the limit found in Reinke's experiment, but this was not the case in the Iizuka and Sato experiment. However, in the latter, a grounded limiter was used at the exit of the source such that the plasma made good contact with ground.

What determines how small the biased electrode must be before gaining ion energy control? This is not a question unique to processing with high-efficiency plasmas, but it is perhaps one that needs more thought when designing such sources with large areas of insulating materials in contact with a magnetized plasma. The voltage division between the biased electrode sheath and the grounded walls is dictated by the relative impedances of the sheaths (Fig. 46), yielding, for the plasma-to-platen dc self-bias voltage V_a:

$$V_a = V_{rf}\left(\frac{Z_a}{Z_a + Z_b}\right). \quad (107)$$

For the dc biasing case, the impedance is simply the sheath resistance, which in turn depends on the electrode area, sheath thickness, and sheath conductivity according to

$$Z_{dc} = R = \frac{s}{\sigma A}. \quad (108)$$

For the more usual rf biasing case, the impedance can be approximated by the capacitive impedance, which for a parallel plate depends only on the area A and sheath thickness s,

$$Z_{rf} = \frac{1}{\omega C} = \frac{s}{\omega \varepsilon_0 A}. \quad (109)$$

Thus, we see that at fixed s/σ in the dc case and fixed s in the rf case, the voltage ratios scale inversely with area ratios. Although these assumptions on s/σ and s are not valid, (107)–(109) illustrate the importance of the area ratio in determining the extent to which ion energies impacting wafer surfaces can be controlled using external bias supplies. Consider two limiting cases,

$$\frac{A_a}{A_b} \ll 1, \quad V_a \to V_{rf}; \tag{i}$$

$$\frac{A_a}{A_b} \gg 1, \quad V_a \to 0, \tag{ii}$$

Clearly, for case (i), the desired result, we have control of ion impact energy. For case (ii), we have no control.[8].

Rf biasing of the wafer platen is generally used in high-efficiency sources. Taking into account the variation of sheath thickness with voltage, let us determine the ratio V_a/V_b of the dc plasma-to-electrode voltages at the powered wafer platen (a) and grounded (b) electrodes in high-efficiency sources having unequal electrode areas A_a and A_b. Referring to Fig. 46, we see that the quantity usually measured, the dc bias voltage of the wafer platen with respect to ground, is given by $V_{bias} = V_b - V_a$. As will be shown, simple one-dimensional arguments (209) for Child law sheaths (18) assuming equal ion densities at the plasma-sheath edges of the platen and grounded electrodes yield the scaling of the voltage ratio on the area ratio as $V_a/V_b = (A_b/A_a)^4$, contrary to measurements (99, 209–215) that indicate a much weaker dependence of V_a/V_b on A_b/A_a for area ratios much different from unity. One-dimensional models incorporating the effects of the dc floating potentials (216) and one-dimensional spherical shell models have also been developed (217, 218), incorporating various assumptions for the sheath and glow physics, and obtaining a scaling more in agreement with measurements. However, rf biasing in high-efficiency sources is generally done in finite cylindrical geometry having two dimensionless parameters: e.g., the powered-to-grounded electrode area ratio and the length-to-radius ratio for a finite length cylinder. In principle, the voltage

---

[8] Note that the measured dc bias voltage in most systems is relative to ground and may bear little relation to the voltage across the sheath that accelerates ions to the wafer. Also, the measured voltage will depend not only on the effective area ratio of the grounded and biased surfaces, but also on the degree to which the electrode is insulated and what value of blocking capacitor is used in the rf circuit.

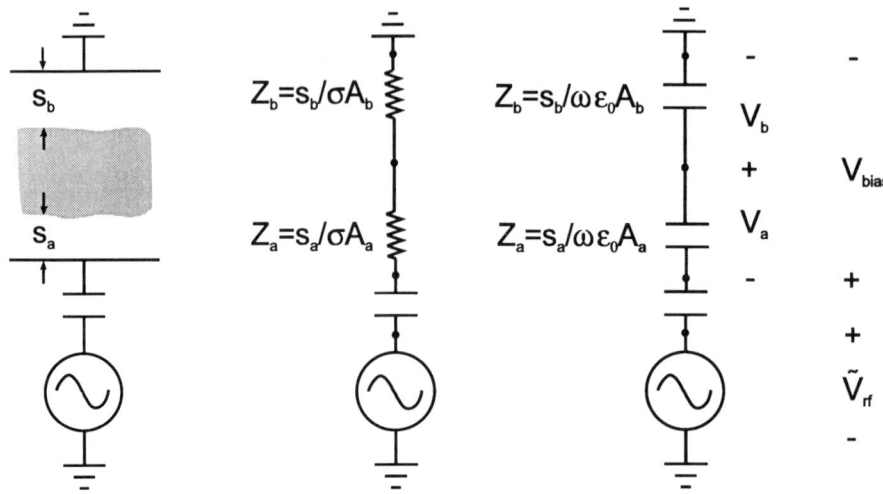

FIG. 46. Equivalent circuit model used to explain distribution of voltages between biased and grounded sheaths for both low and high frequency.

ratio depends on both parameters, and therefore there is no simple scaling with the area ratio alone (219).

To determine the fundamental scaling formula in the limit of high-voltage rf-driven sheaths (see Section II.A.2), we let $\bar{x}$ be a two-dimensional vector that specifies the position on the electrode surface, $n_s(\bar{x})$ be the density at the plasma-sheath edge, $V(\bar{x})$ be the dc plasma-to-electrode (sheath) voltage, $s(\bar{x})$ be the sheath thickness, and $\tilde{J}(\bar{x})$ be the rf current density normal to the electrode surface. A key observation is that the plasma, being highly conducting, cannot support a potential difference greater that a few T_e's. Since $V \gg T_e$, the plasma-sheath edge is an equipotential surface. Since each electrode a and b is also an equipotential surface, the plasma-to-electrode voltage across each sheath is a constant, independent of the position $\bar{x}$ along the electrode surface. For the powered electrode, we therefore have $V_a(\bar{x}) = V_a =$ constant.

For high-voltage capacitive sheaths, the rf voltage amplitude $\tilde{V}_a$ and the dc voltage drop V_a across the sheath are approximately equal: $\tilde{V}_a \approx V_a$. In this limit, using (109), we can relate the rf current density to the dc sheath thickness using the capacitive sheath assumption:

$$\tilde{J}_a(\bar{x}) \propto V_a/s_a(\bar{x}). \tag{110}$$

For a collisionless (Child law) sheath, we have from (18) that

$$n_{sa}(\bar{x}) \propto V_a^{3/2}/s_a^2(\bar{x}). \tag{111}$$

The total rf current $\tilde{I}_a$ flowing to the powered electrode is

$$\tilde{I}_a \propto \int_{a_a} \tilde{J}_a(\bar{x}) \, d^2x. \tag{112}$$

Inserting (110) into (112) and using the collisionless sheath scaling to eliminate s_a, we obtain

$$\tilde{I}_a \propto V_a^{1/2} \int_{A_a} n_{sa}^{1/2}(\bar{x}) \, d^2x. \tag{113}$$

A similar expression to (113) is obtained for the grounded electrode b. Equating $\tilde{I}_a$ to $\tilde{I}_b$, the current to the ground electrode, by continuity of rf current, we obtain the scaling formula for high voltage sheaths,

$$\frac{V_a}{V_b} = \left[\frac{\int_{A_b} n_{sb}^{1/2}(\bar{x}) \, d^2x}{\int_{A_a} n_{sa}^{1/2}(\bar{x}) \, d^2x} \right]^4. \tag{114}$$

We see that the voltage ratio is independent of the rf driving voltage and the electron temperature for high voltage sheaths. The voltage ratio is determined by integrals over the powered and grounded areas that depend on the density at the plasma-sheath edge. This density and its variation along the powered and grounded surface areas are determined by the generation and loss processes for ions in the bulk plasma (*219*).

We note that making the simple assumption $n_{sb} = n_{sa} =$ constant yields the scaling $V_a/V_b = (A_b/A_a)^4$. However, the density variations along the surfaces due to the plasma transport in finite cylindrical geometry strongly modify this scaling (*219*). Further, for large area ratios, the voltage across the grounded electrode is low and the dc floating potentials cannot be neglected (*216, 219*). A further issue for bias control in high-efficiency source design stems from the use of magnetic confinement and insulating liners. Magnetic confinement effectively limits cross-field electron transport and thereby limits the grounded surface in contact with the plasma. To make matters worse, the use of insulating liners to reduce metallic wall sputtering (*87, 101, 112, 113*), further reduces the grounded surface area, and the plasma potential can easily be decoupled from ground. For rf biasing, a thick liner effectively makes s very large and has the same effect as making the grounded surface area very small. In other high-density sources, such as the helicon, inductively coupled plasma, and helical resonator, the power is purposely coupled to the plasma through a dielectric wall and there is an inherent ground reference problem. In the helicon, for this reason, it is common to use a grounded plate at the top of the source. In the TCP and helical resonator, the problem may be less severe because magnetic confinement is not required and the plasma might "find" ground below the wafer surface. Unfortunately, few iedf measurements as a function of dc or rf bias have been made except for ECR sources.

1. Plasma Anodization

In the preceding discussion we have focused on tuning ion energies to optimize selectivity and linewidth control while minimizing atomic displacement damage during etching. However, ion energy control is critical for controlling the properties, such as stress, index of refraction, and stoichiometry, of deposited thin films (*78*). Recently, positive biases have been used for oxidizing single crystalline Si (*61,62*), and the process has been referred to as plasma anodization. As expected, Shirai and Gonda (*207*) and Iizuka and Sato (*208*) both found that the plasma potential floats above the bias potential when the dc bias voltage is made positive with respect to ground and a large electron current is drawn to the wafer platen. So, although the electron flux is increased to the wafer surface during these so-called anodization processes, electrons still diffuse against the electric field and are slowed as they impact the surface. There is also a low-energy, positive ion flux that is less than the electron flux, since net current is drawn to the wafer.

IX. Device Damage

A. ATOMIC DISPLACEMENT DAMAGE

If conventional rf diodes did not produce excessive damage when operated at high power and low pressure, there might not be such a concerted effort toward developing high-efficiency sources. For minimizing atomic displacement damage caused by energetic ion and/or neutral impact, the advantages of using high-efficiency sources with independent ion energy and flux control seem clear. By reducing the plasma potential and decoupling ion energy and flux control, the extent to which atomic displacement occurs can be kept to a minimum. This is clearly evident in the experiments by Yapsir *et al.* (*111*), where Si was etched using a CF_4 diverging-field ECR and the displaced atom density, ΔD_{da}, was measured using Rutherford backscattering spectroscopy (RBS). Hara *et al.* made similar measurements using SF_6 ECR plasmas to treat GaAs

(*220*), where etching should be limited to physical sputtering, and to etch Si (*221*). In these experiments, comparisons to RIE treatments showed that ΔD_{da} can be reduced by typically a factor of two using ECR treatments. Of course, such comparisons must be examined carefully to understand their meaning. In principle, if the same ion and neutral fluxes and energy distributions are obtained in both systems, the same damage should result. There is no inherent difference between RIE and ECR except that the latter affords superior control over these key parameters. Hara *et al.* made such a comparison at constant current density (Fig. 47) suggesting that the improvement with the ECR system stems from a narrowing of the iedf and/or a reduction in the mean ion energy; however, neither parameter was measured in either system.

Interestingly, Yapsir *et al.* (*111*) found that the displaced atom density was slightly reduced by application of an rf bias, which, as discussed earlier, should increase the ion energy. They noted similar improvements in other damage metrics when the rf bias was applied: smaller Schottky diode leakage currents, reduced heavy metal contamination, lower MOS generation current, and fewer defect-induced etch pits. Over the same range of rf bias power, the Si etch rate increased, suggesting that the

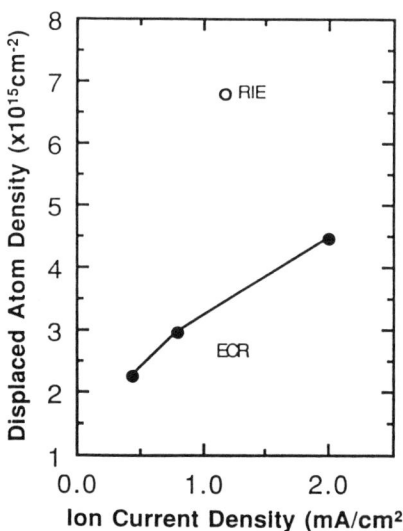

FIG. 47. Displaced atom density, ΔD_{da}, for a GaAs surface as a function of ion current density in SF_6 ECR and RIE plasmas. For the same ion flux, damage in the ECR is significantly reduced presumably because of a reduction in ion energy. Reprinted with permission from Ref. 220.

reduction in damage was correlated with a faster removal of the damage layer. This is supported by the observation that ΔD_{da} increased with rf bias power when argon was substituted for the CF_4 gas and physical sputtering replaced ion-assisted chemical etching. On the other hand, x-ray photoelectron spectroscopy (XPS) measurements show a residual contamination layer containing Si, F, C, and O that is about 0.5 nm thicker when an rf bias is applied during ECR processing (*222*). Apparently, this added contamination has little effect on the electrical properties and RBS is probably too insensitive to detect the difference in lattice displacement. Clearly, more experiments along these lines are needed to understand how best to minimize plasma-induced damage as the bias voltage is changed. Simultaneous measurements of ion flux and energy are also needed.

B. Contamination

Contamination by heavy metals, as documented by Yapsir *et al.* (*111*), raises important design issues for high-density sources. Because the electron temperature can be relatively large compared to rf diode reactors, the potential difference between the plasma and the grounded or floating walls can exceed the sputtering threshold for the wall material (see Section II.A.2). Thus, while the ion energies are reduced relative to those impacting the electrodes in an rf diode, sputtering can still be a problem. This has been observed by several groups (*87, 101, 111–113*), and a common solution is to use a dielectric liner with low sputtering yield, such as quartz or alumina, in the source. However, sputtering of these materials may still occur, compromising etching selectivity (*223*) and leading to particulate formation. For example, if SiO_2 is sputtered from a liner during poly-Si etching, it can be difficult to etch without applying a bias voltage to the wafer; with no bias, net deposition can occur. It is critical that source design take into account the ion energy and flux not only to the device wafer but also to the walls of the reactor.

C. Charging

Another mechanism by which microelectronic devices can be damaged and manufacturing yields deleteriously affected is charging of insulators during plasma treatment. An excellent, short review has recently been presented by Gabriel and McVittie (*224*), who point out that charging damage has become more apparent in recent years as gate oxide

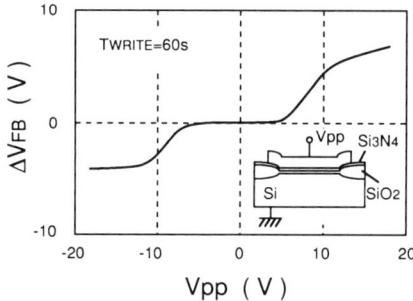

FIG. 48. Metal nitride oxide silicon (MNOS) device used to measure charging voltage in high-density plasmas. The flat band voltage shift ΔV_{FB} depends on the peak voltage, created by charging in the plasma, across the dielectric. Reprinted with permission from Ref. 232.

thicknesses have shrunk to 10 nm or less. This problem is not unique to processing with high-efficiency plasma sources and in fact was initially recognized as a problem in the stripping of photoresist using an rf barrel etcher (225). However, the larger current densities to the wafer and the possibility for large current nonuniformities resulting from magnetic confinement are cause for concern when designing high-efficiency sources for microelectronic device fabrication.

Some of the metrics used for evaluating charging damage include the flat-band voltage shift (ΔV_{FB}) of a metal–nitride–oxide–silicon (MNOS) nonvolatile memory device (226) (Fig. 48), the frequency of dielectric breakdown as a function of field strength across the dielectric (224, 227–229), the leakage current in CMOS inverter circuits (230), and the threshold voltage shift in an MOS transistor (230).

1. Plasma Uniformity

While there may be many factors influencing charging damage, plasma uniformity is clearly one of the most important. For this reason, the damage is readily observed in rf magnetrons (88). Achieving adequate uniformity at the wafer surface is a central issue for high efficiency sources. There is no simple solution because of the many interrelated physics issues that affect uniformity. These include

(a) incident (microwave or rf) power flux profile;
(b) wave refraction during propagation to the absorption zone;
(c) absorption zone size, shape, and location;
(d) wave absorption profile;

(e) transport of heated electrons and their subsequent ionization profile; and

(f) transport and diffusion of bulk plasma (and free radicals) to the wafer surface.

We have touched on all these issues in the preceeding sections. To some extent, these issues are alleviated in low-profile or close-coupled geometries, thus prompting recent interest in those configurations.

The effect of plasma uniformity has been most extensively studied by Samukawa (230, 231), who used a Faraday cup to measure the ion current density as a function of axial and radial position in a diverging-field type ECR (Fig. 12a). The extent of oxide damage clearly correlated with the degree of radial nonuniformity in the ion current to the wafer. This can be seen in Fig. 49 where the leakage current measured from a CMOS inverter circuit is plotted vs. the ion current density difference, ΔJ_i, from the wafer center to the wafer edge. By varying the magnetic field configuration and the position of the wafer with respect to the

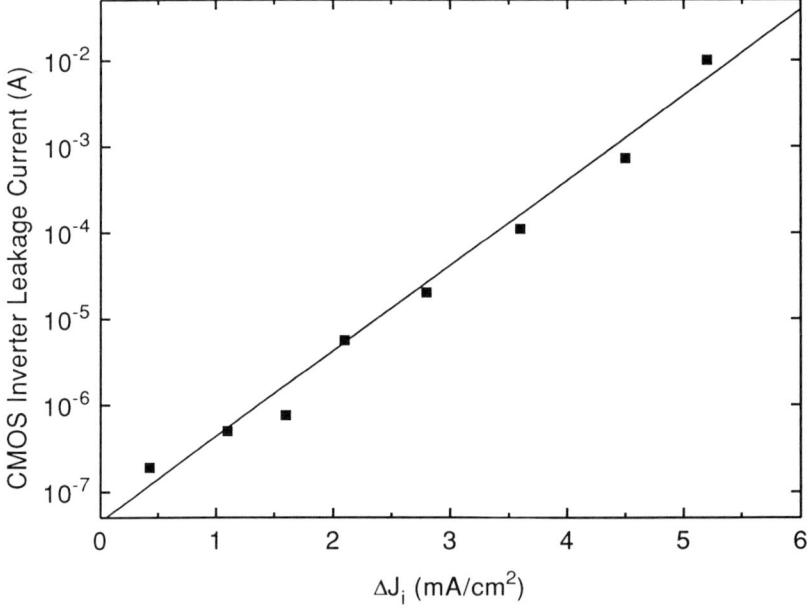

FIG. 49. Leakage current from CMOS inverter circuit as a function of difference in current density from the center to the edge of an ECR plasma. Adapted from Figs. 5 and 6 in Ref. 230.

resonance zone, ΔJ_i could be varied over the range shown, producing a five order-of-magnitude variation in leakage current. By collimating the field and placing the wafer near the ECR resonance zone where the ion current density was more uniform (Fig. 12d), Samukawa showed that the extent of damage could be dramatically reduced.

Why should the uniformity of ion current density affect gate oxide damage? It is safe to say that this remains an open question, but the simple explanation offered in Fig. 50 is worth considering. Since the top surface of the wafer exposed to the plasma is isolated from the bulk of the wafer and substrate holder by, for example, gate and field oxide, the top surface of the wafer will charge to a potential such that the time-averaged ion and electron fluxes are equal. Providing the top surface is electrically conducting, for example, before etching is completed, any nonuniformity in the current density to the wafer will not result in a surface potential difference across the wafer. However, once the pattern clears at the etching endpoint so that the top surface forms isolated regions, a surface potential can exist when the current density is nonuniform. The surface potential, in turn, is a driving force for

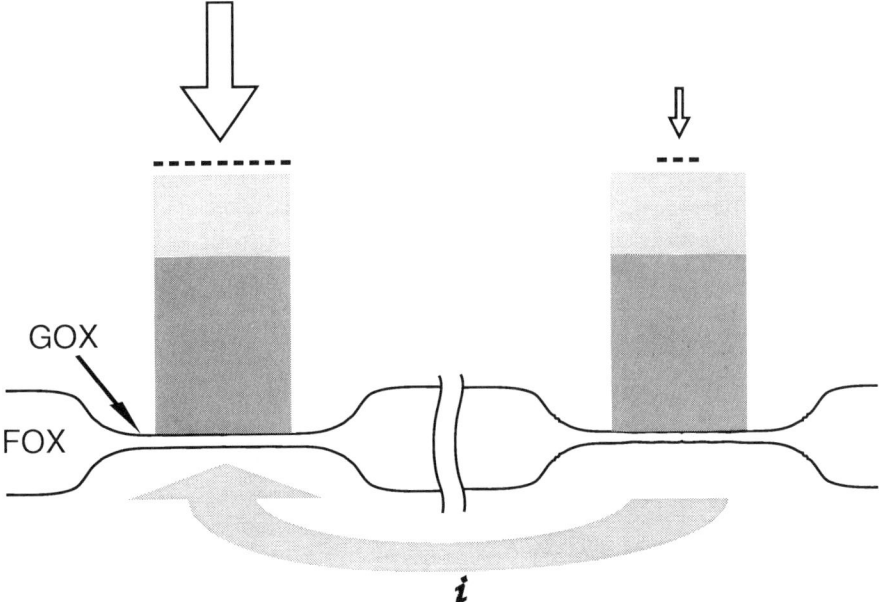

FIG. 50. Schematic illustration of how uniformity of plasma current density can induce charging damage.

Fowler–Nordheim tunneling current (226) through the insulating layer. The largest current will occur where the oxide is thinnest, i.e., the gate oxide, and if the current is large enough it can produce premature breakdown, ΔV_{FB}, leakage current, etc. Processing in high-efficiency plasmas leaves one particularly prone to this problem as the current densities, and therefore the surface charge densities, are much larger than in conventional rf diode processing. Coupled with the difficulty of obtaining uniform plasma density in magnetized systems and the unprecedented thinness of gate oxides, the problem has recently become a matter for alarm (228).

2. Biasing

Application of a bias voltage to the wafer also effects the extent of gate oxide damage in high-density plasma processing. Namura et al. (232) showed that the ΔV_{FB} from an MNOS device increased, albeit nonuniformly across the wafer, as an rf bias was applied to the wafer exposed to a close-coupled ECR source (Fig. 12d). Thus, the advantage of putting the wafer into the source, as discussed by Samukawa, appears to be mitigated when a 13 MHz bias is applied to the wafer. Namura et al. suggest that the application of rf bias produces a nonuniform current density to the wafer, which again leads to a surface potential difference, presumably because the rf current to the walls is larger on the edges nearest to the walls. They offer an equivalent circuit model that includes the MNOS devices on the wafer and the capacitive coupling of the wafer to the plasma through the sheath. A key ingredient of the circuit is the anisotropic conductivity of the plasma: conduction along the magnetic field being much larger than conduction across the field owing to magnetization of the electrons. Although this simple model appears to give good agreement with the observed radial dependence of ΔV_{FB} across the wafer, the model parameters are *ad hoc* and the equivalent circuit model used to describe coupling of the plasma to the substrate of the device is not clearly explained.

Samukawa (233) also suggests that the 13 MHz bias results in a nonuniform plasma above the wafer and reports a "local discharge between the chamber wall and the substrate holder." To alleviate this problem, Samukawa reduced the bias frequency and found that both the plasma uniformity improved (Fig. 51) and the incidence of gate oxide breakdown was dramatically reduced (Fig. 52). At the same time, Tsunokuni et al. (234) reported similar observations. Samukawa says that use of such low frequencies eliminates the local discharge. Since the

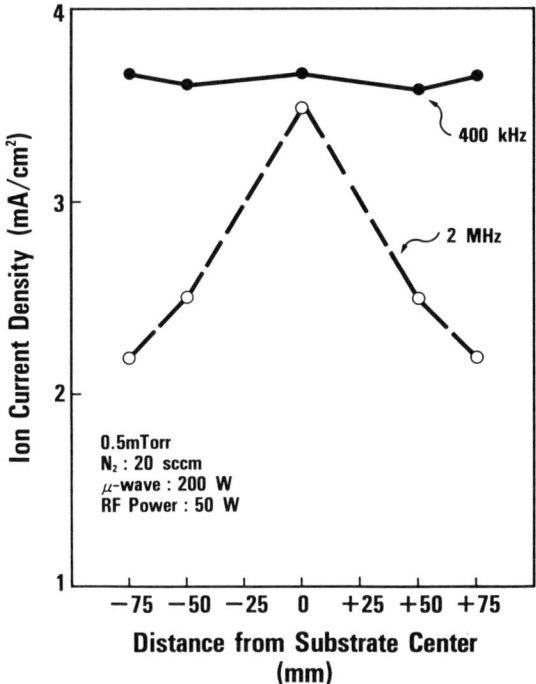

FIG. 51. Dependence of ion current density uniformity on bias frequency in N_2 ECR plasma. Reprinted with permission from Ref. 233.

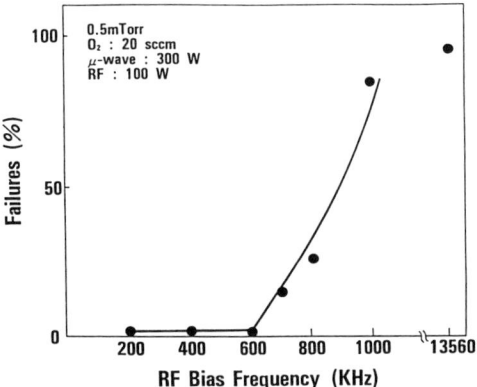

FIG. 52. Cumulative failure rate caused by gate oxide breakdown as a function of bias frequency in an O_2 plasma. Reprinted with permission from Ref. 233.

capacitive impedance of the sheath increases with decreasing frequency, the sheath becomes predominantly resistive at frequencies below 1 MHz with the sheath current being predominantly ion conduction current (*64, 78, 236*). If the plasma is uniform initially, applying a low frequency bias will serve primarily to accelerate ions to the surface without altering the plasma potential profile and producing large, localized currents to the walls.

The bias frequency dependence of the gate oxide damage illustrates once more the importance of plasma source design. Surely, the extent to which local discharges are created will depend on the geometry of the source and, in particular, the proximity of grounded surfaces to the wafer edge. We have already discussed how ΔD_{da} can increase with increasing ion energy so it does not appear desirable to solve the gate oxide damage problem by resorting to low-frequency rf bias. In many ways this defeats the purpose behind using the high-density plasma in the first place. Nonetheless, we note that Hitachi currently employs low frequency bias (typically < 2 MHz) in their ECR machines sold for etching of poly-Si, metals, and oxides.

The sequence by which the plasma process is terminated apparently also has an effect on gate oxide damage. Samukawa (*230*) reports that when the rf (13 MHz) bias is extinguished prior to turning off the microwave power supply, the CMOS leakage current is reduced by more than five orders of magnitude and the gate oxide breakdown voltage is increased by two times. The origin of this effect is not well understood and needs further investigation. Similarly, the extent of gate oxide damage produced by helicons, helical resonators, and TCPs has not been reported to our knowledge. Typically, 13 MHz bias frequencies have been used with these sources, and it is reasonable to expect similar problems to arise.

D. Radiation

Photon irradiation is another mechanism by which microelectronic devices can be damaged during high-efficiency plasma processing. In this regard, the difference between high-density sources and conventional rf diodes stems from the efficiency of electron heating at low pressure that can lead to production of large excited state densities (*143*) and energetic photon irradiation (*236*). Damage to SiO_2 by absorption of above–bandgap-energy photons has been studied most extensively (*237–239*). Depending on the photon energy (*240, 241*), the free electrons and/or holes created by the absorption process may find their way to traps within the oxide or at the oxide-silicon interface. In both cases, device performance can be affected by shifting threshold voltages and by creating leakage and generation currents. Although in some instances it is possible to anneal out such damage (237,238), it is preferable to prevent damage formation in the first place and thereby preserve the thermal processing budget (1). One means for doing this was suggested by Yunogami et al. (241), who showed that radiation damage could be minimized by maintaining the substrate at temperatures as low as 120 K during plasma irradiation. The improvement was ascribed to a reduction in the hole mobility at low temperature that enhanced the probability for recombination before trapping.

In addition to VUV photon production, Castagna et al. (242) have shown that runaway electrons in ECR plasmas can generate significant fluxes of x-rays when they impact the reactor walls. The runaway electrons are created primarily at lower pressures and higher microwave powers where the electrons can gain energy (Section III.B) faster than they dissipate it by collisions with neutrals. After acquiring more than ~ 100 eV of energy, the collision cross-sections begin to decrease with increasing electron energy, permitting the electrons to "run away" to energies approaching 1 keV before impacting the source walls and producing x-rays.

Unfortunately, there are virtually no measurements of *absolute* photon fluxes to device wafers in high-density plasmas, and therefore it is difficult to judge the severity of the radiation problem. Such measurements are essential if one wants to distinguish among x-ray, VUV, ion, electron, and fast neutral bombardment damage mechanisms and distinguish these in turn from the charging damage already discussed.

X. Summary and Remaining Questions

The results of any plasma process depend on more than just plasma properties and plasma source design. Choice of chemistry—both gas-phase and on the surface—plays a paramount role in the outcome of etching, deposition, and passivation treatments. Nonetheless, plasma source design can influence the materials process. The plasma governs the generation and transport of reactive species to and from the surface, and many surface processes are limited by this transport. We have seen that plasma source design can have a profound effect on charged particle transport. For example, high profile ($L \gg R$), diverging magnetic field designs induce a distributed sheath that accelerates ions to higher energies and limits ion energy control.

While most measurements and simulations have been concerned with the transport of ions and electrons, there have been but a few (*201,243,244*) measurements of neutral, reactive species transport in high-density plasmas. In only one case are we aware of a measurement of the degree of dissociation in an ECR plasma, and this was made on N_2 gas (*243*). We have seen how sources differ in the mechanism by which electrons are heated, and although it has been useful to view the electron energy distribution function as Maxwellian, subtle deviations may affect the generation rate of key reactive intermediates. Clearly, if we are to understand the subtle interplay between source design and surface modification, we must measure the flux of reactive species to the surface, and this means measuring the absolute concentrations and "temperatures" of gas-phase reactive intermediates.

Another factor influencing the transport of reactive neutrals is the composition of wall material in plasma sources. Consider that for each of the sources reviewed here, a dielectric, usually fused silica or alumina, is used either as a barrier through which wave energy is coupled or as a liner to protect against sputtering of metallic wall materials. Regardless of electron heating and ion transport mechanisms, the extensive use of these materials marks a sharp distinction between the new generation of high-efficiency sources and their conventional rf diode predecessors. These materials can alter surface chemistry by reducing reactant fluxes, by the loading effect (*223*), or by generating small concentrations of species such as oxygen that can dramatically alter the rates of both gas-phase and surface reactions (*245*). This area, again, is virtually unexplored.

We have seen that source design can be critical for minimizing the extent of damage and yield loss. In particular, plasma uniformity has been empirically correlated with charging damage and, in turn, plasma uniformity depends on such design features as the magnetic field profile, source dimensions, wave propagation, and the electron heating mechanism. Much work remains before we can control plasma uniformity and relate uniformity to charging damage. Similarly, we find it ironic that low-frequency bias voltages are used to minimize charging damage during anisotropic etching, since the ion bombarding energy and the extent of atomic displacement damage is clearly higher at these lower frequencies. Again, the benefits of low frequency bias appear traceable to an improvement in plasma uniformity, but we suspect that this effect is strongly dependent on the details of plasma source design: for example, how near the walls lie to the edge of the wafer. It is also curious that the application of a bias voltage that increases the ion bombarding energy can reduce the extent of electrical damage. Is this simply a question of the faster etching rate leading to faster removal of damaged material or is there a more subtle explanation?

While plasma source design is primarily concerned with efficient plasma generation and ion energy control at low pressures to maintain critical dimensional control and high throughput while minimizing device damage, there are clearly fundamental limits to the control available. Generation of intense plasma necessarily implies larger currents to device wafers, leading to an increased probability for charging damage. Intense plasma generation also implies large densities of excited electronic states that will radiate and expose wafers to above-bandgap radiation and thereby produce trapped charge and interface states. In electron cyclotron resonance plasmas, electron heating can be so efficient that "runaway" occurs and electrons can gain more than 1 kV before impacting walls and generating potentially damaging x-rays. Finally, the control of ion bombardment energy is limited by the extent to which a broad distribution of internal energy states are created.

Plasma stability is an area of concern that we have not reviewed because of a paucity of measurements and theory. The consequences of plasma instability should be obvious: In large-volume manufacturing, one can little afford a process that inexplicably deviates from its normal behavior. Plasma are notoriously nonlinear and, as such, are a rich medium for instabilities and turbulence. Arguably, low-pressure, magnetized plasmas are more prone to unstable operation than their unmagnetized counterparts since more modes can be excited and mode switching is prevalent (*112,114,208,246-248*). But, even for unmagnetized, conventional rf diodes, bistability and hysteresis have been observed and correlated with changes in etching rate (*249,250*). So unmagnetized sources such as the inductively coupled plasma, helical resonator, or surface wave excited plasma can be prone to instability. This is an area in need of careful experimental investigation and theoretical analysis.

To facilitate comparisons between plasma sources, we have relied on simple scaling relationships. Clearly these have limited validity and serve no more than to provide crude estimates. What we really need is a computer-aided design tool with which plasma generation, transport, and stability in two or three dimensions can be simulated. In this manner, reactors and processes could be modified rapidly in software and a preliminary design selected prior to making a hardware prototype. While recent progress along these lines has been impressive, the field is too young to warrant an in-depth review. With rapid improvements in computational power, we look forward to increasingly sophisticated codes in the next few years and a continuing progression toward plasma sources that provide the level of process control needed to meet the demands of future materials processing applications.

XI. Symbol Definitions

A	Loss area
A_{eff}	Effective loss area
$\mathbf{B}$	Magnetic field
$\mathbf{B}_0$	Dc magnetic field
b	Inductive or helical coil radius
C_θ	Normalized, energy-flux-weighted, cumulative angular distribution function
c	Shield radius for helical resonator; speed of light
ΔD_{da}	Displaced atom density
d_{eff}	Effective plasma source size
ER	Etching rate
$\mathbf{E}$	Electric field
E_{rhp}	Amplitude of RHP electric field
E_{lhp}	Amplitude of LHP electric field
$\tilde{E}$	Complex amplitude of rf electric field
ε_{c}	Collisional energy lost per electron–ion pair created
ε_{exc}	Effective excitation energy
ε_{i}	Ion energy
ε_{L}	Total energy lost per electron–ion pair created
e	Electronic charge
eedf	Electron energy distribution function
I	Current
H_z	Magnetic field induced by skin current in inductively coupled plasma
h_{L}	Ratio of plasma density at axial sheath edge to density in plasma center
h_{R}	Ratio of plasma density at radial sheath edge to density in plasma center
I_0	Modified Bessel function of the first kind
iadf	Ion angular distribution function

iedf	Ion energy distribution function
ivdf	Ion velocity distribution function
J_m	Bessel function of order m
J	Current density
J_{iL}	Ion current density at axial wall
$\tilde{J}$	Complex amplitude of rf current density
J_a	Dc ion current density at the wafer platen
$\tilde{J}_a$	Rf current density at the wafer platen
J_b	Dc ion current density at the grounded surface
$\tilde{J}_b$	Rf current density at the grounded surface
K	Second-order rate constant
K_{el}	Elastic scattering rate constant
K_{exc}	Excitation rate constant
K_{iz}	Ionization rate constant
K_0	Modified Bessel function of the second kind
k_0	Free space wavenumber, ω/c
k_r	Spatially varying propagation constant, $k_0 \kappa_r^{1/2}$
k_z	Wavenumber along $\hat{z}$
$k_\perp$	Wavenumber perpendicular to $\hat{z}$
$k_{\perp d}$	Transverse wavenumber for surface wave inside dielectric tube
$k_{\perp p}$	Transverse wavenumber for surface wave inside plasma
k_{zh}	Helical pitch wavenumber
L	Reactor length; inductance
L_d	Characteristic length over which V_d occurs
ℓ	Helicon antenna length
LHP	Left-hand circularly polarized
M	Ion mass
m	Electron mass; azimuthal mode number
N	Neutral gas density
η	Number of turns in inductive coil
n	Plasma density
n_p	Plasma density downstream from source
n_s	Plasma density at sheath edge
n_c	Critical density
P_{abs}	Absorbed power
P_{inc}	Incident power
P_{refl}	Reflected power
P_{trans}	Transmitted power
$P_{\mu w}$	Microwave power
P_w	Power carried by wave
p	Neutral gas pressure

p_{min}	Minimum pressure needed to sustain ECR discharge
P_{ohm}	Ohmic power density
R	Reactor radius; resistance
RHP	Right-hand circularly polarized
S_e	Stochastic energy flux
S_{ECR}	ECR energy flux
S_{ion}	Ion energy flux
S_r	RHP energy flux
SC_0	Reactive sticking coefficient on bare surface
s	Sheath thickness
T	Neutral temperature
T_e	Electron temperature
T_i	Ion temperature
$T_{i\perp}$	Ion temperature perpendicular to the magnetic field
T_{iz}	Ion temperature parallel to the magnetic field
TCP	Transformer coupled plasma
t_{res}	Time electron spends in resonance zone
v_i	Volume removed per unit ion bombarding energy
v_n	Volume removed per reacting neutral
u_B	Bohm velocity
u_i	Ion speed
u_{is}	Ion energy gained by acceleration through sheath potential, V_s
u_{iz}	Ion speed normal to the surface
$u_{i\perp}$	Radial ion speed (parallel to the surface)
$u_{i\perp}^{(0)}$	Directed ion radial speed
u_{ph}	Wave phase velocity, ω/k_z
u_{res}	Parallel electron velocity at resonance zone
u_{Ti}	Ion thermal speed
u_{Te}	Electron thermal speed
V_a	Dc bias voltage of plasma with respect to wafer platen
$\tilde{V}_a$	Rf voltage amplitude of plasma with respect to wafer platen
V_b	Dc bias voltage of plasma with respect to ground
$\tilde{V}_b$	Rf voltage amplitude of plasma with respect to ground
V_{bias}	Dc voltage of wafer platen with respect to ground
V_d	Distributed sheath voltage
V_{rf}	Rf voltage amplitude between powered electrode and ground
V_s	Voltage between plasma and floating surface
V_T	Helical voltage at tap position
$\tilde{V}_{pw}$	Rf voltage amplitude between plasma and wafer
$\bar{V}_{pw}$	Dc voltage between plasma and wafer resulting from rf bias
ΔV_{FB}	Flat-band voltage shift

W_{ECR}	Transverse energy gained from one pass through ECR resonance zone
Z	Impedance
z_{res}	Axial ECR resonance position
Δz_{res}	Width of ECR resonance zone
z'	Distance from resonance zone
z_T	Tap position on helical resonator coil
α	Inverse scale length of cyclotron frequency near resonant zone, $\omega_{ce}^{-1} \partial \omega_{ce}/\partial z'$
α_z	Inverse of decay length along $\hat{z}$
β	Ratio of sheath potential to perpendicular ion temperature $(V_s/T_{i\perp})$
δ_c	Collisional skin depth
δ_p	Collisionless skin depth
$\gamma(\theta, \varepsilon_i)$	Differential ion flux
Γ_e	Electron flux
Γ_i	Ion flux
Γ_{iL}	Ion flux at axial wall
Γ_n	Neutral flux
κ_d	Relative dielectric constant for insulating tube
κ_r	Relative dielectric constant of plasma
λ_{De}	Electron Debye length
λ_i	Combined mean free path for ion–neutral momentum and charge exchange
λ_r	Wavelength for RHP wave in plasma, $2\pi/k_r$
λ_z	Wavelength along $\hat{z}$
μ_i	Ion mobility
η	Microwave power attenuation coefficient, $\omega_{pe}^2/(\omega c |\alpha|)$
ω	Angular frequency
ω_{ce}	Electron cyclotron frequency
ω_{ci}	Ion cyclotron frequency
ω_h	Upper hybrid frequency
ω_L	Lower hybrid frequency
ω_{pe}	Electron plasma frequency
ω_{pi}	Ion plasma frequency
ω_R	RHP cutoff frequency
Φ	Magnetic flux in inductive plasma; spatially varying plasma potential
ψ	Pitch angle of helical coil in helical resonator
σ	Dc conductivity of plasma

σ_{ac} Ac plasma conductivity
σ_i Combined ion momentum and charge transfer cross-section
$\tilde{\sigma}$ Absorption or emission frequency
$\tilde{\sigma}_0$ Rest absorption or emission frequency
Θ Surface coverage
ν_c Electron–neutral collision frequency
ν_{cT} Sum of electron–neutral and electron–ion collision frequencies
ν_{LD} Effective collision frequency for Landau damping
ν_T Total effective collision frequency
ξ Ratio of directed velocity to sheath velocity, $2u_{i\perp}^{(0)}/u_{is}$
ζ Ratio of helicon phase velocity to electron thermal speed, $\omega/(k_z u_{Te})$

Acknowledgments

The authors gratefully acknowledge A. J. Lichtenberg for his reading and comments on a draft of this manuscript, F. F. Chen for his comments on a draft of Section IV, C. W. Jurgensen for helpful discussions on ion energy control and etching anisotropy, and E. S. Aydil, J. Keller, M. Moisan, A. Kornblit, and J. Margot for their critical reading of this manuscript. We thank V. Vahedi for generating Figs. 4, 5, and 7. The support of MAL by National Science Foundation Grant ECS-85-17363 and Department of Energy Grant DE-FG03-87-ER13727 is gratefully acknowledged.

References

1. R. B. Fair, *Proceedings of the IEEE* **78**, 1687 (1990).
2. J. M. Cook and K. G. Donohoe, *Solid State Tech.* **34**, 119 (1991).
3. J. Asmussen, *J. Vac. Sci. Technol.* **A7**, 883 (1989).
4. R. R. Burke and C. Pomot, *Solid State Technol.* **31**, 67 (1988).
5. K.-H. Kretschmer, K. Matl, G. Lorenz, I. Kessler, and B. Dumbacher, *Solid State Technol.* **33**, 53 (1990).
6. W. M. Holber, in "Handbook of Ion Beam Processing Technology," (J. J. Cuomo, S. M. Rossnagel, and H. R. Kaufman, eds.), Noyes Publications, Park Ridge, New Jersey, p. 21.
7. D. Dane, P. Gadgil, T. D. Mantei, M. A. Carlson, and M. E. Weber, *J. Vac. Sci. Technol.* **B10**, 1312 (1992).
8. A. Hall and K. Nojiri, *Solid State Technol.*, May 1991, p. 107.
9. Y. H. Lee, Z. H. Zhou, D. A. Danner, P. M. Fryer, and J. M. Harper, *J. Appl. Phys.* **68**, 5329 (1990).

10. J. Pelletier and M. J. Cooke, *J. Vac. Sci. Technol.* **7**, 59 (1989).
11. T. Ono, M. Oda, C. Takahashi, and S. Matsuo, *J. Vac. Sci. Technol.* **B4**, 696 (1986).
12. H. Shindo, T. Hashimoto, F. Amasaki, and Y. Horiike, *Jpn. J. Appl. Phys.* **29**, 2641 (1990).
13. H. Uetake, T. Matsuura, T. Ohmi, J. Murota, K. Fukuda, and N. Mikoshiba, *Appl. Phys. Lett.* **57**, 596 (1990).
14. G. Fortuno-Wiltshire, *J. Vac. Sci. Technol.* **A9**, 2356 (1991).
15. C. Constantine, D. Johnson, S. J. Pearton, U. K. Chakrabarti, A. B. Emerson, W. S. Hobson, and A. P. Kinsella, *J. Vac. Sci. Technol.* **B8**, 596 (1990).
16. S. J. Pearton, F. Ren, J. R. Lothian, T. R. Fullowan, R. F. Kopf, U. K. Chakrabarti, S. P. Hui, A. B. Emerson, R. L. Kostelak, and S. S. Pei, *J. Vac. Sci. Technol.* **B9**, 2487 (1991).
17. K. D. Choquette, M. Hong, R. S. Freund, J. P. Mannaerts, R. C. Wetzel, and R. E. Leibenguth, *IEEE Photon. Technol. Lett.* **5**, 284 (1993).
18. J. M. Cook, D. E. Ibbotson, P. D. Foo, and D. L. Flamm, *J. Vac. Sci. Technol.* **A8**, 1820 (1990).
19. S. Tachi, K. Tsujimoto, S. Arai, and T. Kure, *J. Vac. Sci. Technol.* **A9**, 796 (1991).
20. D. X. Ma, T. A. Lin, and C. H. Chen, *J. Vac. Sci. Technol.* **A10**, 1217 (1992).
21. W. F. Marx, D. X. Ma, and C. H. Chen, *J. Vac. Sci. Technol.* **A10**, 1232 (1992).
22. S. M. Gorbatkin, L. A. Berry, and J. Swyers, *J. Vac. Sci. Technol.* **A10**, 1295 (1992).
23. P. K. Gadgil, D. Dane, and T. D. Mantei, *J. Vac. Sci. Technol.* **A10**, 1303 (1992).
24. K. W. Whang, S. H. Lee, and H. J. Lee, *J. Vac. Sci. Technol.* **A10**, 1307 (1992).
25. S. J. Pearton, U. K. Chakrabarti, W. S. Hobson, C. R. Abernathy, A. Katz, F. Ren, T. R. Fullowan, and A. P. Perley, *J. Electrochem. Soc.* **139**, 1763 (1992).
26. K. Suzuki, K. Ninomiya, S. Nishimatsu, and S. Okudaira, *J. Vac. Sci. Technol.* **B3**, 1025 (1985).
27. K. Suzuki, S. Okudaira, and I. Kanomata, *J. Electrochem. Soc.* **126**, 1024 (1979).
28. S. Samukawa, M. Sasaki, and Y. Suzuki, *J. Vac. Sci. Technol.* **8**, 1062 (1990).
29. S. Samukawa, S. Mori, and M. Sasaki, *Jpn. J. Appl. Phys.* **29**, 792 (1990).
30. S. Samukawa, Y. Suzuki, and M. Sasaki, *Appl. Phys. Lett.* **57**, 403 (1990).
31. S. Samukawa, M. Sasaki, and Y. Suzuki, *J. Vac. Sci. Technol.* **8**, 1192 (1990).
32. S. Samukawa, T. Toyosato, and E. Wani, *J. Vac. Sci. Technol.* **B9**, 1471 (1991).
33. D. A. Carl, D. W. Hess, and M. A. Lieberman, *J. Appl. Phys.* **68**, 1859 (1990).
34. F. Plais, B. Agius, F. Abel, J. Siejka, M. Puech, G. Ravel, P. Alnot, and N. Proust, *J. Electrochem. Soc.* **139**, 1489 (1992).
35. M. Kitagawa, K. Setsune, Y. Manabe, and T. Hirao, *Jpn. J. Appl. Phys.* **27**, 2026 (1988).
36. T. Hirao, K. Setsune, M. Kitagawa, Y. Manabe, K. Wasa, and S. Kohiki, *Jpn. J. Appl. Phys.* **26**, L544 (1987).
37. Y. Manabe and T. Mitsuyu, *J. Appl. Phys.* **66**, 2475 (1989).
38. C. S. Pai, J. F. Miner, and P. D. Foo, *J. Electrochem. Soc.* **139**, 850 (1992).
39. R. D. Knox, V. L. Dalal, and O. A. Popov, *J. Vac. Sci. Technol.* **A9**, 474 (1991).
40. J. C. Barbour, H. J. Stein, O. A. Popov, M. Yoder, and C. A. Outten, *J. Vac. Sci. Technol.* **A9**, 480 (1991).
41. M. Kitagawa, N. Matsuo, G. Fuse, H. Iwasaki, A. Yoshida, and T. Hirao, *Jpn. J. Appl. Phys.* **27**, L2139 (1988).
42. N. Cheung, *Nucl. Instr. Methods Phys. Res.* **B55**, 811 (1991).
43. X. Y. Qian, N. W. Cheung, M. A. Lieberman, S. B. Felch, R. Brennan, and M. I. Current, *Appl. Phys. Lett.* **59**, 348 (1991).

44. X. Y. Qian, D. Carl, J. Benasso, N. W. Cheung, M. A. Lieberman, I. G. Brown, J. E. Galvin, R. A. MacGill, and M. I. Current, *Nucl. Instr. Methods Phys. Res.* **B55**, 844 (1991).
45. M. H. Kiang, M. A. Lieberman, N. W. Cheung, and X. Y. Qian, *Appl. Phys. Lett.* **60**, 2767 (1992).
46. N. Kondo and Y. Nanishi, *Jpn. J. Appl. Phys.* **28**, L7 (1989).
47. S. Sugata, A. Takamori, N. Takado, K. Asakawa, E. Miyauchi, and H. Hashimoto, *H. Vac. Sci. Technol.*, **B6**, 1087 (1988).
48. Z. Lu, M. T. Schmidt, D. Chen, R. M. Osgood, Jr., W. M. Holber, D. V. Podlesnik, and J. Foster, *Appl. Phys. Lett.* **58**, 1143 (1991).
49. M. Hong, R. S. Freund, K. D. Choquette, H. S. Luftman, J. P. Mannaerts, and R. C. Wetzel, *Appl. Phys. Lett.* **62**, 2658 (1993).
50. G. M. Mikhailov, P. V. Bulkin, S. A. Khudobin, A. A. Chumakov, and S. Y. Shapoval, *Vacuum* **43**, 199 (1992).
51. M. Delfino, S. Salimian, and D. Hodul, *J. Appl. Phys.* **70**, 1712 (1991).
52. M. Delfino, S. Salimian, D. Hodul, A. Ellingboe, and W. Tsai, *J. Appl. Phys.* **71**, 1001 (1992).
53. Z. Lu, M. T. Schmidt, R. M. Osgood, W. M. Holber, and D. V. Podlesnik, *J. Vac. Sci. Technol.* **A9**, 1040 (1991).
54. A. Kishimoto, I. Suemune, K. Hamaoka, T. Koui, Y. Honda, and M. Yamanishi, *Jpn. J. Appl. Phys.* **29**, 2273 (1990).
55. H. Yamada, *J. Appl. Phys.* **65**, 775 (1989).
56. K. Nakashima, M. Ishii, I. Tajima, and M. Yamamoto, *Appl. Phys. Lett.* **58**, 2663 (1991).
57. T. Shibata, Y. Nanishi, and M. Fujimoto, *Jpn. J. Appl. Phys.* **29**, L1181 (1990).
58. H. Hirayama and T. Tatsumi, *J. Appl. Phys.* **66**, 629 (1989).
59. M. Ishii, K. Nakashima, I. Tajima, and M. Yamamoto, *Appl. Phys. Lett.* **58**, 1378 (1991).
60. T. Takeshita, T. Unagami, and O. Kogure, *Jpn. J. Appl. Phys.* **27**, L2118 (1988).
61. K. Miyake, S. Kimura, T. Warabisako, H. Sunami, and T. Tokuyama, *J. Vac. Sci. Technol.* **A2**, 496 (1984).
62. D. A. Carl, D. W. Hess, and M. A. Lieberman, *J. Vac. Sci. Technol.* **A8**, 2924 (1990).
63. D. A. Carl, D. W. Hess, M. A. Lieberman, T. D. Nguyen, and R. Gronsky, *J. Appl. Phys.* **70**, 3301 (1991).
64. R. A. Gottscho, *Phys. Rev. A* **36**, 2233 (1987).
65. A. von Engel, "Ionized Gases." Clarendon Press, 1965.
66. C. W. Jurgensen, *J. Appl. Phys.* **64**, 590 (1988).
67. R. Warren, *Phys. Rev.* **98**, 1658 (1955).
68. B. Chapman, "Glow Discharge Processes." John Wiley and Sons, New York, 1980.
69. R. T. Farouki, M. Dalvie, and L. F. Pavarino, *J. Appl. Phys.* **68**, 6106 (1990).
70. D. H. Economou, D. R. Evans, and R. C. Alkire, *J. Electrochem. Soc.* **135**, 756 (1988).
71. M. A. Lieberman, *IEEE Trans. Plasma Sci.* **16**, 638 (1988).
72. M. A. Lieberman, *IEEE Trans. Plasma Sci.* **17**, 338 (1989).
73. V. A. Godyak and N. Sternberg, *Phys. Rev.* **A42**, 2299 (1990).
74. V. A. Godyak and R. B. Piejak, *Phys. Rev. Lett.* **65**, 996 (1990).
75. G. A. Hebner, J. T. Verdeyen, and M. J. Kushner, *J. Appl. Phys.* **63**, 2226 (1988).
76. V. M. Donnelly, D. L. Flamm, W. C. Dautremont-Smith, and D. J. Werder, *J. Appl. Phys.* **55**, 242 (1984).
77. C. P. Chang, D. L. Flamm, D. E. Ibbotson, and J. A. Mucha, *J. Vac. Sci. Technol.* **B6**, 524 (1988).

78. W. C. Dautremont-Smith, R. A. Gottscho, and R. J. Schutz, "Plasma processing: Mechanisms and applications," *in* "Semiconductor Materials and Process Technolog Handbook" (G. E. McGuire, ed.). Noyes, Park Ridge, New Jersey, 1988, Chapter 5.
79. A. S. Bergendahl, D. V. Horak, P. E. Bakeman, and D. J. Miller, *Semicond. Int.*, Sept. 1990, p. 94.
80. M. A. Lieberman, A. J. Lichtenberg, and S. E. Savas, *IEEE Trans. Plasma Sci.* **19**, 189 (1990).
81. A. V. Lukyanova, A. T. Rakhimov, and N. V. Suetin, *IEEE Trans. on Plasma Sci.* **19**, 197 (1991).
82. R. W. Boswell, "A Study of Waves in Gaseous Plasma." Ph.D. Thesis, Flinders University of South Australia, July 1970.
83. F. F. Chen, *Laser and Part. Beams* **7**, 551 (1989).
84. A. Komori, T. Shoji, K. Miyamoto, J. Kawai, and K. Kawai, *Phys. Fluids* **B3**, 893 (1991).
85. P. K. Lowenhardt, B. D. Blackwell, R. W. Boswell, G. D. Conway, and S. M. Hamberger, *Phys. Rev. Lett.* **67**, 2792 (1991).
86. F. F. Chen, *Plasma Phys. Controlled Fusion* **33**, 339 (1991).
87. S. M. Gorbatkin and L. A. Berry, *J. Vac. Sci. Technol.* **A10**, 3104 (1992).
88. H. Hoga, T. Orita, T. Yokoyama, and T. Hayashi, *Jpn. J. Appl. Phys.* **30**, 3169 (1991).
89. J. H. Ingold, *in* "Gaseous Electronics" (M. N. Hirsh and H. J. Oskam, eds.). Academic Press, Orlando, Florida, 1978, p. 19.
90. T. Nakano, N. Sadeghi, and R. A. Gottscho, *Appl. Phys. Lett.* **58**, 458 (1991).
91. J. E. Stevens, Y. C. Huang, R. L. Jarecki, and J. L. Cecchi, *J. Vac. Sci. Technol.* **A10**, 1270 (1992).
92. M. D. Bowden, T. Okamoto, F. Kimura, H. Muta, T. Sakoda, K. Muraoka, M. Maeda, Y. Manabe, M. Kitagawa, and T. Kimura, *J. Appl. Phys.* **73**, 2732 (1993).
93. E. Eggarter, *J. Chem. Phys.* **62**, 833 (1975).
94. L. R. Peterson and J. E. Allen, Jr., *J. Chem. Phys.* **56**, 6068 (1972).
95. M. Hayashi, Report IPPJ-AM-19, Inst. Nucl. Phys., Nagoya U. (1981).
96. L. Tonks and I. Langmuir, *Phys. Rev.* **34**, 876 (1929).
97. W. Schottky, *Phys. Zeits.* **25**, 635 (1924).
98. Smirnov, B. M. "Physics of Weakly Ionized Gases." Mir, Moscow, 1981, pp. 216–217.
99. V. A. Godyak, "Soviet Radio Frequency Discharge Research." Delphic Associates, Inc., Falls Church, Virginia, 1986, pp. 86–90.
100. V. A. Godyak and V. N. Maximov, *Vestnik Moskovskoy Universiteta, ser. Fiz. Astr.* **18**, 51 (1977).
101. N. Sadeghi, T. Nakano, D. J. Trevor, and R. A. Gottscho, *J. Appl. Phys.* **70**, 2552 (1991); N. Sadeghi, T. Nakano, D. J. Trevor, and R. A. Gottscho, *J. Appl. Phys.* **71**, 3648 (1992).
102. J. A. Thornton and A. S. Penfold, *in* "Thin Film Processes" (J. L. Vossen and W. Kern, eds.). Academic Press, New York, 1978.
103. S. M. Rossnagel and H. R. Kaufman, *J. Vac. Sci. Technol.* **4**, 1822 (1986).
104. A. R. Wendt and M. A. Lieberman, *J. Vac. Sci. Technol.* **A8**, 902 (1990).
105. J. Goree and T. E. Sheridan, *Appl. Phys. Lett.* **59**, 1052 (1991).
106. G. R. Misium, A. J. Lichtenberg, and M. A. Lieberman, *J. Vac. Sci. Technol.* **A7**, 1007 (1989).
107. V. A. Godyak, R. B. Piejak, and B. M. Alexandrovich. To appear in *Plasma Sources Sci. Technol.* (1994). See also M. M. Turner, *Phys. Rev. Lett.* **71**, 1844 (1993).

108. S. C. Brown, "Basic Data of Plasma Physics." MIT Press and John Wiley & Sons, Inc., New York, 1959.
109. F. F. Chen, "Introduction to Plasma Physics." Plenum, New York, 1974.
110. J. Spencer, "Recent advances in ECR plasma sources," as reported at the Plasma Etch Users' Group Monthly Meeting, Varian Research Center, Palo Alto, California, April 9, 1992.
111. A. S. Yapsir, G. Fortuno-Wiltshire, J. P. Gambino, R. H. Kastl, and C. C. Parks, *J. Vac. Sci. Technol.* **A8,** 2939 (1990).
112. S. M. Gorbatkin, L. A. Berry, and J. B. Roberto, *J. Vac. Sci. Technol.* A8, 2893 (1990).
113. S. M. Rossnagel, S. J. Whitehair, C. R. Guarnieri, and J. J. Cuomo, *J. Vac. Sci. Technol.* **A8,** 3113 (1990).
114. S. Samukawa, T. Nakamura, and A. Ishitani, *Jpn. J. Appl. Phys.* **31,** L594 (1992).
115. C. G. Montgomery, R. H. Dicke, and E. M. Purcell, "Principles of Microwave Circuits." Dover, New York, 1965.
116. M. Mauel, private communication (1992).
117. M. C. Williamson, A. J. Lichtenberg, and M. A. Lieberman, *J. Appl. Phys.* **72,** 3924 (1992).
118. T. D. Mantai and T. E. Ryle, *J. Vac. Sci. Technol.* **B9,** 29 (1991).
119. T. D. Mantai and S. Dhole, *J. Vac. Sci. Technol.* **B9,** 26 (1991).
120. S. Samukawa, S. Mori, and M. Sasaki, *J. Vac. Sci. Technol.* **A9,** 85 (1991).
121. M. Pichot, A. Durandet, J. Pelletier, Y. Arnal, and L. Vallier, *Rev. Sci. Instrum.* **59,** 1072 (1988).
122. J. Root and J. Asmussen, *Rev. Sci. Instrum.* **56,** 154 (1985).
123. F. Jaeger, A. J. Lichtenberg, and M. A. Lieberman, *Plasma Phys.* **14,** 1073 (1972).
124. V. L. Ginzburg, "The Propagation of Electromagnetic Waves in Plasmas." Pergamon Press, Oxford, 1964.
125. K. G. Budden, "Radio Waves in the Ionsphere." Cambridge University Press, Cambridge, U.K., 1966, pp. 474–479.
126. T. Stix, "The Theory of Plasma Waves." McGraw-Hill, New York, 1962.
127. M. Born and E. Wolf, "Principles of Optics," Sixth Edition. Pergamon Press, Oxford, 1980, Appendix II.
128. W. P. Allis, S. J. Buchsbaum, and A. Bers, "Waves in Anisotropic Plasmas." MIT Press, Cambridge, Massachusetts, 1963.
129. A. J. Lichtenberg and M. A. Lieberman, "Regular and Stochastic Motion." Springer-Verlag, New York, 1981.
130. M. Tanaka, R. Nishimoto, S. Higashi, N. Harada, T. Ohi, A. Komori, and Y. Kawai, *J. Phys. Soc. Japan* **60,** 1600 (1991).
131. R. W. Boswell, *Phys. Lett.* **33A,** 457 (1970).
132. R. Bowers, C. Legendy, and F. E. Rose, *Phys. Rev. Lett.* **7,** 339 (1961).
133. G. N. Harding and P. C. Thonemann, *Proc. Phys. Soc.* **85,** 301 (1965).
134. B. J. Davies, *Plasma Phys.* **4,** 43 (1970).
135. F. F. Chen, *J. Vac. Sci. Technol.* **A10,** 1389 (1992); F. F. Chen and F. Chevalier, *J. Vac. Sci. Technol.* **A10,** 3599 (1992).
136. F. F. Chen and C. D. Decker, *Plasma Phys. Controlled Fusion* **34,** 635 (1992).
137. R. W. Boswell and D. Henry, *Appl. Phys. Lett.* **47,** 1095 (1985).
138. A. J. Perry and R. W. Boswell, *Appl. Phys. Lett.* **55,** 148 (1989).
139. A. J. Perry, D. Vender, and R. W. Boswell, *J. Vac. Sci. Technol.* **B9,** 310 (1991).
140. P. K. Loewenhardt, B. Zhang, B. D. Blackwell, and R. W. Boswell, *Rev. Sci. Instrum.* **64,** 3334 (1993).
141. R. H. Lovberg, in "Plasma Diagnostic Techniques" (R. H. Huddleston and S. L. Leonard, eds.). Academic Press, New York, 1965, p. 69.

142. R. W. Boswell, *Plasma Phys. Controlled Fusion* **26,** 1147 (1984).
143. K. P. Giapis, N. Sadeghi, J. Margot, R. A. Gottscho, and T. C. Lee, *J. Appl. Phys.* **73,** 7188 (1993).
144. R. W. Boswell and R. K. Porteous, *Appl. Phys. Lett.* **50,** 1130 (1987).
145. F. F. Chen, private communication (1992).
146. R. Conn, as reported at the Plasma Etch Users' Group Monthly Meeting, Varian Research Center, Palo Alto, California, April 9, 1992.
147. W. Hittorf, *Wiedemanns Ann. d. Physik.* **21,** 90 (1884).
148. J. J. Thomson, *Philos. Mag.* **4,** 1128 (1927).
149. J. S. Townsend and R. H. Donaldson, *Philos. Mag.* **8,** 605 (1929).
150. K. A. McKinnon, *Philos. Mag.* **8,** 605 (1929).
151. H. U. Eckert, in "Proc. 2nd Intl. Conf. on Plasma Chem. and Technol." (H. Boenig, Ed). Technomic Publ., Lancaster, Pennsylvania, 1986.
152. J. H. Keller, M. S. Barnes, and J. C. Forster, 43rd Annual Gaseous Electronics Conference, Champaign-Urbana, Illinois, 16–19 Oct. 1990, Abstract NA-5.
153. Lam Research Corporation, Technical Note TN-003 (1992).
154. W. G. Oldham and S. E. Schwarz, "An Introduction to Electronics." Holt, Reinhart and Winston, New York, 1972.
155. J. W. Denneman, *J. Phys. D: Appl. Phys.* **23,** 293 (1990).
156. J. Amorim, H. S. Maciel, and J. P. Sudano, *J. Vac. Sci. Technol.* **B9,** 362 (1991).
157. W. W. Macalpine and R. O. Schildknecht, *Proc. Inst. Radio Engrs.*, December, p. 2099 (1959).
158. G. N. Steinberg and A. R. Steinberg, U.S. Patent No. 4,368,092, issued Jan. 11, 1983.
159. D. L. Flamm, D. E. Ibbotson, and W. L. Johnson, U.S. Patent No. 4,918,031, issued April 17, 1990.
160. D. L. Flamm, as presented at the International Union for Pure and Applied Chemistry, 9th International Symposium on Plasma Chemistry, Pungochiuso, Italy, Sept. 7–15, 1989.
161. J. M. Cook, D. E. Ibbotson, and D. L. Flamm, *J. Vac. Sci. Technol.* **B8,** 1 (1990).
162. M. A. Lieberman, A. J. Lichtenberg, and D. L. Flamm, Memorandum No. UCB/ERL M90/10, Electronics Research Laboratory, College of Engineering, University of California, Berkeley, California, Feb. 5, 1990.
163. P. Bletzinger, 44th Annual Gaseous Electronics Conference, 22–25 October, 1991, Albuquerque, New Mexico, Abstract RB-1, p. 201.
164. Prototech Research, as presented at the Plasma Etch Users' Group Monthly Meeting, Varian Research Center, Palo Alto, California, April 9, 1992.
165. J. R. Pierce, "Traveling Wave Tubes." D. Van Nostrand, New York, 1950, Chapter 3.
166. R. F. Welton, E. W. Thomas, R. K. Feeney, and T. F. Moran, *Measurement Sci. Technol.* **2,** 242 (1991).
167. L. D. Smullin and P. Chorney, *Proc. Inst. Radio Engrs.* **46,** 360 (1958).
168. A. W. Trivelpiece and R. W. Gould, *J. Appl. Phys.* **30,** 1784 (1959).
169. D. R. Tuma, *Rev. Sci. Instrum.* **41,** 1519 (1970).
170. M. Moisan and Z. Zakrzewski, *J. Phys. D: Appl. Phys.* **24,** 1025 (1991).
171. L. Paquin, D. Masson, M. Wertheimer, and M. Moisan, *Can. J. Phys.* **63,** 831 (1985).
172. R. Claude, M. Moisan, M. R. Wertheimer, and Z. Zakrzewski, *Appl. Phys. Lett.* **50,** 1797 (1987).
173. G. Sauve, M. Moisan, J. Paraszczak, and J. Heindenreich, *Appl. Phys. Lett.* **53,** 470 (1988).
174. J. Margot and M. Moisan, *J. Phys. D: Appl. Phys.* **24,** 1765 (1991).
175. T. M. Mayer and R. A. Barker, *J. Vac. Sci. Technol.* **21,** 757 (1982).
176. J. W. Coburn and H. F. Winters, *J. Appl. Phys.* **50,** 3189 (1979).
177. H. F. Winters, J. W. Coburn, and T. J. Chuang, *J. Vac. Sci. Technol.* **B1,** 469 (1983).

178. U. Gerlach-Meyer, J. W. Coburn, and E. Kay, *Surf. Sci.* **103**, 177 (1981).
179. T. M. Mayer, R. A. Barker, and L. J. Whitman, *J. Vac. Sci. Technol.* **18**, 349 (1981).
180. R. A. Barker, T. M. Mayer, and W. C. Pearson, *J. Vac. Sci. Technol. B* **1**, 37 (1983).
181. T. M. Mayer and R. A. Barker, *J. Electrochem. Soc.* **129**, 585 (1982).
182. H. Gokan and S. Esho, *J. Electrochem. Soc.* **131**, 131 (1984).
183. E. S. G. Shaqfeh and C. W. Jurgensen, *J. Appl. Phys.* **66**, 4664 (1989).
184. H. Nogami, Y. Ogahara, S. Mashiro, T. Tsukada, and D. I. C. Pearson, 39th Spring Meeting of the Japanese Society of Applied Physics, 30p-ZS-8 (1992); S. Mashiro, H. Nogami, Y. Ogahara, T. Tsukada, and D. I. C. Pearson, *loc. cit.*, 29-NC-3 (1992).
185. W. M. Holber and J. Forster, *J. Vac. Sci. Technol.* **A8**, 3720 (1990).
186. J. Liu, G. L. Huppert, and H. H. Sawin, *J. Appl. Phys.* **68**, 3916 (1990).
187. R. Koslover and R. McWilliams, *Rev. Sci. Instr.* **57**, 2441 (1986).
188. E. A. den Hartog, H. Persing, J. S. Hamers, and R. C. Woods, 44th Annual Gaseous Electronics Conference, Albuquerque, New Mexico, October 1991, Abstract DA-1.
189. T. Nakano, N. Sadeghi, D. J. Trevor, R. A. Gottscho, and R. W. Boswell, *J. Appl. Phys.*, **72**, 3384 (1992).
190. T. Nakano, R. A. Gottscho, N. Sadeghi, D. J. Trevor, R. W. Boswell, A. J. Perry, T. C. Lee, K. P. Giapis, and J. Margot, *Jpn. Soc. Appl. Phys.* **61**, 711 (1992).
191. F. F. Chen, in "Plasma Diagnostic Techniques" (R. H. Huddlestone and S. L. Leonard, Eds.). Academic Press, New York, 1965, p. 113.
192. N. Hershkowitz, in "Plasma Diagnostics," Vol. 1 (O. Auciello and D. L. Flamm, Eds.). Academic Press, New York, 1989.
193. J. D. Swift and M. J. R. Schwar, "Electrical Probes for Plasma Diagnostics." Elsevier, New York, 1969.
194. V. A. Godyak, R. B. Piejak, and B. M. Alexandrovich, *Plasma Sources Sci. Technol.* **1**, 36 (1992).
195. R. K. Porteous, H.-M. Wu, and D. B. Graves, *Plasma Sources Sci. Technol.* **3**, 25 (1994).
196. M. Matsuoka and K. Ono, *Appl. Phys. Lett.* **50**, 1864 (1987).
197. M. Matsuoka and K. Ono, *J. Vac. Sci. Technol.* **A6**, 25 (1988).
198. P. Reinke, W. Jacob, and W. Moller, in "Proc. NATO-ASI on Diamond and Diamond-Like Films and Coatings" (R. E. Clausing, L. L. Harton, F. C. Augus, and P. Koidl, eds.). NATO-ASI Series B: Physics, Vol. 266. Plenum, New York, 1991.
199. P. Reinke, S. Schetz, W. Jacob, and W. Moller, *J. Vac. Sci. Technol.* **A10**, 434 (1992).
200. E. A. den Hartog, H. Persing, and R. C. Woods, *Appl. Phys. Lett.* **57**, 661 (1990).
201. R. C. Woods, R. L. McClain, L. J. Mahoney, E. A. den Hartog, H. Persing, and J. S. Hamers, *SPIE* **1594**, 366 (1991).
202. S. Samukawa, Y. Nakagawa, and K. Ikeda, *Jpn. J. Appl. Phys.* **29**, 88 (1990).
203. C. Charles, R. W. Boswell, and R. K. Porteous, *J. Vac. Sci. Technol.* **A10**, 398 (1992).
204. M. A. Hussein, G. A. Emmert, N. Hershkowitz, and R. C. Woods, *J. Appl. Phys.* **72**, 1770 (1992).
205. H. Fujita, Y. Okuno, Y. Ohtsu, and S. Yagura, *J. Appl. Phys.* **67**, 6114 (1990).
206. V. E. Golant, A. P. Zhilinsky, and I. E. Sakharov, "Fundamentals of Plasma Physics." John Wiley & Sons, New York, 1977.
207. K. Shirai and S. Gonda, *J. Appl. Phys.* **68**, 4258 (1990).
208. S. Iizuka and N. Sato, *J. Appl. Phys.* **70**, 4165 (1991).
209. H. R. Koenig and L. I. Maissel, *IBM J. Res. Dev.* **14**, 168 (1970).
210. J. L. Vossen, *J. Electrochem. Soc.* **126**, 2345 (1979).
211. V. A. Godyak, *Sov. J. Plasma Phys.* **2**, 78 (1976).
212. K. Kohler, J. W. Coburn, D. E. Horne, E. Kay, and J. H. Keller, *J. Appl. Phys.* **57**, 59 (1985).

213. J. W. Coburn and E. Kay, *J. Appl. Phys.* **43,** 4965 (1972).
214. C. M. Horwitz, *J. Vac. Sci. Technol.* **A1,** 60 (1983).
215. G. Y. Yeom, J. A. Thornton, and M. J. Kushner, *J. Appl. Phys.* **65,** 3825 (1989).
216. A. M. Pointu, *Appl. Phys. Lett.* **50,** 1047 (1987).
217. M. A. Lieberman, *J. Appl. Phys.* **65,** 4186 (1989).
218. M. V. Alves, M. A. Lieberman, V. Vahedi, and C. K. Birdsall, *J. Appl. Phys.* **69,** 3823 (1991).
219. M. A. Lieberman and S. E. Savas, *J. Vac. Sci. Technol.* **A8,** 1632 (1990).
220. T. Hara, J. Hiyoshi, and H. Hamanaka, M. Sasaki, F. Kobayashi, K. Ukai, and T. Okada, *J. Appl. Phys.* **67,** 2836 (1990).
221. G. Washidzu, T. Hara, J. Hiyoshi, M. Sasaki, Y. Suzuki, and K. Ukai, *Jpn. J. Appl. Phys.* **30,** 1045 (1991).
222. A. S. Yapsir, G. S. Oehrlein, G. Fortuno-Wiltshire, and J. C. Tsang, *Appl. Phys. Lett.* **57,** 590 (1990).
223. A. J. Watts and W. J. Varhue, *Appl. Phys. Lett.* **61,** 549 (1992).
224. C. T. Gabriel and J. P. McVittie, *Solid State Technol.*, June 1992, p. 81.
225. K. Tsunokuni, K. Nojiri, S. Kuboshima, K. Hirobe, in "Ext. Abstr. 19th Conf. Solid State Dev. and Materials, Tokyo 1987," p. 195.
226. S. M. Sze, "Semiconductor Devices." John Wiley, New York, 1985.
227. S. B. Felch, S. Saimian, and D. T. Hodul, *J. Vac. Sci. Technol.* **B10,** 1320 (1992).
228. P. H. Singer, *Semiconductor Int.*, May 1982, p. 78.
229. C. T. Gabriel, *J. Vac. Sci. Technol.* **B9,** 370 (1991).
230. S. Samukawa, *Jpn. J. Appl. Phys.* **29,** 980 (1990).
231. S. Samukawa, *Jpn. J. Appl. Phys.* **30,** L1902 (1991).
232. T. Namura, H. Okada, Y. Naitoh, Y. Todokoro, and M. Inoue, *Jpn. J. Appl. Phys.* **29,** 2251 (1990).
233. S. Samukawa, *Jpn. J. Appl. Phys.* **30,** 3154 (1991).
234. K. Tsunokuni, K. Nojiri, and M. Nawata, in "Ext. Abstr. 38th Mtg. Jpn. Soc. Appl. Phys. 1991," p. 499.
235. C. B. Zarowin, *J. Vac. Sci. Technol.* **A2,** 1537 (1984).
236. S. Miyake, W. Chen, and T. Ariyasu, *Jpn. J. Appl. Phys.* **29,** 2491 (1990).
237. D. J. DiMaria, L. M. Ephrath, and D. R. Young, *J. Appl. Phys.* **50,** 4015 (1979).
238. L. M. Ephrath and D. J. DiMaria, *Solid State Technol.*, April 1981.
239. D. A. Buchanan and G. Fortuno-Wiltshire, *J. Vac. Sci. Technol.* **A9,** 804 (1991).
240. T. Yunogami, T. Mizutani, K. Suzuki, and S. Nishimatsu, *Jpn. J. Appl. Phys.* **28,** 2172 (1989).
241. T. Yunogami, T. Mizutani, K. Tsujimoto, and K. Suzuki, *Jpn. J. Appl. Phys.* **29,** 2269 (1990).
242. T. J. Castagna, J. L. Shohet, K. A. Ashtiani, and N. Hershkowitz, *Appl. Phys. Lett.* **23,** 2856 (1992); *J. Vac. Sci. Technol.* **A10,** 1325 (1992).
243. S. Meikle and Y. Hatanaka, *Appl. Phys. Lett.* **54,** 1648 (1989).
244. J. L. Cecchi, J. E. Stevens, R. L. Jarecki, Jr., and Y. C. Huang, *J. Vac. Sci. Technol.* **B9,** 318 (1991).
245. J. P. Booth and N. Sadeghi, *J. Appl. Phys.* **70,** 611 (1992).
246. D. A. Carl, M. C. Williamson, M. A. Lieberman, and A. J. Lichtenberg, *J. Vac. Sci. Technol.* **B9,** 339 (1991).
247. P. K. Shufflebotham and D. J. Thomson, *J. Vac. Sci. Technol.* **A8,** 3713 (1990).
248. R. A. Dandl and G. E. Guest, *J. Vac. Sci. Technol.* **A9,** 3119 (1991).
249. E. S. Aydil and D. J. Economou, *J. Appl. Phys.* **69,** 109 (1991).
250. P. A. Miller and K. E. Greenberg, *Appl. Phys. Lett.* **60,** 2859 (1992).

Electron Cyclotron Resonance Plasma Sources and Their Use in Plasma-Assisted Chemical Vapor Deposition of Thin Films

OLEG A. POPOV

Matsushita Electric Works, R&D Laboratory
Woburn, Massachusetts

I. Introduction	122
II. ECR Fundamentals and Microwave Power Absorption	126
A. Principles of Electron Cyclotron Resonance	126
B. Microwave Power Absorption in ECR Plasmas	129
1. "Pure" Electron Cyclotron Resonance	131
2. Doppler Broadened ECR	132
3. Whistler Wave Absorption at $B \gg B_{ce}$	134
4. Left-Hand Polarized Wave Absorption	136
5. Upper Hybrid Resonance	139
6. Second Harmonic Resonance	142
7. Geometrical Resonances	145
8. Stochastic Heating	146
9. Nonlinear Microwave Dissipative Process	148
III. ECR Plasma Sources: Designs and Characteristics	152
A. Basic Types of ECR Plasma Sources	152
1. ECR Plasma Source Classification	152
2. Major ECR Source and Plasma Characteristics and Parameters	155
B. Hitachi/NTT ECR Plasma Source	157
1. Magnetic Field Structure and Coil Arrangement	157
2. Microwave Introduction Window	165
3. Microwave Mode Converters	169
4. ECR Plasma Chamber: Resonant Cavity, Cutoff, Scaling	178
C. ECR Plasma Source Characteristics and Plasma Parameters	183
1. Microwave Plasma Modes	184
2. Plasma Parameters as Function of Microwave Power Absorbed	188
3. Axial Distribution of ECR Plasma and Plasma Stream	193
4. ECR Plasma Parameter Pressure Dependencies	202
IV. ECR Plasma Sources in PA CVD of Thin Films	208
A. Silicon Nitride Film ECR CVD	209
1. Review of Previous Work	209
2. Silicon Nitride Films Deposited in ECR and SHR Microwave Plasmas	210
B. Silicon Oxide Films	216
1. Review of Previous Work	216
2. ECR CVD of Silicon Dioxide Film in the Microscience 9200 system	218

C. ECR Plasma Planarization with Silicon Dioxide Films 221
    1. Silane/Oxygen Planarization 221
    2. Metalorganic ECR CVD and Planarization of SiO_2 Films 222
V. Conclusion 226
    Acknowledgments 227
    References 227

I. Introduction

In recent years electron cyclotron resonance (ECR) plasma sources have been widely explored in various plasma and ion beam technologies. Initially ECR was used in the late 1940s by Lax, Brown and co-workers as a method for bounded plasma diagnostics (*1,2*). Later, S. Brown with colleagues studied extensively the fundamental properties and characteristics of microwave plasmas inserted in static magnetic fields, including ECR: breakdown, plasma confinement and ionization balance, microwave power absorption, and resonant phenomena (*3,4*).

The idea of using ECR for the plasma generation was at first implemented in the early 1960s by Dandl et al. for plasma fusion research (*5*), and by Miller with co-workers (*6,7*) and Consoli et al. (*8,9*) for plasma acceleration. However, during the next 20 years electron cyclotron resonance was used mainly for fusion plasma electron heating (*10–13*), and for generation of multiply charged heavy ions (*14,15*).

The first attempts to apply ECR for plasma technology were done in the mid-1970s by Musil et al. for silicon oxidation (*16,17*), Suzuki et al. for plasma etching (*18*), and Sakudo et al. for ion implantation (*19*). In 1983 Matsuo introduced an ECR source for low-temperature plasma-assisted chemical vapor deposition (PA CVD) of dielectric thin films (*20*). Since then, ECR plasma has been applied to many other PA CVD technologies: silicon nitride (*21–23*) and silicon dioxide (*24–28*) thin films; intermetal planarization with silicon dioxide 1–2 micron thick films using rf bias (*29–31*); silicon carbide (*32,33*) and boron nitride (*34,35*) coatings; diamond (*36–38*) and diamondlike-carbon (DLC) (*39–42*) coatings; amorphous silicon (*43–45*), and silicon germanium (*46,47*) films; epitaxial growth of silicon and GaAs (*48–50*); MO CVD of doped and undoped silicon oxide films (*51–53*); ECR sputtered CVD of metal oxides such as Ti_2O_5, Al_2O_3, and ZnO (*54–56*); and silicon oxidation (*57,58*). Recently new technologies of ECR CVD of tungsten layers on low-temperature substrate were developed (*59*) and ECR hydrogen passivation of semi-conductor films (*60, 61*) were developed. Other metal and metal nitride and oxide coatings have been recently reported.

The growing interest in ECR CVD technology is motivated by the unique features of ECR plasma, which are superior to those of the conventional rf plasma source used in PA CVD. Some of most promising ECR plasma CVD characteristics and features are briefly listed here.

- Low gas pressure. Typical ECR operational pressures are 10^{-4}–10^{-3} Torr, while rf plasma pressures are 10^{-2}–5 Torr. Such low pressures allow undesirable homogeneous reactions to be avoided in volume. At pressures below 1 mTorr, the ion mean free path is larger than the source and stream dimensions, so by controlling the ECR plasma potential and magnetic field profile, one can control the energy of ions oncoming to the substrate.
- Low process temperature, 50–200°C. Conventional rf CVD systems operate at temperatures $T > 300°C$. This unique feature opened the possibility of depositing films on different materials that cannot stand high temperature.
- Low ion energy: $V_i = 10$–50 eV. The ion bombardment with this energy is extremely important for the formation of the deposited film with good physical and mechanical properties.
- High ion density at wafer location, $N_i > 10^{11}$ cm^{-3}. At the same power level, rf plasmas have densities 10 times lower, $N_i \simeq 10^{10}$ cm^{-3}.
- High degree of gas decomposition and high concentration of excited species and radicals involved in film formation. This improves the gas utilization and dramatically increases the film growth rate.
- Separation of plasma generation zone from the sample (wafer) stage. This allows independent control of the energy and flux of charged particles (ions and electrons) bombarding the sample.
- Controllability of energy of ions bombarding the substrate by varying of magnetic field and its structure.
- Low ion energy spread.

Though industrial ECR PA CVD equipment has been on the market for more than six years (mainly in Japan), ECR technology is still the "state-of-the-art" (62).

The slow pat in the "industrialization" of ECR technology is associated with some factors that are not "purely" technological: (i) the complexity of the system (often two vacuum chambers, two or three magnetic coils and/or permanent magnets, bulky microwave transmission line); (ii) the high equipment cost compared with that of rf CVD and LPCVD systems (expensive turomolecular pumps, coil and microwave power supplies); (iii) the huge size and weight of the whole system compared with those of rf plasma systems.

All disadvantages would probably be forgiven if ECR plasma tools provided process and film with properties needed by the industry. Unfortunately, ECR CVD technology is far from being an "engineered tool." There are many technological problems that are common for ECR plasma systems and have not been solved yet. Among them are (i) control of film uniformity, in particular across large-area wafers of 15–20 cm diameter; (ii) x-ray radiation from ECR plasma which could be a source of film structural damage; (iii) relatively low (compared with that of rf CVD and high-temperature LPCVD) film deposition rate limited by low operational pressures, 0.1–30 mTorr; (iv) particulation from flaky film deposited on the process chamber walls. To solve these and other problems that we will discuss in Section III, extensive research on ECR plasmas and deposition process chemistry is needed.

EC heating physics was traditionally the field of plasma fusion and space plasma related sciences and was extensively studied (theoretically and experimentally) in the 1960s and 1970s (*8–13, 63–75*). The results of that work cannot be always applied directly to ECR plasma sources used in plasma processing. Only six or seven years ago, when ECR plasma became an emerging technology in silicon and silicon oxide etching and PA CVD of dielectric films, the number of publications related to ECR plasma and ECR processes began to grow very fast.

A considerable contribution and effort to attract attention to ECR plasma source capability in plasma processing were made in NTT Labs, Japan, by Matsuo *et al.* (*20, 54, 55, 140*), of Michigan State University by J Asmussen *et al.* (*76, 81, 87, 92, 98, 101, 102, 148, 151*), at the Plasma Physics Institute, Prague, Czechoslovakia, by J. Musil with colleagues (*11, 16, 17, 71, 77, 157*), and at C.N.E.T. (France) by J. Pelletier *et al.* (*85, 103, 141*).

Extensive research into ECR plasmas, which is important for the better understanding of the basic phenomenon and is critical for successful source design and performance, has been done in the last three or four years. Mechanisms of microwave power absorption in low-pressure ECR plasma sources were studied by Dandl *et al.* (*100, 110, 139*), by Lichtenberg and Lieberman with co-workers (*107, 123, 136, 167, 206, 235*), and by J. E. Stevens *et al.* (*127, 149, 236*). Electron energy distribution was studied by Kushner *et al.* (*153*), Amemiya with co-workers (*82, 84, 99*) and Hershkowitz with colleagues *230, 240*). Ion energetics were studied by Matsuoka *et al.* (*83, 126*), Holber with colleagues (*94, 117, 134*), and Gottscho et al. (*119, 142, 238*). ECR plasma modes were studied by Popov (*106, 114, 159*), Gorbatkin (*104, 105, 219*), and Carl (*107, 123*). ECR plasma UV and x-ray radiation was studied by Buchanan et al. (*132*) and T. J. Castagna et al. (*152*). Extensive plasma characterization (electron density and temperature, plasma and floating potentials, ion current density) were done by Carl et al. (*107, 123*), Gorbatkin et al. (*104, 105*), Popov (*86, 93, 114, 130*), Samukawa et al. (*120, 135, 220, 239*), Hopwood et al. (*101, 102*), Rossnagel et al. (*115, 129*), and Nihei et al. (*109, 146*).

However, in spite of the extensive research on processes in ECR plasmas, many phenomena that have been observed in ECR plasma sources are still not well understood. So, much more effort needs to be applied to make ECR plasma sources and PACVD technology "engineered" objects.

Though the number of ECR papers exceeded many hundreds, there are only two or three ECR PA CVD review papers. The first review was written by S. Matsuo in 1986 (*156*). Matsuo described only the "NTT-type" ECR source and presented some results of ECR CVD of Si_3N_4 and SiO_2 films and ECR sputtered deposition of Ta_2O_5 and Al_2O_3 films, which he accomplished in the early 1980s. Only 15 references dealing with ECR CVD, all performed in Japan, were cited.

The other review paper, written by Dusek and Musil (*157*), was published in 1990. The authors cited more than 60 ECR PACVD works made in Japan, USA, Great Britain, France, Germany, Czechoslovakia, Canada, and China.

This paper is an attempt to review and analyze the main types of ECR plasma sources suitable for ECR PACVD of thin films, mainly ECR sources using magnet coils, which are often called NTT/Hitachi type. In Section II we discuss principles of electron cyclotron resonance and possible mechanisms of microwave power absorption, which is a "core" of the ECR plasma source. Section III is devoted to a brief analysis of the most widely used ECR plasma source. In this section we present major plasma parameters and their dependencies on external source parameters (pressure, microwave power, geometry). In Section IV we present and discuss the most "established" ECR CVD technologies: silicon nitride and silicon oxide film deposition. This section deals more with process characterization than with film properties. The author of this paper is a plasma physicist whose expertise is plasma research and plasma and ion source design. Therefore, he asks to be excused if the reader finds too much about plasma and too little about thin films.

II. ECR Fundamentals and Microwave Power Absorption

A. Principles of Electron Cyclotron Resonance

Electron cyclotron resonance (ECR) source is a nickname for the class of plasma and ion sources where high-density plasmas, typically $N_e = 10^{11} - 10^{13}$ cm^{-3}, are generated by microwave power at frequencies $f = \omega/2\pi = 916$–$3{,}500$ MHz, in the presence of a static magnetic field, $B_z = B_{ce} = \omega c m / e$. Here m and e are electron charge and mass, respectively.

Several types of ECR plasma sources are discussed later, and schematics of the five most-used sources are given in Section III. The typical ECR plasma source consists of the microwave introduction technique (window, antennae), magnetic field arrangement (coil, permanent magnets), plasma chamber (cavity, waveguide), and the substrate zone. The key element in an ECR plasma source is the microwave absorption zone where a plasma is generated.

The absorption zone is usually conceived as a site or a layer where the static magnetic field has a value of B_{ce} and the microwave electric field is not zero.

The requirement for the "resonance" came from the expression for time-averaged microwave power density, P, which is transferred to "cold" electrons ($T_e = 0$) at the microwave breakdown (1,2):

$$P = \mathrm{Re}\{J \times E\} = \sigma E_r^2 = \frac{e N_e E_r^2}{4m} \left\{ \frac{v_e^2}{v_e^2 + (\omega - \omega_{ce})^2} + \frac{v_e^2}{v_e^2 + (\omega + \omega_{ce})^2} \right\}. \quad (1)$$

Here σ is the plasma conductivity; E_r is a component of the microwave electric field, which is perpendicular to the static magnetic field, $E_r \perp B_z$; and v_e is the electron–neutral elastic collision frequency. It can be seen that there are two resonances for the microwave power: (i) $\omega = \omega_{ce}$, and (ii) $\omega = v_e$. The first term in the right-hand part of Eq. (1) represents the resonance between microwave and gyrotron frequencies when the electron rotates around the static magnetic field line in the same direction as the vector of the microwave electric field. These waves are commonly called right-hand circularly polarized waves (R).

It is important to note here that this resonance is irrelevant to wave phase velocity, v_{ph}. Since Eq. (1) was derived for $N_e \ll N_{\mathrm{cr}}$, wave phase velocity is always much higher than electron velocity, v_e: $v_{\mathrm{ph}} = \omega/k = c/n \gg v_e$ (68, 70). Here, k is wavenumber, n is the refractive index (at $N_e \ll N_{\mathrm{cr}}$, $n \simeq 1$), and c is the velocity of light.

The second term in Eq. (1) is the absorbed microwave power of so-called left-hand polarized waves (L), whose vector of the electric field rotates in the direction opposite to that of the electron rotation in the magnetic field. It can be seen that the presence of the magnetic field even reduces the level of L-wave power absorption. Thus, "pure" electron cyclotron resonance can be only for right-hand polarized waves.

Equation (1) looks very simple and convenient to use. Unfortunately, it does not represent real conditions that exist in ECR plasma sources. Indeed, Eq. (1) was derived with many assumptions: zero electron temperature; low plasma density, $N_e < N_{cr}$ (in industrial ECR sources $N_e \gg N_{cr}$); homogeneous plasma density, $dN_e/dr = dN_e/dz = 0$; homogeneous magnetic field, $dB_z/dr = dB_z/dz = 0$; diffusion regime for electrons, which is not the case for low gas pressures, $p < 1\,\text{mTorr}$. But probably the most "controversial" assumption is that of the approximation of the linear electrodynamics, i.e., the electron drift velocity under effect of microwave electric field, v_E, is assumed to be much smaller than the electron thermal velocity, v_{Te} (64). The threshold microwave electric field, E_{th}, which causes the deviation from the linear electrodynamics $v_E = 0.1\,v_{Te}$, was defined by Ginzburg as (64)

$$E_{th} = 4 \times 10^{10} \left(\frac{2m}{M} T_e(\omega^2 + v_{eff}^2)\right)^{1/2}. \quad (2)$$

For typical electron temperatures in the ECR absorption zone of 15–20 eV, the microwave threshold electric field $E_{th} = 200\text{–}300\,\text{V/cm}$. In ECR plasma sources, having plasma chamber diameters of 15–25 cm, such a microwave "vacuum field" corresponds to an incident microwave power of 200–300 W, which is minimal operational power in commercial ECR reactors. Thus, the approximation of the linear electrodynamics is not valid for processes in ECR plasmas, and Eq. (1) is not correct for the calculation of microwave power absorbed in ECR heating.

The approximation of the liner electrodynamic also implies that the dissipation of the microwave power (decreases of the electric field) does not change plasma conductivity, σ, i.e., plasma parameters: electron temperature, plasma density. As a result, according to Eq. (1), the absorbed microwave power, $P = \vec{J} \times \vec{E} = \sigma E^2$, is linearly proportional to plasma density and to the power square of the electric field.

The approach made in Ref. 2 implied that electron temperature $T_e = 0$ (cold plasma approximation), which is not the case in real ECR plasmas. This leads to a "pure" electron cyclotron resonance for R waves: $\omega = \omega_{ce}$. The account of the electron velocities and electron energy distribution, f_{ee} (as it is shown later), broadens the absorption zone and shifts the electron cyclotron resonance frequency to larger values (*13, 63, 66, 68, 75, 110, 139, 149*).

The width of the resonant region in Eq. (1) is determined by the value of the electron collision frequency, v_e. At pressures of 0.1–10 mTorr, and microwave frequencies of 1–10 GHz, $v_e \ll \omega$. Therefore, the half-width of the resonance curve $P = f(\omega - \omega_{ce})$ is expected to be very small, and the ratio v_e/ω is equal to 10^{-4}–10^{-5}.

Meantime, numerous experiments showed that even at pressures of 0.1–1.0,Torr the value of the half-width was ≈ 0.5–0.7, which is several orders higher than the ratio v_e/ω. This indicates the collisionless nature of the microwave power absorption mechanism at pressures up to 10 mTorr.

The authors who derived Eq. (1) realized the nature of their assumptions. That is why they tested their model only for the microwave breakdown where Eq. (1) seemed to be valid (*2*). But even at breakdown conditions, where plasma density is much lower than its critical value, $N_{cr} = \omega^2 m/4\pi e^2$, the experimental value of v_{eff} was found to be equal or close to v_e only at pressures higher than 10 mTorr where the electron mean free path λ_e is smaller than the plasma chamber dimensions. At pressures below 10 mTorr, the experimental half-width of the breakdown resonance curve was found to be practically constant and much larger than the electron–neutral collision frequency v_e.

Many experiments made in the 1970s showed that even at $N_e \ll N_{cr}$, the effective electron collision frequency, v_{eff}, could be a function of microwave electric field, plasma density, electron temperature, and magnetic field (*160–165*). This means that microwave power absorption mechanism can be nonlinear even at the low plasma density, $N_e = 0.1 N_{cr}$, that makes Eq. (1) irrelevant for underdense magnetoactive plasmas, $N_e < N_{cr}$.

Thus, "pure" ECR microwave power absorption occurs only at microwave breakdown and a very low microwave power absorption level where the collective plasma effects do not play any role and where absorption mechanisms other than ECR cannot be effective at $B \approx B_{ce}$. In industrial ECR plasma sources operated at plasma densities much higher than N_{cr}, and at magnetic fields higher than ECR field, the actual microwave power absorption process is more complicated and can be caused by other wave dissipative mechanisms that are not necessarily associated with electron cyclotron resonance.

B. Microwave Power Absorption in ECR Plasmas

When electromagnetic waves enter magnetoactive plasma (i.e., plasma in a static magnetic field), they can be completely or partially damped by various dissipative processes. The dissipative mechanism may result either in heating of electrons and ions with the successive loss of their energy in elastic and inelastic collisions or their loss on the plasma boundaries (walls, window, substrate holder, etc.), or in the transformation (conversion) into other waves that are absorbed in the plasma or reflected back or that propagate to the substrate area.

In the case of electron cyclotron resonance where the microwave frequency ω is of the same order as the electron gyrotron frequency, $\omega = \omega_{ce}$, waves' dissipative processes go via electron heating. In order to heat plasma ions directly, the magnetic fields must be increased in $\omega_{ce}/\omega_{ci} = M/m$ times. For the commonly used microwave frequency of 2.45 GHz, the ion cyclotron resonance (ICR) requires the magnetic field $B_{ei} = 1{,}604{,}150\,G$ for the hydrogen ion, H^+, which is 1,834 times higher than ECR magnetic field, $B_{ce} = 875\,G$. Which mechanism of microwave power absorption will be dominant depends on electromagnetic waves mode (TE_{nm} or TM_{nm}), magnetic field configuration and magnitude, plasma density, chamber geometry, and gas pressure.

Next we briefly describe some dissipative mechanisms of microwave power that can occur in ECR plasma sources used in plasma processings:

(1) "Pure" EC heating ($\omega = \omega_{ce}$) (*1, 4, 20, 74, 81, 92–98*).
(2) Doppler shifted ECR (*5, 6, 7, 13, 63–71, 100, 104, 106, 110, 123, 149, 154*).
(3) Whistler waves absorption in overdense magnetoactive plasmas (*11, 17, 63, 64, 68, 71, 113, 127, 130, 131, 149, 154, 168, 169, 194, 236, 239*).
(4) Left-hand polarized wave absorption (*11, 64, 65, 68, 71, 130, 170, 171*).
(5) Upper hybrid resonance (UHR), $\omega_{UHR} = \omega_{ce} + \omega_{pl}$ (*72, 145, 175–178*).
(6) Higher harmonic resonances, $\omega = 2\omega_{ce}$, $3\omega_{ce}$ (*11, 71, 75, 126, 145, 179–181*).
(7) Plasma geometrical resonances (*182–185, 201*).
(8) Stochastic heating (*172–174, 186, 202–211, 235*).

(9) Nonlinear power absorption at high microwave electric field, $E > E_{th}$, (*10–12, 69–71, 160–166, 168, 176–178, 192*).
(10) Nonresonant collisional absorption of left-hand polarized waves (*1, 2, 63–65, 71, 89*).

We do not consider here other mechanisms of microwave power absorption that can be efficient at low driving frequencies, such as low hybrid resonance, $\omega = (\omega_{ce}\omega_{ci})^{1/2}$ (*72, 184, 187–190*) or ion cyclotron resonance (*191–193*).

All effective mechanisms of microwave power absorption at low pressures, $v_e \ll \omega$, can be divided in two groups: cyclotron resonances and plasma resonances.

Group A: Cyclotron resonances. An electron gains energy when one of its characteristic frequencies associated with magnetic field (gyrotron, hybrid, etc.) is equal to the frequency of the microwave electric field (or its harmonics). An electron rotates in phase with the rotating electric field E, which is perpendicular to the magnetic field B_z and to the wave propagation vector k. Plasma density N_e, and hence, plasma frequency ω_{pl} and electron temperature T_e, do not affect the mechanism of microwave power absorption. The major mechanism of electron momentum transfer, which determines the plasma conductivity, σ, is electron–neutral collisions. Plasma density grown linearly with microwave power absorbed, P_{abs}, and the microwave electric field is assumed to be uniform in the absorption zone.

Such a process of absorption occurs at breakdown in a resonant cavity with microwave standing waves that oscillate with a frequency $\omega = 2\pi f$, as (*1, 167, 186, 206*)

$$E(r, t) = E(r) \exp(j\omega t), \qquad (3)$$

and absorbed microwave power density, P/S, is given by Eq. (1).

However, as soon as plasma density in the resonant cavity approaches the critical value, N_{cr}, the cavity is no longer the resonant one, and the approach of the standing waves with zero phase velocity is not valid. The "replacement" of standing waves with travelling waves (transverse or longitudinal) raises the issue of the wave phase velocity with respect to those of electrons, and the plasma effect on the wave characteristics: polarization, wavelength, propagation, reflection, absorption, etc.

"Real" ECR plasma sources deal with travelling waves in overdense magnetoactive plasmas where mechanism(s) and level of microwave power absorption are controlled not only by the microwave electric field and static magnetic field and their spatial distribution, but also by electron energy and plasma density.

Group B: Plasma resonances. In magnetoactive plasmas, the transverse electromagnetic wave, which has vacuum electric fields E_x and E_y perpendicular to the wave propagation direction, "acquires" a longitudinal component, E_z, due to different phase velocities of right-hand (R) and left-hand (L) circularly polarized waves. The formation of the electric field E_z, which is parallel to wave propagation, causes the possibility of the transformation of the predominantly transverse wave, say TE_{11}, to a longitudinal plasma wave. It can occur at plasma densities N_e close to N_{cr} when the wave complex refractive index n approaches infinity and the plasma wave phase velocity v_{ph} approaches the electron thermal velocity v_{te}. Then, an electron moves in the wave propagation direction along and in phase with the electric field ($E_z \| K$) and absorbs microwave energy (*63–65*).

This mechanism of wave attenuation is called "Landau damping" (*194*). It was experimentally observed by many researchers at low microwave power level: from $1-2\,\text{mW/cm}^2$ to a few watts per square centimeter (*10, 71, 144*). In other words, when the microwave frequency becomes equal to the plasma frequency, the energy of electromagnetic waves is transformed into plasma electron oscillations.

The role of the magnetic field is to provide conditions at which the wavelength of the wave entering an overdense plasma reduces to the dimensions of the Debye radius, $D_e = v_{te}/\omega_{pl}$, which is the characteristic length of plasma electron oscillations. Indeed, the implementation of conditions $v_{te} = v_{ph}$ and $\omega = \omega_{pl}$ is equivalent to the requirement $D_e = 2\pi/K$.

Let's review the most probable mechanisms that can exist in ECR plasmas and might be "responsible" for microwave power absorption.

1. "Pure" Electron Cyclotron Resonance

Microwave power absorption occurs when a single electron or a group of electrons in a plasma with density below N_{cr} are accelerated by the electric field, E_r and E_θ, of the right-hand polarized wave which rotates in phase with electron(s) under the effect of the magnetic field B_z ($B_z \perp E_r, E_\theta$). Accelerated electrons lose their energy in collisions with heavy neutral particles that result in ionization and electron momentum transfer. In "pure" ECR, $v_{\text{eff}} \equiv v_e$. Thermal electron velocity is assumed to be zero, and thus the wave space dispersion is ignored, i.e., an electron passes during the wave period a distance much smaller than the wavelength, $Kv_{te} \ll \omega$ (64). In the presence of the static magnetic field the cold plasma approximation gives (63)

$$Kv_{te} \ll |\omega - \omega_{ce}|. \tag{4}$$

In other words, the wave phase velocity, v_{ph}, is much larger than electron thermal velocity, v_{te} (cold plasma approximation). Such a model is correct only at very low pressure where plasma density is much lower than N_{cr}, and, of course, at ECR microwave breakdown. The electron momentum transfer process (phase randomization) that is needed for microwave power absorption is electron collisions with neutral particles, molecules, and electrons, which is the case at relatively high pressures, $p > 1$ mTorr. At low pressures, $p < 1$ mTorr, phase randomization mechanisms other than collisions are needed (wave instabilities, stochastic heating), which we will discuss briefly later (172, 173, 186, 206).

2. Doppler Broadened ECR

With account of the electron finite velocity and moreover, the electron energy distribution, f_{ee}, the process of microwave power absorption becomes more complicated because of the Doppler shift and broadening in the resonance frequency. The resonant frequency broadens to include higher magnetic fields and different groups of electrons in the electron energy distribution function satisfy the resonances over a wide range of magnetic fields (6, 13, 63–66, 75, 149).

The increase of the plasma density to values close to N_{cr} results in a decrease of the wavelength so it becomes equal or smaller than the resonance zone, where $\omega = \omega_{ce}$. This creates the possibility of microwave attenuation by the energetic tail in electron energy distribution, f_{ee}, at magnetic fields even before the ECR resonance zone (13, 66). This results in the broadening of the range of magnetic fields "participating" in the microwave power absorption. Mathematically, the effects of both plasma density and electron velocity on the resonant magnetic fields are written as (63, 64)

$$|\omega - \omega_{ce}| \simeq Kv_{te}. \tag{5}$$

The broadened ECR magnetic field at different plasma density N_e and at electron thermal velocity (electron temperature, T_e) was studied in Refs. 13, 100, 110, and 149. The relationship between "resonant" magnetic field range, B/B_{ce}, and N_e and T_e in the Doppler broadened absorption ECR zone was given in Ref. 149:

$$B/B_{res} \simeq 1 + \{N_e/N_{cr}(v_{te}/c)^2\}^{1/3}. \qquad (6)$$

The modeling and experimental results obtained in Ref. 149 did not give the resonant values, but gave the range of magnetic fields where microwaves are completely absorbed. At plasma densities of 10^{12} cm^{-3} and electron temperatures $T_e = 6\text{-}10$ eV, this range is between 1,000 G and 925 G. It can be seen from (6) that the range of the absorption magnetic field decreases with N_e and T_e.

The author of this paper performed a series of experiments in ECR plasmas excited in chambers of different diameters at a frequency of 2.45 GHz in radially and axially uniform magnetic fields at microwave power of 300–1,500 W (*106, 113, 145*). No mode converter was used. The author found two major resonant magnetic fields: (i) at $B_z = 870\text{-}880$ G ("pure" ECR), and (ii) at a magnetic field within a range of $B_z = 915\text{-}940$ G. The magnitude of the resonant field, B_{res}, grew with plasma density and electron temperature. Typical dependencies of the plane probe ion saturation current, J_{io}, measured several centimeters downstream from the ECR zone are given in Fig. 1. It can be seen that the observed ion current peak at $B_z = 925$ G is higher than that at an ECR field of 875 G. The increase of the plasma density shifts the "resonant" magnetic field from 925 G to higher values. The maximum resonant magnetic field corresponding to the plasma density of $N_e = (2\text{-}3) \times 10^{12}$ cm^{-3} was 940 G. The same values were reported in Ref. 149 where a circular polarized TE$_{11}$ mode converter was used.

The thermal spread in electron velocities in the wave propagation direction causes not only the broadening of electron cyclotron resonance, but also an effective mechanism of noncollisional electron momentum scattering. The physics of this mechanism were discussed in many publications (*7, 13, 63, 66, 75, 195*). With the assumption of Maxwellian electron velocity distribution and uniform wave electric field and static magnetic field, Ben Daniel et al. (*75*) obtained an expression for microwave plasma conductivity σ_{res} ar electron cyclotron resonance $\omega = \omega_{ce}$:

$$\sigma_{res} = \frac{N_e e^2 (\pi m)^{1/2}}{m K_r (2k T_e)^{1/2}} = \varepsilon_0 \omega_{pl}^2 / v_{eff}, \qquad (7)$$

where the effective collision frequency, v_{eff} (*75*), is

$$v_{eff} = K_r (2k T_e/m)^{1/2}/\pi^{1/2}, \qquad (8)$$

and K_r is the real part of the complex wave number K (*64*).

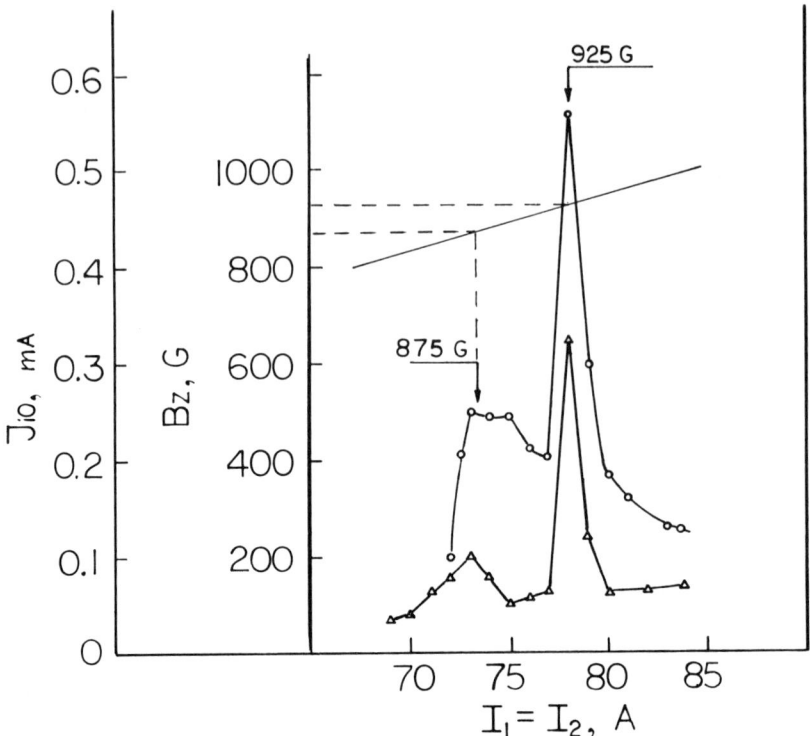

FIG. 1. Probe ion saturation current on the plasma axis and magnetic field as functions of magnetic coil currents, $I_1 = I_2 = I$. After Ref. 53.

3. Whistler Wave Absorption at $B \gg B_{ce}$

Many authors generated very high plasma density in microwave plasmas inserted in a magnetic field much higher than Doppler shifted ECR, $B/B_{ce} > 1.2$ (*19, 71, 130, 163, 168*). The most effective microwave power absorption was observed when microwaves entered a magnetoactive plasma from the "high-field side" (Whistler wave), provided the plasma density was higher than N_{cr} (*130, 168, 193, 100, 139, 149, 236*). Few absorption mechanisms were suggested to explain the observed results. One of them was based on the experimentally observed transformation of transverse electromagnetic waves into longitudinal plasma waves near N_{cr}. Longitudinal plasma waves have very small wavelength, and hence small phase velocity, v_{ph}, at $N_e \simeq N_{cr}$, close to the velocities of electrons on the tail of the electron energy distribution, f_{ee}. Under some conditions v_{ph} is even close to v_{te}, the electron thermal velocity. The electric field of a plasma longitudinal wave is parallel to the wave propagation. Therefore, plasma electrons that move in phase with the wave very efficiently acquire microwave power in Landau damping (194).

The wave transformation is illustrated in Fig. 2 where the dispersion curves for complex refractive indexes of extraordinary and ordinary waves, n_1^2 and n_2^2, propagating without losses ($v_{eff} = 0$) in a cold magnetoactive plasma ($Kv_{te} \ll |\omega - \omega_{ce}|$) at small angle α, are plotted as functions of plasma density normalized to its critical density, $N_e/N_{cr} = \omega_{pl}/\omega$ (63, 64, 71, 130).

The denominator in the equation for n_1, n_2 has sign ($-$) for R waves and sign ($+$) for L wave (63–65):

$$n_{1,2}^2 = 1 - \frac{2\frac{N_e}{N_{cr}}\left(1 - \frac{N_e}{N_{cr}}\right)}{2\left(1 - \frac{N_e}{N_{cr}}\right) - \frac{B}{B_{ce}}\sin^2\alpha \pm \left[\left(\frac{B}{B_{ce}}\sin^2\alpha\right)^2 + 4\frac{B}{B_{ce}}\left(1 - \frac{N_e}{N_{cr}}\right)^2\cos^2\alpha\right]^{1/2}}. \tag{9}$$

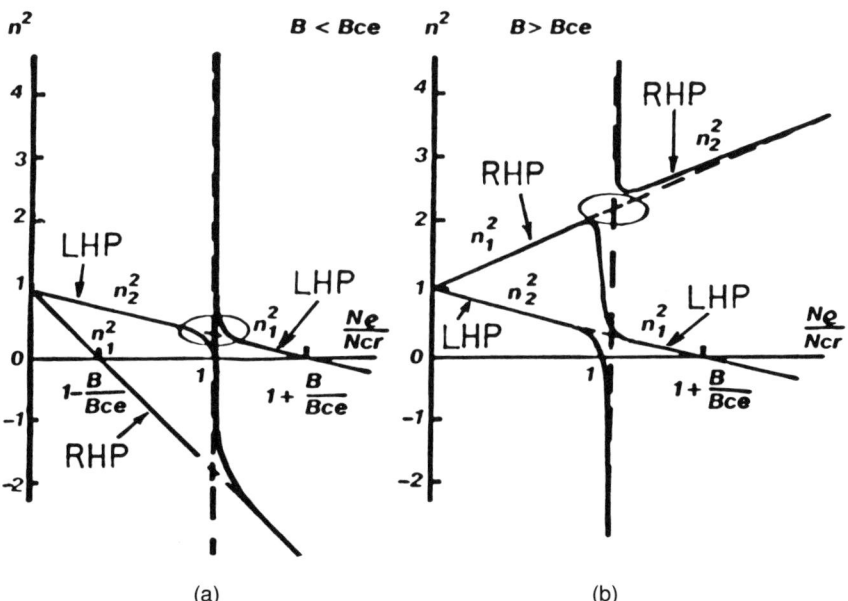

FIG. 2. Complex refractive index, n^2, as a function of the ratio N_e/N_{cr}, for two magnetic fields: (a) $B < B_{ce}$, and (b) $B > B_{ce}$. $v_{eff} = 0$.

It can be seen from Eq. (9) and Fig. 2 that each wave changes its polarization (the sign in the denominator) when passing a zone $N_e = N_{cr}$. Also, at $N_e \simeq N_{cr}$, the refractive index of each wave approaches infinity, which indicates that the zone of wave transformation ($N_e \simeq N_{cr}$) is a resonant zone. The remarkable feature of these two resonances is that they can occur at magnetic fields that are much higher and lower than the ECR (or Doppler shifted ECR) field.

From the practical point of view it means that a high level of microwave power absorption of transverse waves can occur at $B_z \ll 875$ G, as well as at $B_z \gg 875$ G, provided the plasma density is not lower than N_{cr}. This was experimentally verified by Musil et al. (71, 168, 193), who used slow wave structure for the generation of Whistler waves propagating in a high magnetic field, $\omega \ll \omega_{ce}$, plasma. The absorption was in sites where plasma density was about N_{cr}, which indicated Landau damping wave dissipation.

It can be noticed also that there is an evanescent zone in the front of the resonant zone where $n \to 0$ (Fig. 2). The width of this zone varies with the magnetic field gradient, angle between magnetic field and wave propagation, and plasma density gradient, dN_e/dz. The portion of microwave power that propagates through the evanescent zone (tunneling) was analyzed by Zheleznyakov (196).

It was shown (64, 71) that when the width of the evanescent zone is much smaller than the wavelength, λ_0, almost all the microwave power "tunnels" through the region between N_{cr} and N_{res} and either is absorbed in Landau damping or propagates downstream and is absorbed in EC heating. (We do not place emphasis on ECR because of the substantial portion of power that could be absorbed before it reaches $B_z = B_{ce}$.)

In commonly used ECR plasma sources operating at microwave power $P_f > 300$ W, the plasma densities near the microwave introduction (window or antenna) grow from 0 to 10^{11} cm^{-3} within a few millimeters, while the wavelength in overdense magnetoactive plasmas with $B > B_{ce}$ is typically 1–3 cm (6, 7, 71, 149). Therefore, one can expect an increase of the wave propagation through the evanescent zone and efficient microwave power absorption in cavities with axially uniform magnetic field and operated at high microwave power (higher plasma density, N_e, and higher density axial gradient, dN_e/dz).

4. Left-Hand Polarized Wave Absorption

Microwaves introduced into a plasma chamber via a rectangular waveguide (typically WR 284 at a frequency of 2.45 GHz) are plane waves with linear polarization. They can be represented as the superposition of right- and left-hand polarized waves (*63, 167*). The R waves as shown earlier can be absorbed at $B > B_{ce}$ either in (Doppler-shifted) ECR or in Landau damping at $N_e \approx N_{cr}$ via transformation into longitudinal plasma waves.

What happens with L waves? They cannot be absorbed directly in EC heating. Therefore, they are expected to be reflected back or propagate downstream and are absorbed inefficiently via collisions with neutral particles. Meanwhile, it was experimentally shown by many authors that linearly polarized waves were almost entirely absorbed (90–95%) at low pressures, $p < 1\,\text{mTorr}$, in cylindrical plasma chambers at magnetic fields much higher and much lower than ECR (*18–20, 26, 35, 71, 78, 79, 100, 120, 123, 126, 127, 130, 135, 137, 145, 149, 196, 236, 239*). As a result, the plasmas with densities much higher than N_{cr} ($N_e > 10^{12}$) were generated at microwave power of 500–1,000 W.

Experiments aimed at temporal study of the absorption of L waves propagating at a small angle to the magnetic field, $B < B_{ce}$, were done by Ferrari *et al.* (*70*) and by Musil *et al.* (*71, 158*). They observed a very high level of microwave power absorption as soon as plasma density N_e reached N_{cr}. The absorption resulted in sharp growth of plasma density to values several times higher than N_{cr}. Musil explained this phenomenon by the transformation of transverse L waves into the longitudinal plasma waves at $N_e \approx N_{cr}$ that was theoretically discussed earlier by Ginzburg (*64*) and Allis *et al.* (*65, 198*) and experimentally observed at low microwave power level by Budnikov *et al.* (*68*) and Arkhipenko *et al.* (*170*).

Ferrari *et al.* also observed absorption of L waves only at $N > N_{cr}$. But the absorption was accompanied by the generation of ion acoustic waves, which are supposed to be dissipated in Landau damping (*70*).

The portion of L waves that is transformed into plasma waves depends on angle, magnetic field and plasma density gradients at $N_e \approx N_{cr}$ and discussed in Refs. 64, 71, and 158.

The indirect experimental evidence of L wave absorption in magnetoactive overdense plasmas, $N_e > N_{cr}$, in magnetic fields higher and lower than ECR was observed by the author of this paper in Refs. 93, 106, 130, and 145. The linearly polarized waves (2.45 GHz) were introduced into 10 cm and 15 cm diameter chambers via a quartz window. The initial plasma was ignited by R waves at ECR ($B_{ce} = 875\,\text{G}$). When the plasma density in the chamber was below cutoff density for L waves (*63*)

$$N_{\text{cutoff}} = N_{cr}(1 + \omega_{ce}/\omega) = N_{cr}(1 + B/B_{ce}), \tag{10}$$

the substantial microwave propagation through the plasma was detected downstream. As soon as plasma density in the chamber (at microwave absorption zone) reaches values of $N_e \simeq 10^{11}\,\text{cm}^{-3}$, the microwave stopped propagating downstream, and plasma density jumped to values much higher than N_{cutoff}.

The numerous experiments carried out by many authors in microwave plasmas generated by nonpolarized TE_{11} waves reported a sharp increase in plasma density (or in Langmuir probe ion saturation current) at forward microwave powers of a few hundred watts (*31, 53, 89, 93, 106, 113, 120, 122, 123, 126, 127, 147*). Some of these data are given in Section III. In Fig. 3 we present two experimental curves J_{io} vs. microwave power taken from Ref. 120. One can see a distinctive surge of ion current density (hence plasma density) at microwave power 300–700 W. Such a large range in the threshold power can be caused by the uncertainty of the actual absorbed microwave power, which was not measured in many experiments. But one can notice that the "threshold" plasma density is around $(5-10) \times 10^{10}\,\text{cm}^{-3}$ in both graphs. These electron densities are close to $N_{\text{cr}} = 7.4 \times 10^{-10}\,\text{cm}^{-3}$, which is the critical plasma density for microwaves of a frequency of 2.45 GHz used in Ref. 120.

The same threshold plasma density was observed in microwave plasmas maintained at low magnetic fields of $B = 100-300\,\text{G}$ (*145*). This observation questions the idea that the complete absorption of L wave in magnetoactive plasma after it is reflected from the surface $N_e = N_{\text{cutoff}}$

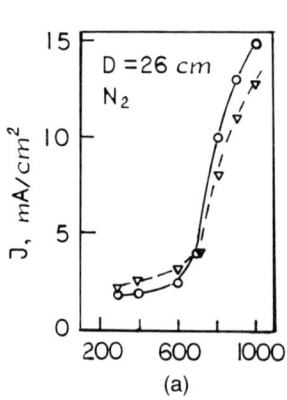
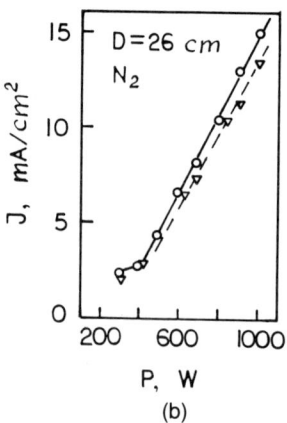

FIG. 3. Ion current density at ECR magnetic field, J_i, as a function of the incident microwave power, P, for two radial magnetic field profiles: (a) nonuniform; (b) uniform. Note that for both cases the "threshold" ion current density is the same, $J_i = 2.5\,\text{mA/cm}^2$. Nitrogen, 0.25 mTorr. After Ref. 120.

is due to the change of wave polarization to the opposite one (R wave) with successive absorption in ECR (*167*). More realistic is the concept of L wave transformation into plasma longitudinal waves (*71*) or into ion acoustic waves (*70*) at $N \simeq N_{cr}$ that can occur at very low magnetic field, $B_z \ll B_{ce}$, provided this field can maintain the plasma ionization balance so the plasma density in the microwave absorption zone is always higher than N_{cr} (*139, 153, 167*).

5. Upper Hybrid Resonance

In the previous section we mentioned that microwave plasmas with densities much higher than N_{cr} can be generated at magnetic fields 2–8 times lower than an ECR field. The efficient attenuation can occur for the transverse electromagnetic wave $K \perp E$, propagating at a right angle to magnetic field, $K \perp B$, with the electric field E perpendicular to B. This wave is often called the "extraordinary wave" (X wave) (*63, 64*). The dispersion relation of this wave is (*63, 64*)

$$n^2 = \frac{c^2 K^2}{\omega^2} = \frac{c^2}{v_{ph}^2} = 1 - \frac{\omega_{pl}^2(\omega^2 - \omega_{pl}^2)}{\omega^2(\omega^2 - \omega_{UH}^2)}. \tag{11}$$

Here ω_{UH} is the upper hybride frequency (*64, 178*):

$$\omega_{UH} = (\omega_{pl}^2 + \omega_{ce}^2)^{1/2}, \tag{12}$$

where ω_{ce} is electron gyratron frequency in a magnetic field B.

When the microwave frequency ω becomes equal to $\omega_{UH} = (\omega_{pl}^2 + \omega_{ce}^2)^{1/2}$, or in other words, when there is a site in a plasma where plasma density and magnetic field have values that satisfy Eq. (12), the complex refractive index, n, approaches infinity. It means that the wave has a resonance at $\omega = \omega_{UH}$ where the phase velocity decreases to the velocity of the electrons, and the energy of the extraordinary wave is transformed into electrostatic oscillations on the upper hybrid frequency (*178*). The electrostatic oscillations are attenuated within a few millimeters in Landau damping (*144*).

There are two cutoff frequencies for the X wave ($n = 0$) (*144*):

$$\omega_{cutoff} = 0.5\{(\omega_{ce}^2 + 4\omega_{pl}^2)^{1/2} \pm \omega_{ce}\}, \tag{13}$$

where the plus sign is for R waves, and the minus sign for L waves.

By substituting ω_{cutoff} for ω, one can derive from (16) two equations for N_{cutoff} for R and L waves. N_{cutoff} for L wave was derived earlier (Eq. (10)). N_{cutoff} for R waves is (*63*)

$$N_{cutoff} = N_{cr}(1 - \omega_{ce}/\omega) = N_{cr}(1 - B/B_{ce}). \tag{14}$$

If a bounded magnetoactive plasma has a maximum plasma density higher than N_{cutoff} for R waves, there is always a site (or layer) in the plasma volume where $N_e = N(\text{UHR})$:

$$N(\text{UHR}) = N_{cr}[1 - (B/B_{ce})^2]. \quad (15)$$

A schematic illustration of Eqs. (10), (14), and (15) is shown in Fig. 4. Minimum initial plasma densities, generated by another method, which were needed for the ignition of self-sustained overdense microwave plasma at $B < B_{ce}$ taken from Ref. 168, are also shown in Fig. 4. It can be seen that these data are in good agreement with Eq. (15).

The plasma densities generated in microwave plasmas with magnetic field below ECR, $B < B_{ce}$, are typically 30–60% lower than in ECR plasmas and decrease as the magnetic field is reduced. This is illustrated by the dependence of ion saturation currents measured in the process chamber on the chamber axis, I_{io}, and off the axis ($R = 6\,\text{cm}$) I_{ir}, as

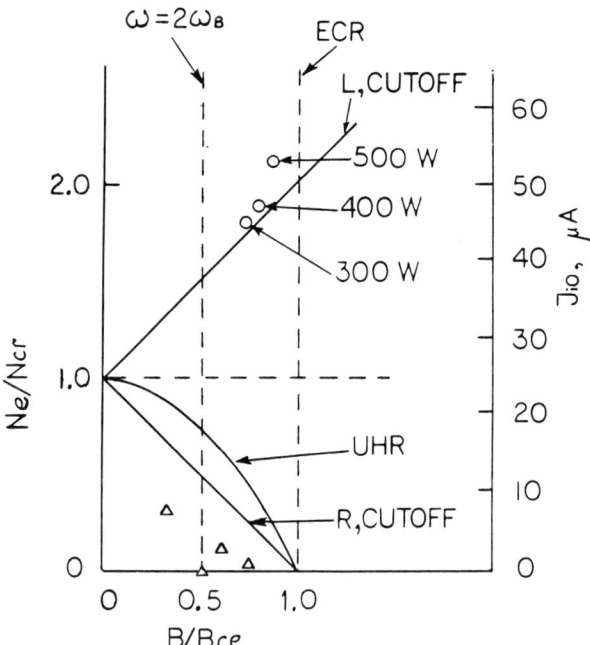

FIG. 4. A schematic illustration of eqs. (13)–(15) as a function of B/B_{ce}. Triangles represent minimal plasma densities generated by other than UHR mechanisms which were required for the ignition of overdense microwave plasma. Circles are the transition ion saturation currents taken from Fig. 6. After Refs. 145 and 168.

functions of magnetic coil currents $I_1 = I_2$ (Fig. 5). These currents generate an axially and radially uniform magnetic field in the plasma generation chamber, so these data are in fact dependencies of ion saturation currents, and hence the plasma density in the stream, from the magnetic field in the microwave absorption zone.

UHR as a method of plasma electron heating has been known and proved in fusion programs for more than 20 years (*175–177, 197*). However, it was not explored by the plasma processing community for plasma generation. Meanwhile, microwave plasmas maintained at UHR required much lower magnetic fields and hence smaller magnetic coils. However, magnetic field confinement is still needed for the reduction of plasma losses to walls and for the maintenance of a microwave plasma at pressures as low as 0.1 mTorr.

There is one substantial disadvantage in UHR: It requires the initial plasma be ignited by another method (*71*). It was shown in Refs. 145

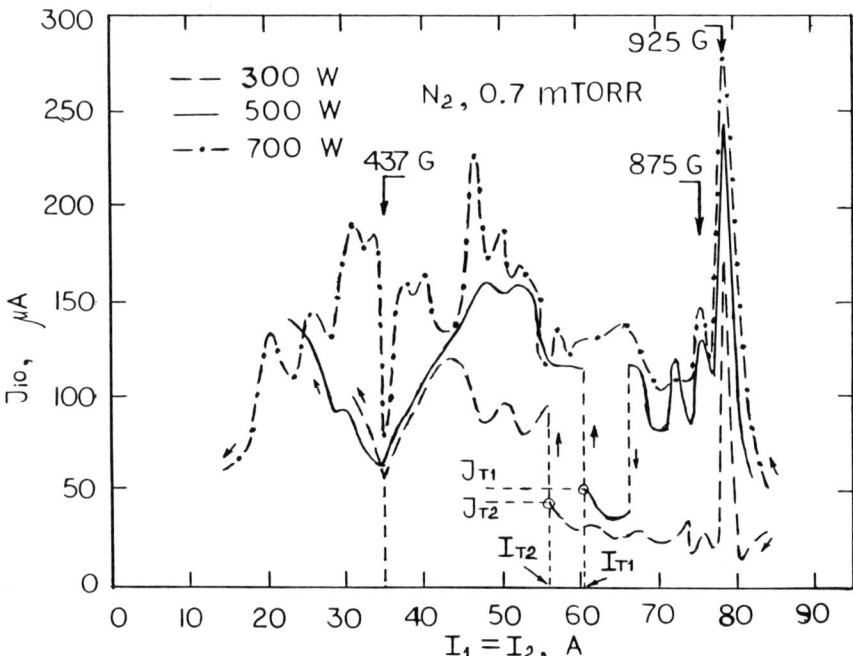

FIG. 5. Probe ion saturation currents on the chamber axis, J_{io}, and at $r = 12$ cm as functions of magnetic coil currents, $I_1 = I_2 = I$. Arrow shows probe currents at "pure" ECR magnetic field of 875 G. After Ref. 145.

and 168 that there are at least two magnetic field where the initial plasma is not required: ECR and second harmonic resonance (SHR).

6. Second Harmonic Resonance

While Studying the ignition of microwave overdense plasma in magnetic fields below ECR, few authors found that there is a peak in plasma density at magnetic fields half of the ECR field (*126, 145, 168*). We present here the experimental data of probe ion saturation current, I_{io}, as function of coil currents, $I_1 = I_2 = I$, measured at three different microwave powers (Fig. 6). It can be seen that there is a peculiar coil current, $I \simeq 35$ A, where I_{io} decreases and then jumps. It was shown in Ref. 145 that this peak corresponds precisely to $B = 437$ G, which is half of the ECR magnetic field of 875 G. In recent work (*181*) made in a magnetic mirror ECR machine (RIKEN), the highest plasma density was found (measured with a double probe) not on the ECR surface but on the surface $B = 0.5 B_{ce}$.

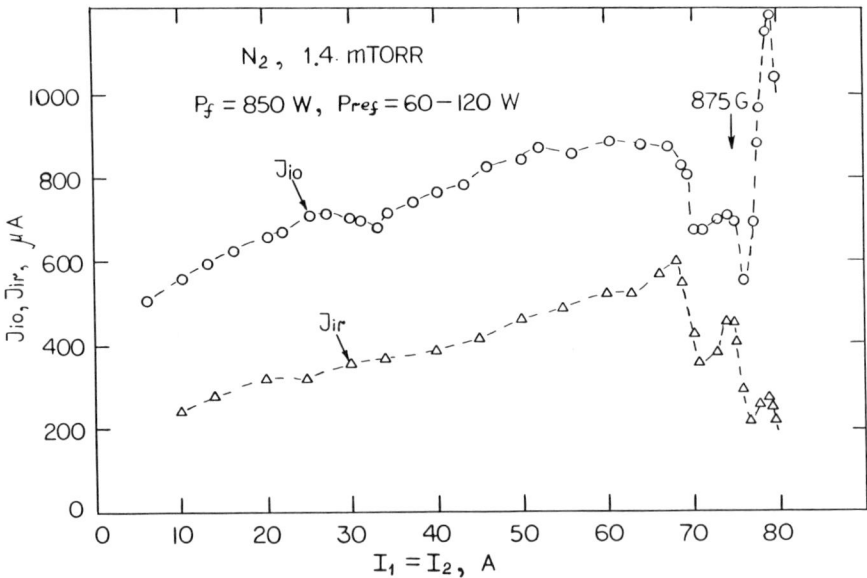

FIG. 6. Probe ion saturation current on the chamber axis, J_{io}, as a function of coil currents, $I_1 = I_2 = I$. Small arrows show the direction of the coil current changes. Large arrows show positions of 437 G (SHR), 875 G, and 925 G. Transition ion currents, J_{t1} and J_{t2} are shown in Fig. 4 as function of the normalized magnetic field, B/B_{ce}. After Ref. 145.

One of the first experimental observations of the resonance on the second gyratron harmonic,

$$\omega = 2\omega_{ce}, \qquad (16)$$

was done in 1963 by Miller and Gibbons, who used right-hand circularly polarized waves for the generation of a plasma accelerator (7). A few years later Datlov et al. (155) observed sharp resonances on second and other higher harmonics: $2\omega = 3\omega_{ce}$, $\omega = 3\omega_{ce}$, in microwave plasmas excited in the TE_{111} resonant cavity. In the following years there were few efforts to generate and heat fusion-related plasmas on the second harmonic (179, 180, 198, 199). In only the last few years, SHR appeared

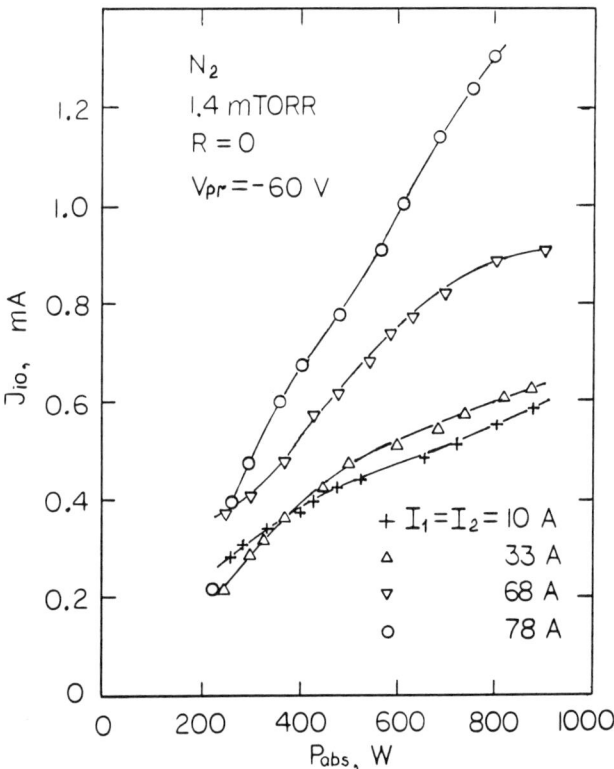

FIG. 7. Probe ion saturation current on the chamber axis as function of the microwave power absorbed. Measurements were done at different coil currents, $I_1 = I_2 = I$.

to be considered as a possible candidate for a high-density microwave plasma source with a low magnetic field (*126, 145, 181*).

It was found that at the same SHR magnetic field near the microwave introduction window ($z = 1-3$ cm), the highest plasma density in the source and downstream was when the magnetic field slightly decreased from the window, $dB_z/dz < 0$ (*145*). The plasma density on the axis, I_{io}, was found to be 30–50% lower than in the ECR plasma, but it was higher than in the ECR plasma at radii of 10–12 cm. This is illustrated in Figs. 7 and 8, where probe ion saturation currents I_{io} and I_{ir}, measured in SHR and ECR plasmas at $z = 35$ cm from the window, are plotted as functions of the microwave power absorbed, $P_{abs} = P_f - P_{ref}$. (No microwave propagation downstream was observed.)

Many authors reported the shift of resonant frequency to higher and to lower magnetic fields (*126, 145, 155, 180*). It was often explained by the electron thermal motion and energy spread (*200*) or by plasma and magnetic field radial inhomogeneity (*126, 145*). But this phenomenon could have another physical nature, which we will discuss in the next section.

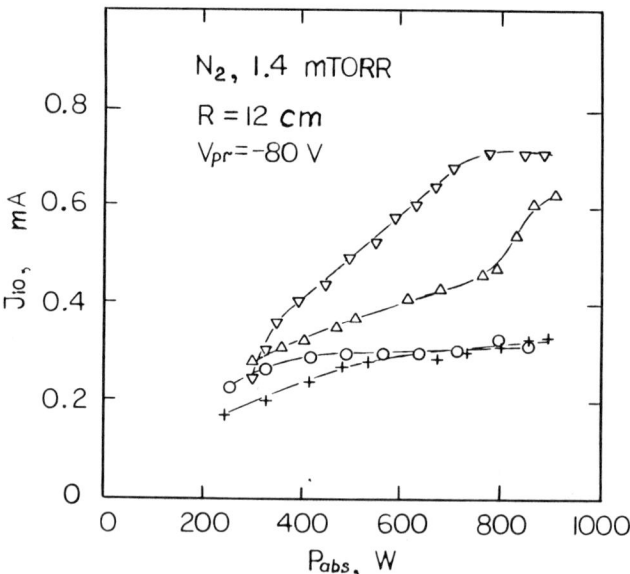

FIG. 8. Ion saturation current at $r = 12$ cm as a function of the microwave power absorbed. Conditions and key as for Fig. 7. After Ref. 145.

7. Geometrical Resonances

In previous sections we discussed microwave power absorption when a resonance has a local character, that is when it was associated either with local growth of the complex refractive index, n, (wavenumber K) to infinity, or with equality of the microwave frequency ω to the gyratron frequency ω_{ce} at a particular point. In industrial ECR sources having overdense plasma density, $N_e > N_{cr}$, the plasma volume is bounded with the conductive or dielectric walls where $N_e = 0$, and gradually increases to its maximum value at the volume axis over the distance comparable with the wavelength in vacuum, λ_0. (For a commercial frequency of 2.45 GHz, $\lambda_0 = 12.24$ cm.) If the wavelength is larger than the plasma diameter ($\lambda_0 > D_{pl}$, the quasi-static approximation), a series of geometrical resonances can occur on the fundamental and higher wave harmonics that can maintain plasma at low microwave (rf) power levels *(182–185, 201)*.

The nature of geometrical resonances can be explained with a circuit diagram. The plasma volume can be represented as a combination of active resistance, inductive impedance caused by the inertia of electrons in the high plasma density region, and capacitive impedances in the low plasma density region and in the sheath near the walls. When inductive and capacitive impedances are equal, the reflected power is zero, and the whole amount of power introduced to the chamber is absorbed.

The actual absorption mechanism could be collisional via electron–neutral and electron–ion collisions, or collisionless: EC heating, Landau damping, or stochastic heating (see the later discussion).

The reported plasma slabs and striations with dimensions of a few centimeters observed in an ECR chamber at a relatively low microwave power level of 2–50 W *(93, 116)* can be explained by geometrical resonances. We observed plasma striations of various structures and shapes at low microwave power of 2–50 W in a 14.6 cm diameter plasma chamber. The shape and location of plasma "slabs" were controlled by microwave power tuning and magnetic field. One pattern consisted of four plasma slabs oriented symmetrically around the chamber axis ("plasma cross"). Another pattern was a typical narrow plasma column with a diameter of 2–3 cm, which is similar to those observed by Brown *et al.* in a TE_{111} resonator *(3, 4)*.

Geometrical resonances were thought to be the cause of the observed shift of plasma density peaks at magnetic field corresponding to higher harmonic resonances: $\omega = 2\omega_{ce}$, $\omega = 3\omega_{ce}$, $2\omega = 3\omega_{ce}$, $\omega = 4\omega_{ce}$ *(145, 184)*.

The further increase of microwave power absorption level that results in the plasma filling of the whole chamber cross section with the growth to values $N_e > N_{cr}$ practically rules out the possibility of the geometrical resonance on the pumping frequency of 2.45 GHz. Therefore, in ECR sources operating at $N_e > N_{cr}$, and $B > B_{ce}$ (high K, small wavelength), the geometrical resonances can be effective only for ion-sound waves propagating within the small gap between chamber walls and critical plasma density, $N_e = N_{cr}$.

The analysis of geometrical resonances is beyond the scope of this paper. I would refer readers to the excellent papers written by A. M. Messiaen and P. E. Vandenplas (*184, 185*).

8. Stochastic Heating

At low gas pressures ($p = 10^{-4}-10^{-3}$ Torr) when electron–neutral and electron–ion collisions are very low ($v_e = 10^6-10^7 \text{ s}^{-1}$), there must be a phase randomization mechanism that is equivalent to electron momentum transfer in its collision with particles (*202, 203*). In the cases of "traveling waves," when the wave phase velocity v_{ph} approaches the velocities of electrons, the wave is attenuated in Landau damping ($E \parallel K$) or in Doppler-broadened electron cyclotron resonance ($E \perp K$). In both cases the interaction between microwave electric field and electrons is noncollisional with the effective phase randomization because of the large magnitude of Kv_{te} ($Kv_{te} \gg |\omega - \omega_{ce}|$ (*13, 63, 64, 66, 75, 172, 195*). (Note that Landau damping is a stochastic mechanism in its nature (*172, 194*).)

The randomization effect appears in the formulae for the plasma conductivity σ, as an effective electron collision frequency, v_{eff}, which includes both collisional terms, v_{e-n}, v_{e-i}, and a noncollisional term, Kv_{te}. Thus, by identifying the effective electron collision frequency, v_{eff}, one can determine the plasma conductivity, σ, and hence the microwave power absorbed. However, at conditions where the wave attenuation in a plasma volume due to Landau damping or collisions with particles is excluded or insignificant (e.g., $N_e < N_{cr}$, or $p < 1$ mTorr), some noncollisional phase randomization mechanism must be effective.

Kawamura *et al.* analyzed electron heating at ECR and second gyratron harmonic in a magnetic mirror by microwaves entering the chamber with $K \perp B$ (*173*). Plasma was underdense for hot electrons ($T_e = 100-150$ keV) and overdense for "cold" electrons, which were generated in collisions of fast electrons with neutrals (hydrogen). The authors (*173*) suggested that electrons are selectively accelerated or decelerated by microwaves dependent on electron velocity and its vector.

The phase randomization was caused by the fluctuation of the density of cold electrons.

It was shown theoretically by Godyak (*174, 205*) and Lieberman (*206*), and experimentally by Godyak and Popov (*207, 208*) and Lieberman et al. (*211*), that in low-pressure rf plasmas, where the conditions

$$\omega_{ci}^2 \ll \omega^2 \ll \omega_{pl}^2 \tag{17}$$

are met, the major mechanism of rf power absorption is noncollisional stochastic electron heating on the border of the overdense plasma and sheath formed by the rf field near the rf electrodes.

Jaeger et al. (*186*) and Lieberman et al. (*136, 209*) analyzed the multiple electrons bouncing in the magnetic mirror between two ECR heating zones. An electron gained energy from the oscillating microwave electric field localized in the ECR zone as a result of numerous passages through the ECR zone where an electron was accelerated or decelerated. Such stochastic heating can occur only if an electron does not "know" the electric field phase.

The author made an assumption in Ref. 145 that a similar mechanism could exist in a low-pressure microwave overdense plasma, where microwaves do not propagate through the plasma but are localized in a "sheath" between the microwave introduction window and the plasma surface, $N_e = N_{cr}$. The approach is similar to that in rf discharge. The microwave window plays the role of an rf electrode with a high microwave electric field localized in the zone between the window and the plasma surface where $N_e = N_{cr}$.

Because of high electron velocity, the window is negatively biased to a potential of 25–60 V, and thus forms a sheath with a negative potential barrier that reflects electrons and accelerates ions. The microwave electric field oscillates with a frequency of $f = (\omega/2\pi) = 2.45\,\mathrm{GHz}$, which is much lower than the plasma frequency $\omega_{pl} = (4\pi N_e^2 e^2/m)^{1/2}$. Therefore, plasma electrons follow microwave electric field oscillations in a manner similar to rf plasma electron oscillations, i.e., $v_E \ll v_{te}$. The stochastic mechanism of the phase randomization and the interaction of the localized oscillating electric field and electrons entering the zone of the oscillations was described by Akhiezer and Bakai (*204*), Godyak (*174, 205*), and Lieberman (*209*). The role of the magnetic field is to confine electrons in the volume and not allow them to be lost on the radial chamber walls. The effective electron collision frequency v_{eff}, which determines the plasma conductivity σ and hence the microwave lower absorption P_{abs}, is controlled by the distance between the microwave introduction window and the plate terminating the chamber (or substrate holder) which is also negatively biased and repels slow electrons back to the plasma.

It was theoretically and experimentally shown by Godyak that the resulting effect of electron acceleration and deceleration by the oscillating rf electric field is the heating of fast electrons on the electron energy distribution function tail (*174, 210, 211*). The main bulk of electrons were found to have unexpected low temperatures of 1–2 eV at pressures as low as 1 mTorr (*210, 211*).

The important feature of the stochastic heating studied in Refs. 174, 208–211 is that it is a linear process. It means that the energy acquired by electrons from the microwave (rf) field is much smaller than the average electron energy [*64*]:

$$W_e = e^2 E^2 / m v_{eff} \ll 3/2(kT_e). \tag{18}$$

In other words, the electron drift velocity in the field direction, v_E, is much smaller than the electron thermal velocity, v_{te}.

9. Nonlinear Microwave Dissipative Process

So far we discussed microwave dissipative processes that have linear character. Rigorously saying it implies that the propagation and absorption of microwaves in a plasma do not affect plasma parameters: plasma density, electron temperature, electron effective collision frequency. In the linear approximation, the plasma conductivity, σ, is not varied with the electric field of the electromagnetic waves, and the high-frequency current density, j, is the linear function of the wave electric field: $j = \sigma E$.

This is the case when the electric field is so weak that the energy that an average electron gains from the electric field between two consequent collisions, W, is much smaller than the electron thermal energy, W_e (*144*):

$$W \ll 3/2 kT_e, \tag{19}$$

where

$$W P_{abs} \tau_h / N_e \tag{20}$$

and

$$P_{abs} = 0.5 J_r * E = 0.5 \sigma E^2. \tag{21}$$

Here τ_h is the characteristic electron heating time. The electron energy balance requires that τ_h be equal to the characteristic electron energy loss time, τ_l:

$$\tau_h = \tau_l = 1/\chi v_{eff}. \tag{22}$$

Here χ is the mean fraction of energy that an electron loses between two collisions. At low pressures when the electron mean free path λ_e is much larger than plasma dimensions, the electron energy balance is an integral plasma characteristic, where the electron heating is the local process, while the losses are integral processes and occur primarily on the plasma boundaries (walls). At typical ECR operational pressures of 10^{-4}–10^{-3} Torr, $\chi \gg m/2M$, and $v_{eff} \gg v_{e-n}$, v_{e-i}.

The requirement of the linear approximation for electron velocity (negligible thermal nonlinear effects) comes from Eq. (19):

$$v_E \ll v_{te}, \tag{23}$$

where the electron drift (oscillation) velocity under the action of the electromagnetic field E is (at ECR)

$$v_E = eE/mv_{eff}. \tag{24}$$

The threshold electromagnetic field, which accelerates electrons to the oscillation velocity of $v_E = 0.1 v_{te}$, was derived by Golant for the transverse wave ($E \perp B$) (including R and L waves) (144):

$$E_{th} = (3\chi mkT_e(\omega^2 + v_{eff}^2)/e \times F(B), \tag{25}$$

where

$$F(B) = \frac{[\{(\omega - \omega_{ce})^2 + v_{eff}^2\}\{(\omega + \omega_{ce})^2 + v_{eff}^2\}]^{1/2}}{\{(\omega^2 + \omega_{ce}^2 + v_{eff}^2)(\omega^2 + v_{eff}^2)\}^{1/2}}. \tag{26}$$

It is important to note that the actual electric field in a plasma which interacts with plasma particles, E_{pl}, can be larger or smaller than the so-called vacuum electric field in the absence of a plasma, E_0. Therefore, in the calculation of the electron oscillation velocity one has to measure the plasma electric field, E_{pl}. Nevertheless, for the rough estimation of the degree of nonlinearity one can use the vacuum electric field of the electromagnetic wave entering the plasma:

$$E_0 = \{(P_{abs}/S_w)Z_w\}^{1/2}. \tag{27}$$

Here, S_w is the plasma chamber cross-section; z_w is the characteristic impedance of the wave entering the chamber.

The theoretical and experimental studies of nonlinear wave–plasma interaction have shown that nonlinearity becomes noticeable when the electron oscillation velocity $v_E \simeq 0.1\,v_{te}$ (*12, 144, 160, 163, 164, 166*). For typical electron temperatures in an ECR absorption zone $T_e = 8$–$15\,\text{eV}$, and a microwave frequency of 2.45 GHz and magnetic field $B \simeq 1{,}000\,\text{G}$, the electron oscillation velocity is 10 times smaller than the electron thermal velocity when the vacuum electromagnetic field is $\simeq 100$–$200\,\text{V/cm}$. To generate such a field by the TE_{11} microwave mode in an "empty" cylindrical waveguide 15 cm in diameter, a microwave power of 3,000 W is needed.

Meanwhile, many authors observed nonlinear microwave power absorption in magnetoactive underdense and overdense plasmas at microwave power ranging from a few milliwatts (*12, 191*) up to a few kilowatts (*71, 161, 165*). They attributed nonlinear wave dissipation to the excitation of plasma instabilities. The physics and conditions of the excitation of various plasma parametric instabilities were done in the excellent book written by Chen (*178*). In the following we present a short description of some processes that might be attributed to nonlinear microwave power absorption observed in ECR plasma sources.

Beginning in the early 1970s, numerous studies were made of nonlinear plasma heating with strong electromagnetic waves in the presence of a static magnetic field. In systems where the magnetic field was parallel to the microwave electric field, $E \parallel B$, the absorption of microwave power was thought to be associated with dissipative parametric instabilities that were excited at the plasma resonance, $\omega = \omega_{pl}$ (*12, 144, 166, 178, 191*). It was found that the threshold microwave electric field had a minimum of $E_{th} \simeq 12\,\text{V/cm}$, in a plasma with $N_e N_{cr}$ and the ratio of v_E/v_{te} was ≈ 0.06–0.07. The electron energy distribution function, f_{ee}, exhibited the large population of hot electrons that grew with electric field and plasma density. No parametric instability was observed at plasma densities below N_{cr}. The effective electron collision frequency, v_{eff}, grew proportionally to E^2, which proved the nonlinear character of electromagnetic wave dissipation. The magnitude of v_{eff} was close to the microwave pumping frequency: $v_{eff} = 0.5$–$0.7\,\omega$.

Porkolab with co-workers studied the parametric excitation and damping of electron plasma waves, electron cyclotron harmonic waves, and ion acoustic waves in dc and microwave (slow wave structure) plasmas inserted in magnetic fields, $B \perp E$, below and higher than the ECR field: $\omega \gg \omega_{ce}$, and $\omega \ll \omega_{ce}$ (*69, 161, 162, 176, 192*). The authors observed the efficient nonlinear microwave power absorption at low and upper hybrid resonances, and at higher electron cyclotron harmonics, $n = 2, 3$. For our discussion it is important to note that the nonlinear absorption became noticeable at microwave electric fields of $10\,\text{V/cm}$, which is the same as that observed by Dreicer et al. (*12*) and by Chu et al. (*191*) in plasmas inserted in microwave cavities with $E \parallel B$.

The other interesting observation was that at the UHR heating even 20–40 W of microwave power heated all plasma electrons including thermal ones in the electron energy distribution function (*176*), while the nonlinear absorption of microwave power of 10–20 times greater magnitude, in magnetic fields higher than the ECR field, $\omega < \omega_{ce}$, caused heating of only the energetic tail of the electron energy distribution (*161*).

Similar results were obtained by Musil et al., who studied the propagation and the absorption of right-hand polarized waves in the plasma generated by a Lisitano gun, in a high magnetic field, $\omega_B = 2.3 \omega$ (*163*). The authors observed the quadratic dependence of plasma density from microwave power, $N_e \propto P_{abs}^2$, beginning from $P = 500$ W that was correspondent to the microwave vacuum electric field, $E_o = 150$ V/cm. The nonlinear growth of plasma density was accompanied by the appearance of fast electrons on the electron energy distribution function, f_{ee} (*163*).

The threshold electric field was found to be dependent on magnetic field and plasma density with the formation of very fast electrons (*160, 164, 191*). In the recent work of Dandl et al., the generation of very energetic electrons was observed as a result of the nonlinear absorption of the Whistler wave launched from "the high-field side," $B > B_{ce}$ (*100, 110, 139*).

Calderon and Perez observed the appearance of the quadratic dependence, $N_e(P_{abs})$, at high microwave power of 3 kW, but they attributed the nonlinear effects to the exceeding of the threshold plasma density, $N_e \simeq 10^{12}$ cm^{-3}, rather than the high microwave electric field (*165*).

The effective electron collision frequency, v_{eff}, which "represents" the nonlinear absorption mechanism, was found to be much higher than that at linear absorption (ECR or Landau damping) (*71*). At a moderate microwave power level, $P_{inc} \simeq 100$ W, the effective electron collision frequency grew as $v_{eff} \sim E^2$, from 1.2×10^{10} s^{-1} at $P = 120$ W to 7.7×10^{10} s^{-1} at $P_{inc} = 150$ W.

Parametrically excited ion acoustic waves ($f = 85$ kHz) can also be "responsible" for the absorption of left-hand polarized waves at a magnetic field equal to the ECR field that was experimentally observed by Ferrari et al. (*70*). The plasma density was about cutoff for L waves with microwave power of 250 W. At microwave incident power below 250 W, the absorption of L waves was not observed. It seems to be obvious that both high microwave electric field, $E > 10$ V/cm, and plasma density, $N_e > N_{cr}$, might play a critical role in providing conditions for a high degree of microwave power absorption. However, this phenomenon has not yet been studied in ECR plasma sources.

III. ECR Plasma Sources: Designs and Characteristics

A. BASIC TYPES OF ECR PLASMA SOURCES

1. ECR Plasma Source Classification

In the 30 years of their history, electron cyclotron resonance plasma sources had periods both of extensive development and "stagnation." In the early 1960s, there were two designs made by Miller et al. (6, 7) (ion propulsion) and R. Dandl et al. (5) (fusion experiments). Today, in the early 1990s, there are a large number of different concepts in ECR plasma source design that are used in PA CVD. Along with them there are also ECR plasma sources used for MBE, ion beam etching, ion implantation, etc., which are out of the scope of this paper.

We classify all ECR plasma sources used in PACVD in five groups based on the method of plasma excitation and magnetic field arrangement. The design concepts of these sources are shown schematically in Fig. 9(a–e).

Group 1: Hitachi/NTT type. This is a "pioneer" ECR source that was designed and explored in 1962 by Miller et al. for ion propulsion technology (6, 7). In the 1970s, a similar design approach was used by Musil et al. for ECR plasma oxidation (16, 17), and by Suzuki et al. for plasma etching (18). In the early 1980s, Matsuo used the divergent part of an ECR plasma for low-pressure CVD (20, 156). Since then, this type of ECR sources has been often called an "NTT ECR source." Its major components are (Fig. 9a)

- Tuned microwave introduction (dielectric) window that couples entering electromagnetic waves with a plasma chamber. Some designs employ a mode converter from the rectangular TE_{10} mode to the circular TE_{11} mode (94, 95, 127, 133) or TM_{01} mode (104, 115, 119, 239).
- Two/three coils that generate an azimuthally uniform magnetic field with controllable axial and radial profile.
- Open-ended plasma chamber, which is actually a section of a circular waveguide.

Group 2: Microwave plasma disk reactor (MPDR). This source was

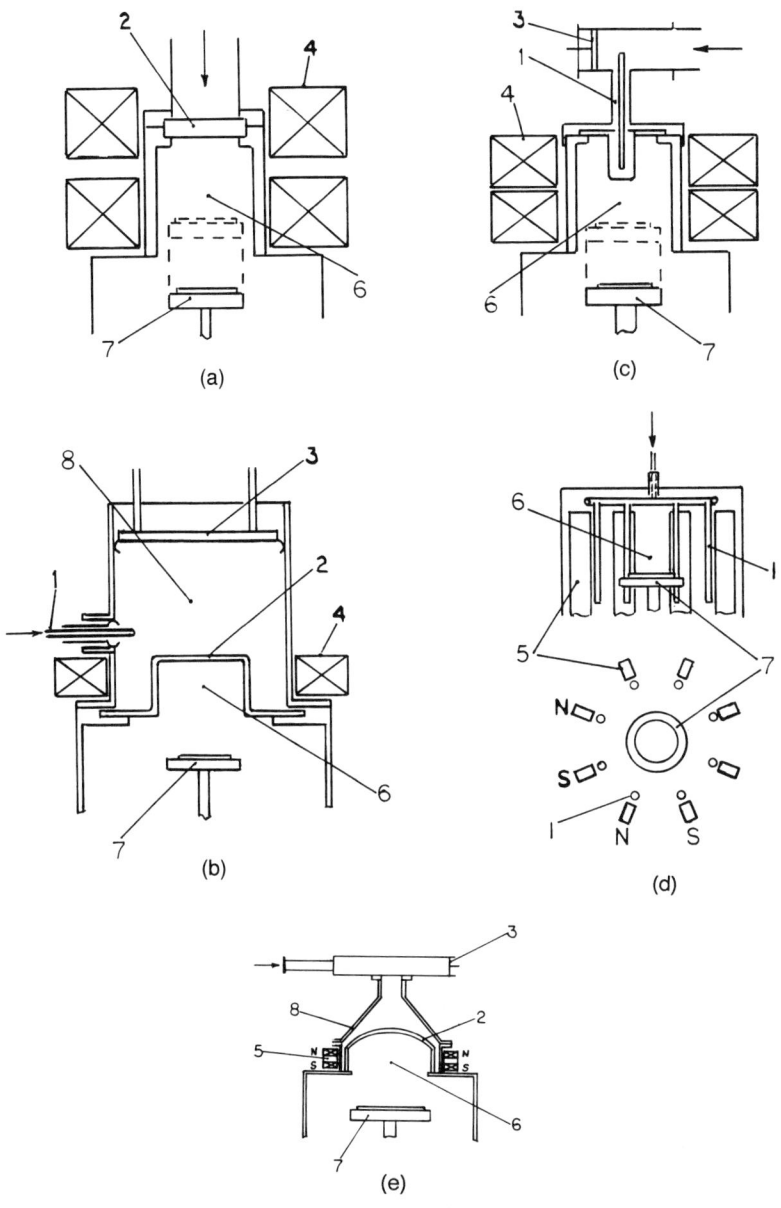

FIG. 9. Schematic conceptual design of five ECR plasma sources: (a) NTT/Hitachi source; (b) Microwave Plasma Disk Reactor (MPDR); (c) antenna/waveguide plasma source; (d) distributed ECR (DECR) source; (e) Leybold type source. Large arrow shows microwave introduction; 1—antenna, 2—microwave window (quartz disk, jar walls), 3—tuning element (plunger), 4—magnetic coil, 5—permanent magnets, 6—plasma generation zone, 7—wafer holder, 8—horn-waveguide.

originally designed in the 1980s by Asmussen and explored in ion propulsion and ion beam etching (*76, 81, 87, 92*). In the last few years the same concept was used for the design of ECR sources for PA CVD of thin films and plasma etching (*102, 213, 214*). MPDR ECR sources comprise three major parts (Fig. 9b):

- Tunable (for selective mode or multimode) resonant cavity with a loop/coaxial antenna.
- Dielectric (usually quartz) plasma reactor located on the bottom of the cavity to minimize the loading effect and to keep cavity's Q high.
- Multipolar permanent magnet (multicusp) arrangement around quartz plasma reactor. It generates a circumferential ECR surface near the quartz walls with zero magnetic field on the axis.

Group 3. Antenna/waveguide ECR source. This type of source was reported in the 1970s and originally employed magnetic fields that were much higher than the ECR field, $B/B_{ce} > 1.5$ (*18, 19, 72, 188*). Its major components are (Fig. 9c):

- Water-cooled antenna axially introduced into a plasma chamber.
- Plasma chamber with diameter above or below cutoff.
- A series of magnetic coils generating a static magnetic field in the chamber.

In some later designs the antenna was isolated from plasma and vacuum by a dielectric part (*215, 216*). Other designs employed permanent magnets instead of magnetic coils (*150, 217, 218*).

Antenna type "ECR-off" sources have not so far found a broad application in PA CVD, probably because of inherent plasma nonuniformity (*19*) and because of some problems associated with antenna cooling at microwave power greater than 500 W.

Group 4. Distributed ECR (DECR) plasma sources. This type of sources was designed in CNET (France) in the 1980s (*80, 103*) and was used in plasma processing equipment built by Alcatel (France) and Electrotech Ltd. (United Kingdom) (*141*). The major components of DECR sources are (Fig. 9d):

- Eight antennas symmetrically distributed inside the plasma chamber near its walls. One microwave power supply with a power divider is used for the even feeding of each antenna through a waveguide/coax converter.
- Eight permanent magnets that form an ECR circular zone in the vicinity of the antenna. The magnetic field diminishes from $B > B_{ce}$ at walls to $B = 0$ on the chamber axis.

- Large diameter plasma chamber with a substrate holder on the axis.

Group 5. Leybold ECR sources (Fig. 9e). They were reported in the late 1980s by Geisler *et al.* (*108*), G. Neumann *et al.* (*122*), and Pongratz *et al.* (*137*). Main components of Leybold ECR sources are:

- Horn antenna with an output aperture of circular shape (*137*) or rectangular shape (*108*). Aperture dimensions are $D = 25$ cm (*137*) and 60×9 cm (*108*).
- Multipolar permanent magnets arrangement at the horn aperture. Magnets form a multicusp (*108, 122*) or "racetrack" configuration (*137*).
- Multimode plasma chamber of a "pyramidal" shape with wafers on its bottom.
- Microwaves are introduced into the wafer location zone via a quartz window.

We do not consider here some ECR sources that were designed and used in plasma fusion research and recently were modified for plasma processing. Among them is an ECR plasma source developed by Dandl *et al.* (*100*). This source comprises two ECR plasmas ignited in resonant cavities in a magnetic mirror, and one process chamber positioned between the cavities. Such a system is large and bulky; it employs few pumps and several magnetic coils, which makes it very impractical and expensive. To my knowledge, this type of ECR plasma source has not so far found applications in plasma processing.

It is impossible to present here and discuss performance and plasma parameters of each ECR source in each group. All authors have their own preferences ("personal biases") as to which sources are more "interesting" and are to be described. The author of this paper spent many years in design, research and application to PA CVD, plasma etching, and ion beam processing of one type of ECR source—the NTT/Hitachi type. Therefore, he will discuss next the design features and source and plasma characteristics of this type of source.

Next we list major ECR source characteristics and plasma parameters that are essential for source performance and plasma processing.

2. *Major ECR Source and Plasma Characteristics and Parameters*

ECR source characteristics:

- Magnetic field spatial distribution near microwave absorption zone (ECR zone, SHR, Doppler-shifted ECR zone).
- Magnetic field and its structure near the wafer.

- Plasma ignition conditions. Standing wave pattern. Minimum microwave power required for ignition, breakdown magnetic fields, effect of gas pressure, location of ignition (breakdown) zone.
- The efficiency of microwave power absorption: absorbed, reflected, and propagated through the plasma microwave power.
- Effects of gas pressure, plasma density, and microwave tuning on the level of microwave power absorption.
- Microwave mode converters and their effect on breakdown, electric field pattern in chamber, and microwave power absorption.
- Microwave power losses in microwave transmission line (waveguide, tuner, walls, window, antennas, etc.).
- Effect of plasma chamber diameter on microwave propagation and absorption.
- Different microwave plasma modes and transitions between them.
- Location and length of the microwave absorption zone in a particular mode.
- Source efficiency: total ion current at the source output (or near substrate holder) as function of microwave power, $\eta = J/P_{inc}$.
- Operational gas pressure range.

ECR plasma characterization:

- Plasma density spatial distribution near the microwave absorption zone. $N_e(r)$, $N_e(z)$.
- Plasma density as function of microwave power absorbed, $N_e(P_{abs})$.
- Plasma density as function of magnetic field.
- Electron temperature and its spatial distribution, $T_e(r)$, $T_e(z)$.
- Electron temperature vs. microwave power, $T_e(P_{abs})$.
- Electron energy distribution function, f_{ee}, and its spatial distribution, $F_{ee}(r)$, $F_{ee}(z)$.
- Electron energy distribution as a function of microwave power absorbed.
- Electron energy distribution, F_{ee}, as a function of gas pressure.
- Two electron temperatures ("slow" and "fast" electrons).
- Plasma potential, V_{pl}, and its spatial distribution, $V_{pl}(z)$, $V_{pl}(r)$.
- Plasma potential, V_{pl}, as a function of microwave power absorbed and gas pressure.
- Ion energy near the substrate holder. Effect of magnetic field and its spatial distribution near the substrate.
- Effects of the substrate (wafer) platen, its location, diameter and (rf) bias on ECR plasma parameters.
- Dielectric (floating) and conductive (grounded or biased) liners' effects on plasma parameters.

It is unrealistic to expect that all the aforementioned characteristics have been studied systematically in one particular ECR source. Typically, in most publications only a few source characteristics and plasma parameters are presented. Using the published data, it is unlikely that a complete comparison could be done even between ECR plasma sources of the same type. Therefore, in the next section we will present and discuss data taken from different sources, including some that have never been published before.

B. Hitachi/NTT ECR Plasma Source

The Hitachi/NTT type of source is probably the most developed and widely used in plasma processing. It has many modifications that differ each from other by magnetic field arrangement, by microwave mode and introduction window, and by plasma chamber design and size.

Next we discuss the major modifications of this ECR source which found (or can find) their applications in PACVD.

1. Magnetic Field Structure and Coil Arrangement

Classification within Hitachi/NTT type ECR plasma sources can be made by the magnetic field configuration. There are four major magnetic field structures: (i) radially and axially uniform magnetic field, (ii) divergent field, (iii) mirror field, and (iv) mirrorlike field with collimating downstream magnet. All four magnetic field configurations have their own magnetic coil arrangements, which are shown schematically in Fig. 10a–d, together with a schematic magnetic field profile.

Magnetic coil arrangements that generate in a plasma chamber radially and axially uniform magnetic fields (Fig. 10a) were used in Refs. 26, 38, 41, 53, 109, 133, 138, and 146. To form a radially uniform ($\pm 1\%$) axial magnetic field, B_z, two coils must be separated by a certain distance (Helmholtz pair) (*138, 159*). When currents in both coils are equal, $I_1 = I_2$, the radial component of the magnetic field, B_r, is close to zero ($B_r = 1-5$ G) in the region between two coils within the distance of $z \simeq H_{\text{coil}}/2$ (H_{coil} is the coil height).

Because of the negligible magnitude of B_r, the ion flux to chamber walls (or liner) is very small, which reduces radial plasma losses but also decreases wall (liner) material sputtering caused by ion bombardment. The latter is especially important because wall sputtering is believed to be a major source of particulation in ECR CVD of dielectric films (*129, 219*).

By varying the coil currents one can generate a magnetic field in the

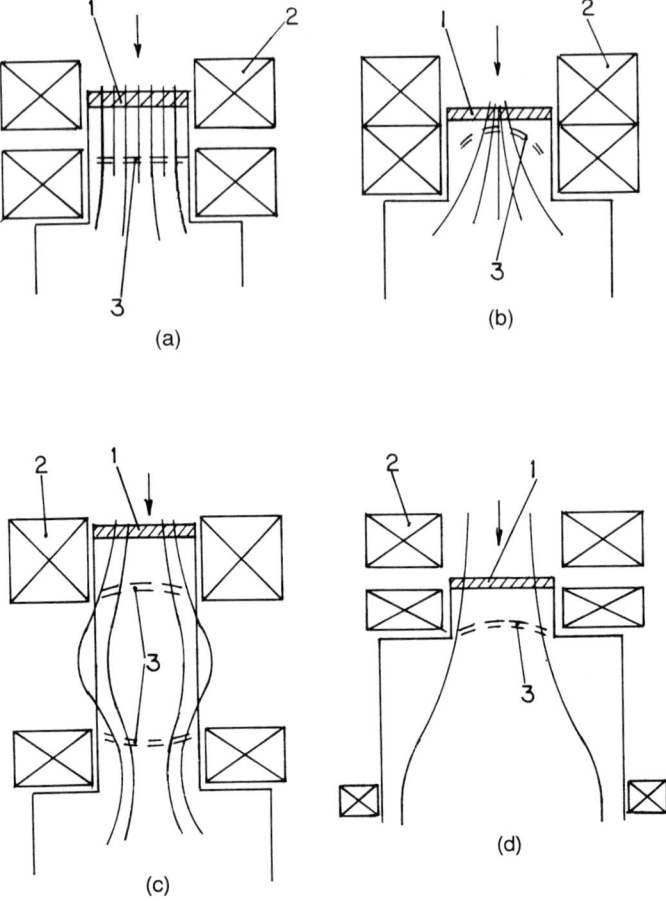

FIG. 10. Four types of magnetic field configuration used in Hitachi/NTT sources. (a) Uniform magnetic field, (b) divergent magnetic field, (c) magnetic mirror, (d) mirrorlike structure with the downstream magnet. 1—microwave introduction window, 2—magnetic coil, 3—ECR magnetic field surface. Large arrow shows the introduction of microwave power.

plasma chamber that decreases with distance from the window, $dB_z/dz < 0$, or increases with z, $dB_z/dz > 0$. But because of Helmholtz separation, the axial component of the magnetic field is always radially uniform within a distance of $z \simeq H_{\text{coil}}$.

ECR plasma sources employing the divergent magnetic field usually have a radially nonuniform magnetic field both in the plasma chamber and in the process chamber. The first ECR plasma source with a divergent magnetic field was designed by Miller *et al.* for the acceleration of ions in a divergent plasma stream for ion propulsion (*6, 7*), and by Consoli *et al.* for high-energy ion beams production (*8, 9*).

In the early 1980s this type of source was applied by Matsuo *et al.* for PA CVD of dielectric films (*20, 55*), and later for plasma etching (*54*). During the next few years the Miller/Matsuo source (now called the NTT source) became a major type of ECR plasma source used for plasma processing. Within this group, ECR sources with two magnetic coils and with an ECR zone placed somewhere in a plasma chamber are probably the most popular ones. They were used for various PA CVD technologies, such as silicon nitride films (*20, 21, 22, 138*), silicon oxide films (*26, 28, 53, 221*), SiO_2 planarization (*29, 30, 31*), boron nitride coatings (*34, 35*), silicon carbide coatings (*32, 33*), silicon oxidation (*58, 107*), amorphous silicon films (*43–46, 222*), amorphous boron films (*47, 223*), diamond and diamondlike carbon (*38–42*), epitaxial silicon (*48, 49*), and epitaxial GaAs (*50*).

A two magnetic coil ECR source with a divergent magnetic field was also very thoroughly studied by Matsuo (*20, 55, 140*), Gorbatkin *et al.* (*104, 219*), Stevens *et al.* (*127, 149*), Carl *et al.* (*107, 123*), Popov *et al.* (*86, 106, 113, 130, 145*), and Shirai *et al.* (*91*).

ECR plasma sources with a divergent plasma stream were found attractive for PA CVD technology because of the enhancement of the deposition process by the ion bombardment of the growing film. Ions are generated in an ECR chamber (primarily) and partially in a plasma stream (*153*) and are accelerated along the magnetic lines towards the substrate by the ambipolar electric field, E_{dc} (*5–9*). The magnitude of E_{dc} can be calculated from the spatial distribution of plasma potential, V_{pl}, and is typically between 0.5 and 2 V/cm (*119, 153*).

The energy of accelerated ions ranges from a few electron volts to 30–40 eV (*20, 54, 55*) depending on microwave power, electron temperature, gas pressure, chamber geometry, and magnetic field structure (*6–9, 20, 54, 83, 99, 112, 117, 119, 142, 153*). (More details can be found in a paper written by Lieberman and Gottscho and published in this book (*167*).)

In many ECR sources using the magnetic field profiles shown in Fig. 10a–d, microwaves are launched from the "high-field" side, i.e., the magnetic field at a microwave introduction window, B_{wv}, is higher than the Doppler-shifted ECR field. (For typical source plasma densities $N_e < 10^{13} \text{ cm}^{-3}$, and electron temperatures $T_e < 20 \text{ eV}$, the magnitude of B_{wv} should be 1,000–1,100 G. See below.) The magnetic mirror field profile has usually two ECR zones that have concave and convex shapes due to the radial nonuniformity of the magnetic field (*123*). The application of the mirror magnetic field for electron heating in ECR plasma sources came from fusion experiments. The microwave power absorption was associated with the heating of a single electron that was bouncing between two mirror ends with many passages through two ECR zones where it was heated with a local microwave electric field (*5, 82, 136, 172, 186, 203*). Meanwhile, some authors who used a mirror magnetic field in ECR plasma sources used in plasma processing reported that one ECR region (zone) was sufficient enough for complete microwave power absorption, provided the plasma density, N_e, exceeded N_{cutoff}. The experiments made in a mirror-type field with ECR underdense plasmas, $N_e < N_{cr}$, have shown that even two ECR surfaces ($B = 875\,G$) did not provide complete microwave power absorption (*123*).

By setting in the plasma chamber a mirrorlike magnetic field with the ratio $B_{wv}/B_{ce} > 1.6$, the designer "moves away" from the window both the microwave absorption zone and the high-density plasma, thus reducing the ion bombardment of the window.

ECR plasma sources employing "high-field" divergent and "high-field" uniform magnetic field profiles have only one ECR zone (surface). It was shown by some authors that when the microwave absorption zone, $B_{abs} = 920\text{--}940\,G$ (dependent on plasma density), is located near the microwave introduction window, the generated plasma has a density with a sharp peak on the axis and with a steep decrease in the radial direction (*114, 149, 159*). Plasmas with such a radial plasma profile were called "narrow" mode plasmas. The location of the microwave power absorption zone can be identified as the intersection of magnetic field profiles "corresponding" to the appearance of high-density plasma on the axis (*106, 145, 159*). This is illustrated in Fig. 11, where "narrow plasma mode" magnetic field axial profiles are shown. It can be seen that all three profiles intersect near the microwave introduction window at $B = 925\,G$. The plasma density radial profile measured at 23 cm from the window is shown in Fig. 12.

To generate a radially uniform plasma, the microwave absorption zone must be moved from the window downstream. It was shown by Samukawa *et al.* (*120, 220, 239*), Popov *et al.* (*114, 138, 159*), and Stevens *et al.* (*127, 149*) that for the formation of a radially uniform plasma, an ECR zone (surface) must be located several centimeters downstream, at $z > 10\,cm$. Popov *et al.* (*114, 138*) and Samukawa *et al.* (*120, 220*) exploring ECR plasma sources with different plasma chamber diameters

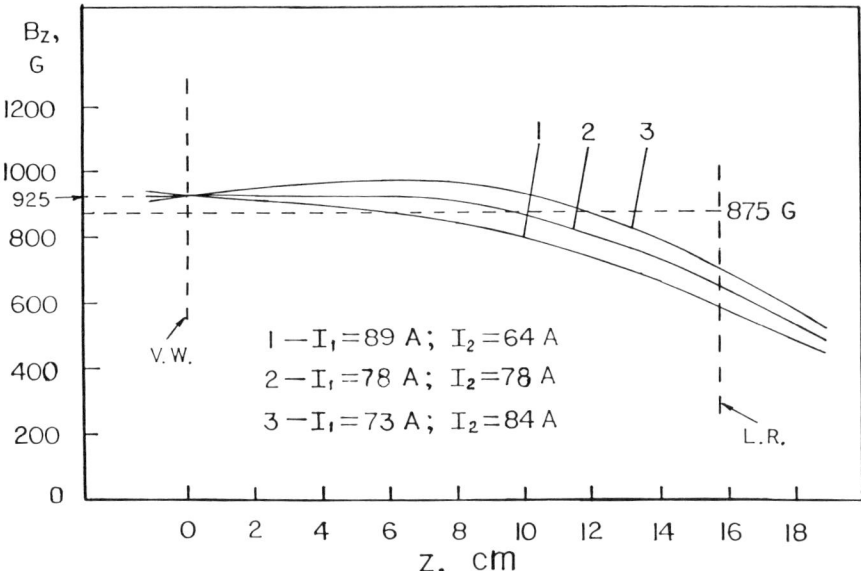

FIG. 11. Magnetic field axial profile $B_z(z)$ for ECR narrow plasma mode. All magnetic lines intersect at $B_z = 925$ G near the microwave introduction window. I_1 is the bottom coil I_2 is the top (window) coil. V.W. is the vacuum (microwave introduction) window. L.R. is the limiting ring.

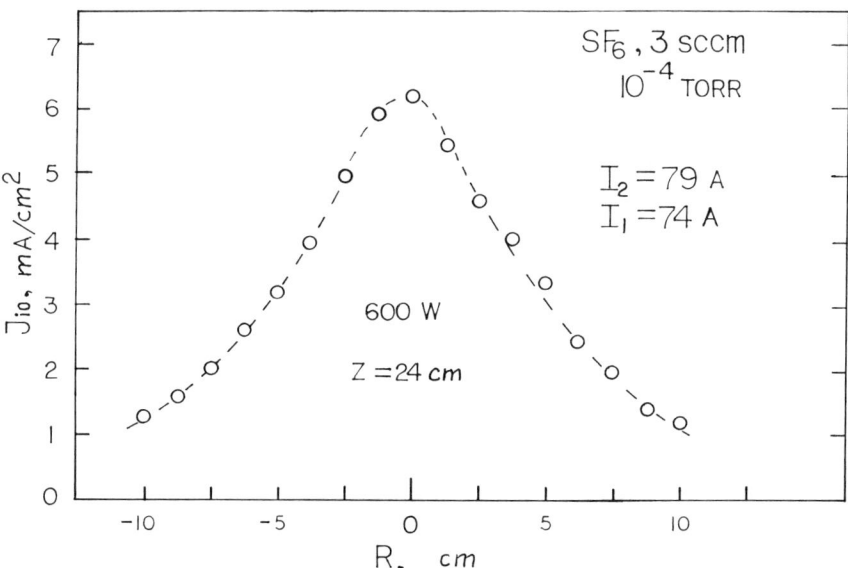

FIG. 12. The radial profile of ion saturation current density in ECR narrow plasma mode measured at 25 cm from the limiting ring aperture.

(15 cm and 26 cm, respectively) and magnetic coils of different sizes have found that a radially uniform plasma is generated if the microwave absorption zone ($B = 875$–930 G) is radially uniform.

In a large three-coil ECR source developed by Samukawa et al. (*120*, *220*), the location of the radially uniform ECR surface was $z = 40$–50 cm (*120*), while in ECR sources of smaller size the radially uniform ECR (and Doppler-shifted ECR) surface can be located much closer to the window, at $z = 10$–16 cm. A magnetic field axial profile with a radially uniform ECR zone located at 15.5 cm from the microwave window is shown in Fig. 13. Plasma density radial distribution generated in this magnetic configuration and measured at 11 cm from the ECR zone is given in Fig. 14.

The physics of microwave power propagation and absorption that results in a radially uniform plasma density is briefly discussed in Section I and in more details in Refs. 138 and 149. Whistler waves entering the overdense plasma from the high-field side must propagate some distance before they form a radially uniform electric field pattern over the plasma cross-section. A very small Whistler-wave wavelength, $\lambda_w = 1$–2 cm, and a radially uniform magnetic field provide a uniform microwave power "deposition" absorption surface.

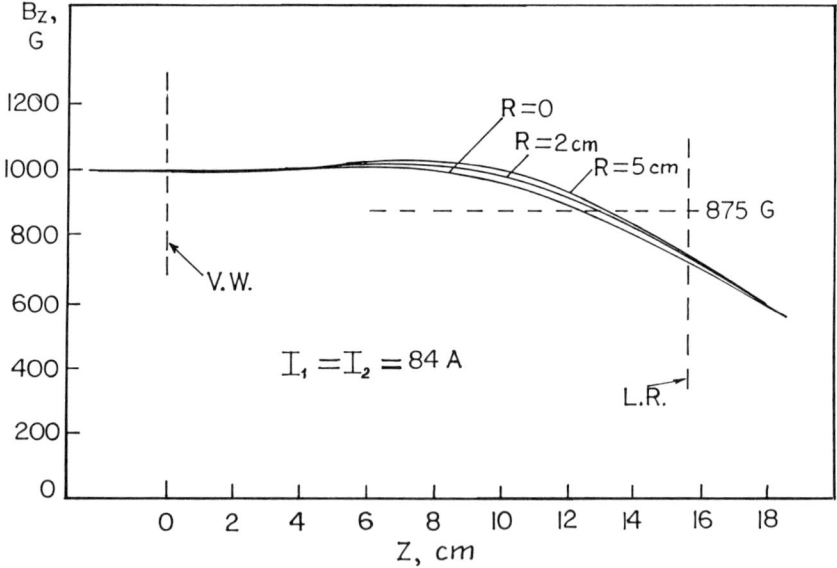

FIG. 13. Axial magnetic field profile, $B_z(x)$, generated by two magnetic coils separated in a Helmholtz pair, and measured at $R = 0$, 2 cm, and 5 cm. After Ref. 138.

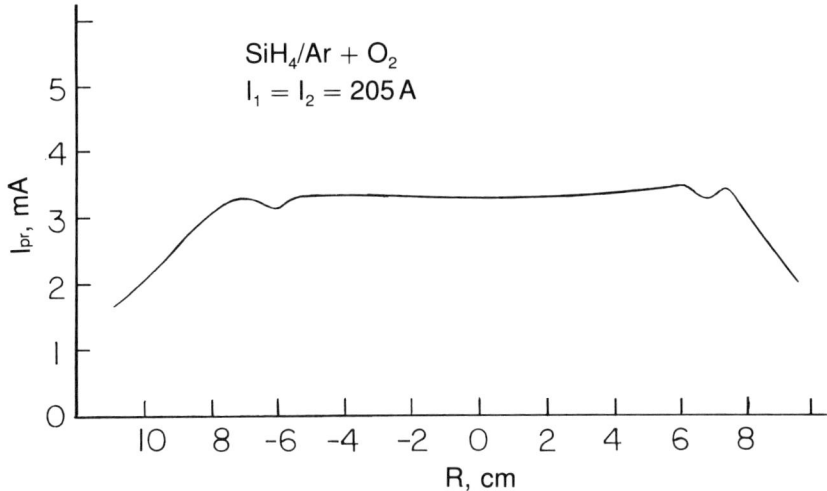

FIG. 14. Probe ion current radial distribution in a mixture of SiH_4/Ar and oxygen measured at $z = 11$ cm from the source output (limiting ring aperture). Whistler uniform mode, $p = 1$ mTorr. Two small dips at $R = 6$ cm are caused by wafer holder edge effect. After Ref. 53.

Consequently, the ionization zone is also radially uniform. (In fact, because of diffusion in the electron velocity phase space, the plasma is more uniform than the microwave electric field even near the absorption surface (*153*).)

The ECR source developed by Samukawa *et al.* (*120, 220, 239*) could be considered as the logical step in the development of the Hitachi ECR source designed by Suzuki *et al.* in 1977 and used for anisotropic plasma etching (*18*) (Fig. 15). It employed a mirrorlike magnetic field with one ECR surface in the plasma chamber and another ECR surface in the process chamber near the substrate. To increase the magnetic field in the process chamber, and hence to reduce plasma losses on chamber walls, a permanent magnet was originally positioned behind the substrate holder. In its latest version the permanent magnet was replaced with a downstream coil (Fig. 15).

The successive research and development by many groups resulted in various modifications of Hitachi-type ECR plasma source with the use of a collimation magnet downstream (*25, 83, 94, 96, 112, 119, 132, 134, 147*). They differ each from other by the number of coils in the ECR chamber and in the downstream (process) chamber. Mantei *et al.* use one coil near the window and two coils in the process chamber so microwave absorption occurs near the source aperture (Fig. 16). Two

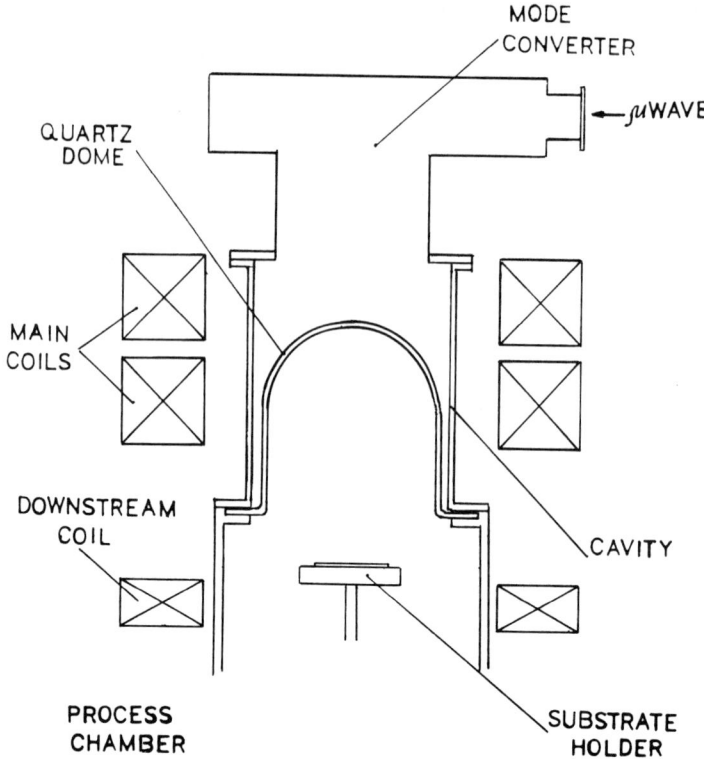

FIG. 15. Schematic of Hitachi-type ECR plasma source. Wafer holder can be moved along the chamber axis.

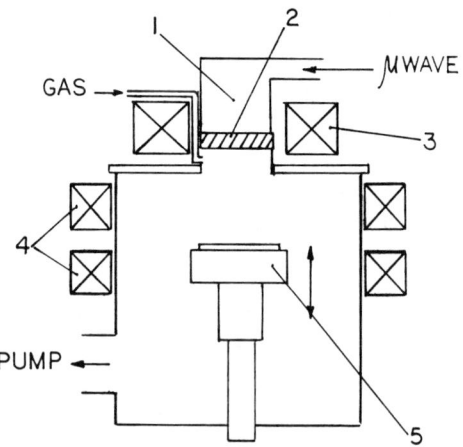

FIG. 16. Mantei ECR plasma source, which does not utilize a special ECR plasma chamber. ECR absorption can occur at any place between the microwave window and the wafer holder. 1—TM_{01} microwave mode converter, 2—quartz window, 3—window coil, 4—process chamber coils, 5—wafer holder. After Ref. 121.

coils placed around an ECR plasma chamber and one coil set downstream is probably the most popular configuration these days. It was used by Holber with co-workers (*94, 134*), Matsuoka et al. (*83*), Buchanan et al. (*132*), Minomo et al. (*147*), Samukawa et al. (*120, 135, 220*), and Gottscho with co-workers (*119, 142, 238*).

Some groups employ three or four magnetic coils around a long ECR plasma chamber (*24, 133*), and the combination of coils in an ECR chamber and coils and permanent magnets in the process chamber (*109, 146*).

2. Microwave Introduction Window

Microwaves propagating from the magnetron via rectangular or circular waveguides enter into a plasma (ECR) chamber through an introduction window. It is usually one (sometimes two or even three) disks made from a dielectric that is transparent to microwaves, i.e., microwaves are not absorbed in the window at any temperature that the window can attain. The microwave introduction window has several functions.

(1) It provides the propagation of microwaves of a particular mode into the plasma without losses in the window itself or in the space between window components, if any.
(2) It should provide matching of the impedance of the microwave transmission waveguides and the ECR plasma impedance within a wide range of microwave power, plasma density, and gas pressure.
(3) The incorporation of the window in the microwave transmission line should provide a radial uniformity of microwaves propagating to the absorption region.
(4) The window also seals a high-vacuum (down to 10^{-8} Torr) plasma chamber.
(5) It should not be a source of particulation during PACVD.

The first ECR plasma stream sources designed for ion propulsion used a ceramic disk with the thickness of half of the wavelength of the TE_{11} circular microwave mode, which was introduced into the ECR plasma chamber (*6, 7*). Such a window was thought to provide propagation of TE_{11} waves without absorption in the overdense plasma.

Musil et al. used a double window for the matching of the circularly polarized TE_{11} mode with a very dense plasma, $N_e > 10^{12}$ cm^{-3} (*71, 227*). It consisted of two mica disks with a thickness of a quarter-wavelength each and separated by a "free" space of length $\frac{1}{4}\lambda_0$. Such an arrangement resulted in more than 95% microwave power absorption (*71, 227*).

In the mid-1980s Sakudo et al. (224) and Torii et al. (78,225) used a double microwave introduction window in ECR ion beam sources. A quartz window that has a low dielectric constant, $\varepsilon = 3.7$, was placed from the waveguide side and also used as a vacuum sealing window. A dielectric block made from material with a high dielectric constant, boron nitride ($\varepsilon = 4.0$) or alumina ($\varepsilon = 9.3$), was placed on the plasma side, which provided a good impedance matching and also protected the quartz vacuum window from plasma (or ion beam) interaction.

The author of this paper used a double quartz window for the matching of impedances of a microwave plasma excited in the 15 cm diameter chamber and nonpolarized TE_{11} mode. The structure of this window is shown in Fig. 17. A quartz disk with a thickness of $\frac{1}{4}\lambda_q = 1.7$ cm was used for the matching purpose as well as for vacuum sealing, while a thin quartz disk (0.4 cm) was placed inside the plasma chamber and protected the vacuum/matching window from plasma heating and ion bombardment. The latter is important because the calorimetric measurements made by the author in an oxygen ECR plasma at pressures of 1–3 mTorr have shown that up to 70–75% of the total microwave power consumed by the ECR plasma goes to the microwave introduction window when the major portion of microwave power is absorbed at a few centimeters from the window (154).

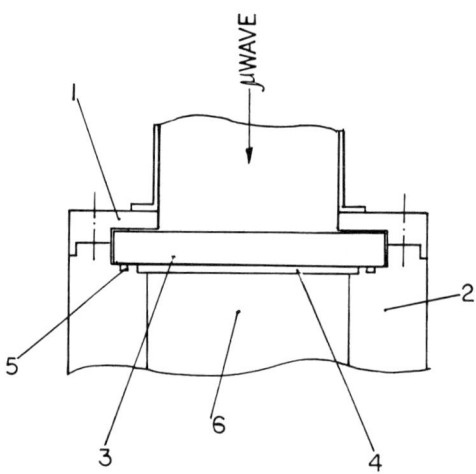

FIG. 17. Microwave introduction double window. 1—top flange, 2—plasma chamber, 3—quartz main window that seals plasma chamber, 4—protection quartz window, 5—plasma generation zone.

With the use of such a microwave window, only 2–10% of the incident microwave power was reflected from the ECR plasma. (Note that low reflected power is not necessarily accompanied by high microwave power absorption. The substantial portion of power could propagate downstream through the ECR absorption zone. We will discuss this issue later.)

The quarter-wave length alumina window for matching of the right-hand circularly polarized TE_{11} mode entering the plasma from the high-field side was designed by Stevens et al. (127, 149). The theoretical and experimental reflected power normalized to the incident power, P_r/P_f, are shown for two cases—tuned alumina window and non-tuned quartz window—and are shown as functions of plasma density in Fig. 18 (127). One can see a good agreement between the model and experimental data.

The diameter of the microwave introduction window was found to be an important characteristic. The original design of the NTT type of source made by Matsuo et al. (20) employed a plasma chamber with dimensions of a TE_{113} microwave resonator ($D = H = 20$ cm). Microwaves were introduced into the cavity from the rectangular waveguide having dimensions 9.2 cm × 2.6 cm × 1 cm. This waveguide was called in Matsuo's patent a "coupling aperture." A quartz window simply separated the cavity from the atmosphere. Such an introduction was considered appropriate because of the resonant character of the chamber. However, as soon as the plasma density in the cavity exceeds N_{cr}, it is

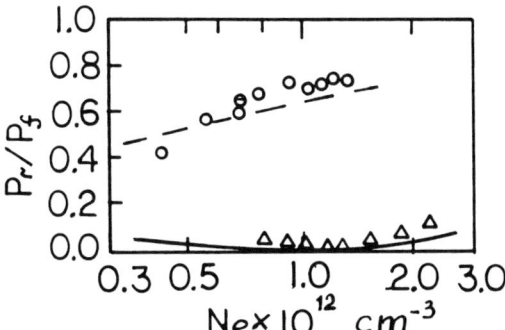

FIG. 18. Reflection coefficient vs. line-averaged plasma density obtained in ECR plasma sustained by R waves (TE_{11} mode). Triangles were obtained with the quarter-wave alumina window; circles, without a tuned window but with a thin quartz window. Solid and dotted curves are theoretical curves. After Ref. 127.

no longer a resonator, but a circular waveguide where other than TE_{11} circular modes can exist and propagate.

The author of this paper found experimentally that the ignition of the microwave plasma in a Matsuo-type ECR source occurred predominantly in the "coupling aperture" where the microwave electric field was the highest in the plasma cavity (*154*). This was obviously because of the small aperture dimensions. Consequently, microwave power was absorbed mainly in this aperture, which led to the generation of radially nonuniform plasma with the peak on the chamber axis. The magnetic field profiles that represented the highest plasma density in the source were found to intersect in the coupling aperture (*154*).

For several years many researchers and designers used the Matsuo concept of the "coupling aperture" for the introduction of microwaves in the ECR plasma chamber (*21, 23, 29, 36, 44, 48, 49, 51, 56, 91, 223*). Research and development of NTT-type ECR plasma sources led to the modification of the coupling aperture. Popov *et al.* introduced microwaves of both TE_{11} and TM_{01} in ECR sources with different plasma chamber diameters, $D_{ch} = 8$ cm, 10 cm, 15 cm, through a quartz window that had a diameter, D_q, slightly larger than the diameter of the plasma chamber, $D_q/D_{ch} = 1.1-1.2$ (*26, 86, 106, 113, 114, 138, 145*).

Some ECR source designers use a single quartz window with a metal coated sealing surface to minimize microwave leakage. The ASTeX ECR source utilized such a window tuned to circular TM_{01} mode and incorporated with a converter from a rectangular waveguide WR 284 to TM_{01} circular mode waveguide (*104, 115, 129, 132, 134, 219*).

Samukawa studied the effects of window diameter on the dependence of plasma density from the microwave power, and on plasma radial uniformity (*120*). He has shown that an increase of the plasma chamber diameter must be accompanied by the enlarging of the microwave introduction window. Otherwise, the plasma density saturates with microwave power at a relatively low power level, $P \simeq 600$ W. For instance, for a plasma chamber diameter of 26 cm, the minimum introduction window diameter must be $\simeq 14-15$ cm. In this case, microwaves are thought to be introduced in the chamber in a radially uniform manner, which results in radially uniform power deposition, and hence uniform plasma density (*120*).

It appears to be reasonable to believe that when microwave tuning is employed (a plunger or a two/three stub tuner) the requirements as to the size, thickness, and diameter of the dielectric window is not so critical. This seems to be the case in Hitachi-type ECR plasma sources that employ a quartz jar instead of a dielectric window (*18, 24, 134, 147*).

This source was found to be very successful in submicron etching technology. The schematic of such a source was given in Fig. 15. The quartz jar was also used for vacuum sealing near the source aperture. The author is aware of only a single publication that dealt with the matching of ECR plasma with the microwave transmission line (147). The authors of that work, using a three-stub tuner, studied the correlation between reflected power, plasma impedance, and plasma density in low- and high-density ECR plasma modes. No reference to dimensions or thickness of the quartz jar were made in this work (147).

3. Microwave Mode Converters

The kind of microwave mode that should be introduced in the plasma chamber is one of the controversial issues in ECR plasma source design. There are two "parties" in the ECR source community: advocates of the introduction of circular mode waves, who claim that without a mode converter a plasma generated in the ECR chamber (or in the process chamber) is azimuthally nonuniform and difficult to tune; and opponents of circular mode waves, who believe that a radially uniform, stable and easy tuned plasma can be achieved without the incorporation of a converter to one of the circular modes.

The author of this paper does not have enough proof to claim who is right. He prefers to present here some experimental data available in the literature and those which he obtained in his studies.

The pioneer of ECR plasma sources, Miller *et al.* (6, 7), used a circular polarizer to convert an originally plane TE_{10} wave to a right-hand circularly polarized wave (R wave) of TE_{11} mode. They did it to eliminate left-hand circularly polarized waves (L waves), and hence to provide conditions for only ECR microwave power absorption. Many ECR fusion-related experiments were also made using only R waves (11, 71, 74, 171, 227). The first ECR plasma and ion beam sources that were designed and developed in Japan did not employ mode converters (18, 19, 20, 24, 36, 54). In the mid-1980s, Forster and Holber reported a characterization of an ECR plasma source with a TE_{11} circular mode converter (94), and ASTeX has brought to the market an ECR source with a circular TM_{01} mode microwave launcher (104, 112, 115).

In the late 1980s, Microscience, Inc., introduced to the market and published some performance and characterization data on an ECR source with a circular TM_{01} mode converter (106). In the early 1990s, a few groups reported the use of TE_{11} mode converters: Princeton University (127, 149, 236) and Mitsubishi Electric Co. (133).

In the meantime, the majority of groups who explore new ECR CVD technologies, or those who develop the established processes such as silicon oxide interlevel films or silicon nitride passivation, still use "converter-free" ECR sources [25, 30, 31]. And there are no obvious signs that future ECR CVD equipment will necessarily include a mode converter (TE_{11} or TM_{01}).

Though one can find in the available literature many experimental results of the characterization of ECR plasma sources that have used both modes, there are only three publications (to my knowledge) where ECR plasma was ignited and studied without a converter and with TE_{11}, and TM_{01} mode converters (106, 127, 239). Here we present some data taken from Refs. 107 and 127 and data recently obtained by the author of this paper.

A typical experimental setup exploring ECR plasma source with two equal magnetic coils and a mode converter is shown in Fig. 19 (107). In Fig. 20 we present a set of minimum and maximum magnetic coil currents needed for a microwave breakdown, I_2 vs. I_1, measured in the ECR source with a TM_{01} mode converter and without a converter. In the latter case, a rectangular waveguide was directly attached to the

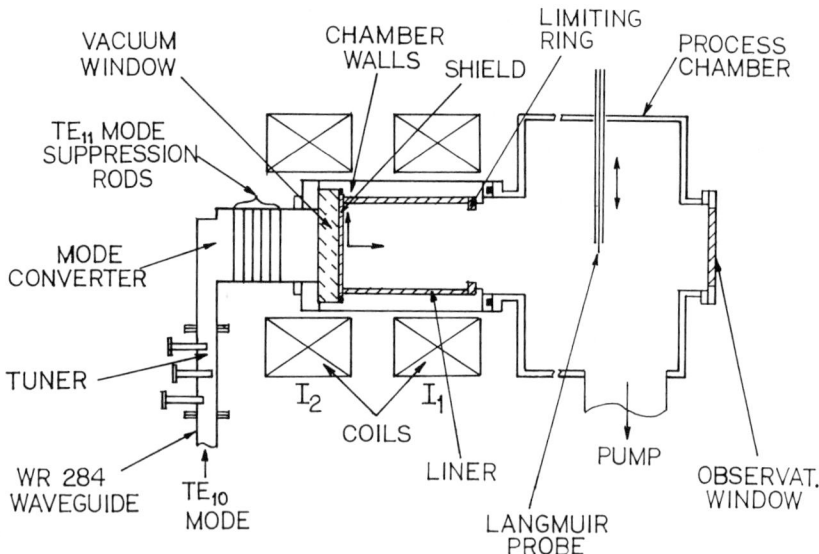

FIG. 19. Schematic of an ECR plasma source with a TM_{01} mode converter. After Ref. 106.

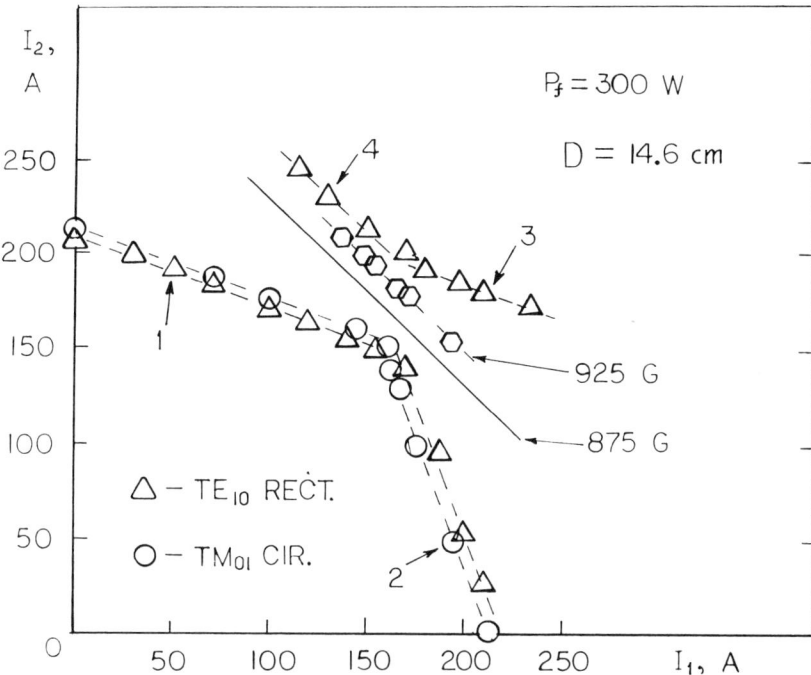

FIG. 20. Minimal and maximal microwave ignition (breakdown) coil currents, I_2 vs. I_1. 1 and 3 are straight lines, I_2 vs. I_1, that represent two breakdown magnetic fields near the microwave introduction window; 2 is a straight line that represents the breakdown magnetic field at $z = 13$ cm; 4 is a straight line that represents the breakdown magnetic field at $z = 6$ cm.

quartz window. One can see that the microwave breakdown occured at the same magnetic field (the same coil currents) in both cases. The dependencies I_2 vs. I_1, which are identical for both microwave modes, form three straight lines that represent three breakdown sites along the chamber. By plotting magnetic field profiles that correspond to each pair of coil currents I_1 and I_2, one can find that all breakdown magnetic field profiles intersect at three sites: (i) near the microwave window, $z \simeq 0$; (ii) at $z = 6$ cm; and (iii) at $z = 13$ cm (Fig. 21). These sites are believed to be the location of the maxima of the microwave electric field standing waves that are formed in the chamber.

The dependencies of plasma density, N_e, and Langmuir probe ion saturation current, I_{i0}, on coil currents, $I_1 = I_2$ (i.e., on the magnetic field in the chamber) have the same character with and without a TM_{01}

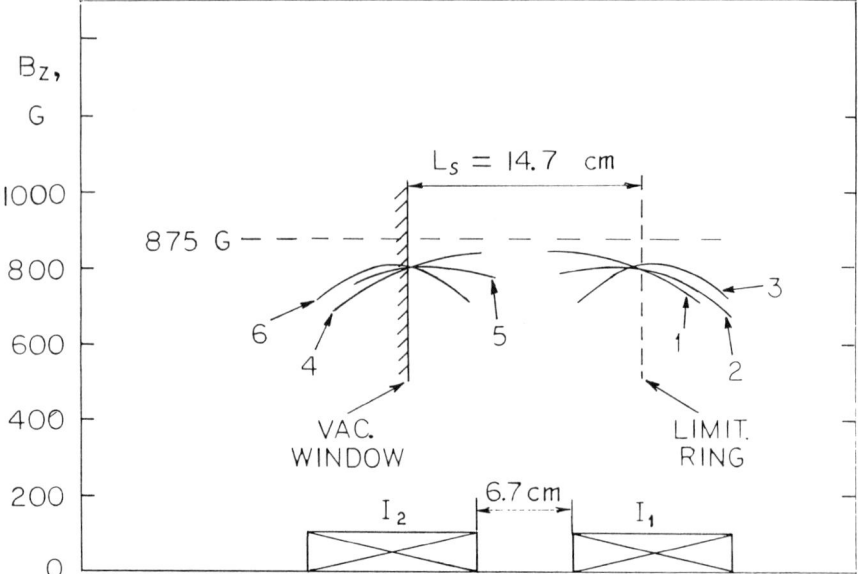

FIG. 21. Magnetic field lines, $B_z(z)$, that are correspondent to minimal microwave ignition coil currents, lines 1 and 2.

mode converter (106). They have two "resonant" magnetic fields, $B = 875\,G$ and $B = 925\,G$, at which the ion saturation current (and plasma density) has the highest magnitudes. The set of the "resonant" coil currents, I_1 vs. I_2, are plotted in Fig. 20. One can see that they form a straight line that represents a particular site (probably a surface) along the chamber where microwave power absorption occurred, $z \simeq 6\,cm$.

Plasma density in the TM_{01} ECR plasma stream was found to be about half of that in the ECR stream without employing a TM_{01} mode converter. This is illustrated in Fig. 22, where ion current density, I_{i0}, is plotted as a function of the microwave power absorbed, $P_{abs} = P_{inc} - P_{ref}$. The measurements were done in argon and in oxygen plasma streams at 47 cm from the vacuum window (31 cm from the source output) at a gas pressure of 0.4 mTorr. It can be seen that at a low microwave power level, $P_{abs} = 150\text{–}400\,W$, ion currents (and hence plasma densities) are the same in both ECR plasmas. As microwave power grows, plasma density increases more slowly in the TM_{01} stream than in TE_{11} streams. The similar difference between plasma densities generated by TM_{01} nonpolarized waves and TE_{11} right-hand circular

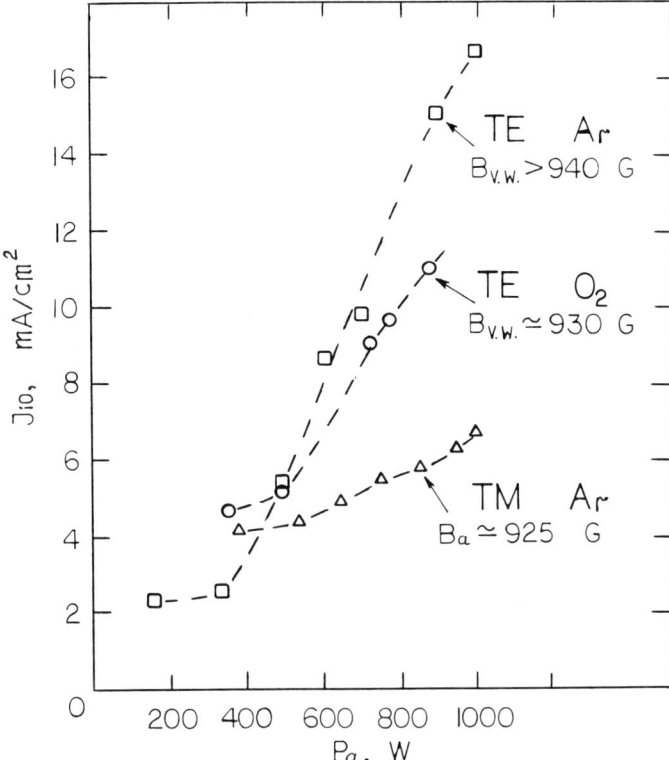

FIG. 22. Ion saturation current density, J_{io}, as a function of microwave power absorbed, measured on the process chamber axis at $z = 31$ cm in two microwave modes, TE and TM.

polarized waves was observed by Stevens et al. (127) (Fig. 23). It is necessary to note that in Figs. 22 and 23, ion current and plasma density are plotted as functions of the absorbed power, P_{abs}. Thus, the difference in plasma densities is believed to be attributed to different microwave absorption mechanisms and/or to different ionization balances.

Langmuir probe diagnostics of argon plasma streams "extracted" from both ECR plasma modes ($B_{abs} = 925$ G), circular TM_{01} and without mode converter, have shown that in the downstream plasma, at $z = 31$ cm from the source limiting ring, electron temperatures T_e and plasma and floating potentials V_{pl} and V_f are essentially the same (154).

We also found that the introduction of a TM_{01} mode converter did not noticeably affect the plasma density radial distribution (Figs. 24a, b).

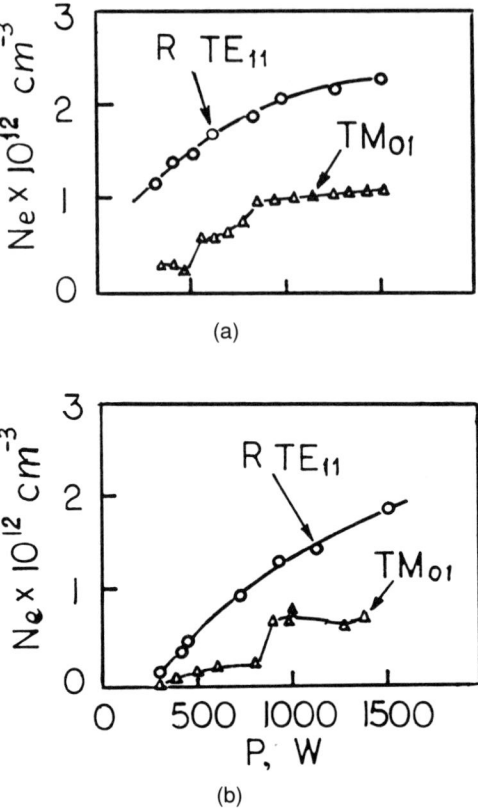

FIG. 23. Line-averaged plasma density measured in ECR plasmas ignited by the right-hand circularly polarized TE_{11} mode using $1/4\lambda$ alumina window (circles) and by the nonpolarized TM_{01} mode using a thin quartz window (triangles). (a) Argon plasma; (b) SF_6 plasma. 1 mTorr. $B_{wv} = 950$ G. After Ref. 127.

In both microwave modes the plasma density has a radially uniform area of $\simeq 13$–14 cm diameter at a distance of 26 cm from the source output. The TM_{01} mode ECR plasma has more distinctive "notch" on the chamber axis. The appearance of this minimum in a plasma excited by the circular TM_{01} microwave mode seems to be obvious: The radial component of the microwave electric field of this mode $E_r = 0$ on the axis and has its maximum off the axis at $r = 0.57R$ (228). However, the nature of the wide "notch" in the $N_e(r)$ distribution (doughnut-shaped

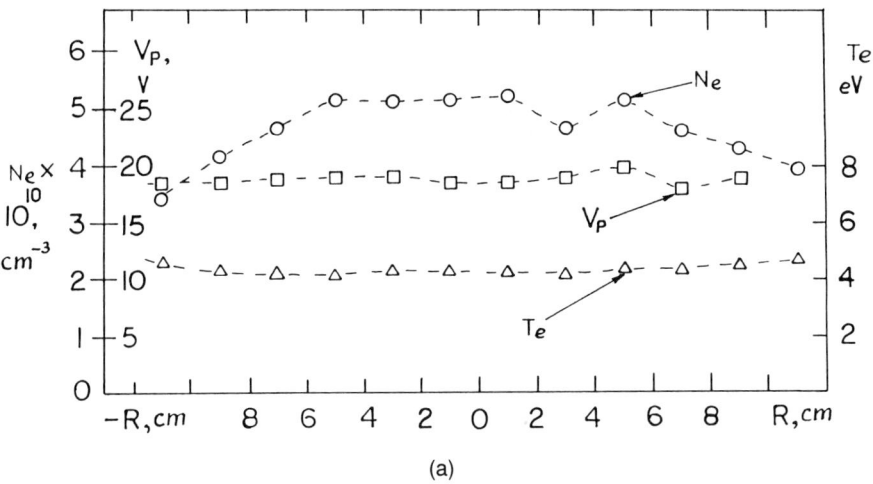

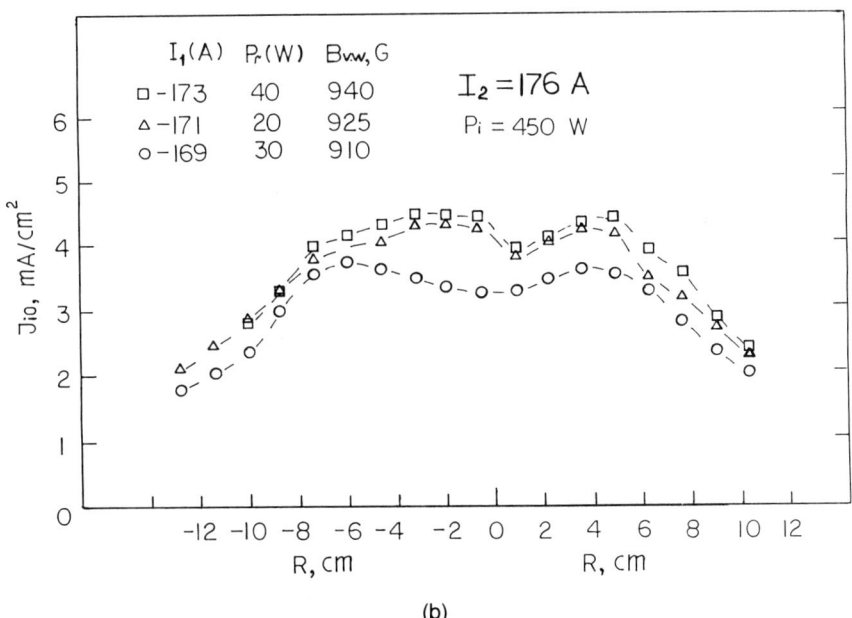

FIG. 24. Plasma parameters radial distribution in stream extracted from ECR plasma ignited by (a) TE mode, and (b) TM_{01} mode. Argon, 0.7 mTorr, 450–500 W. After Refs. 86 and 106.

plasma) in ECR plasma generated by the TE_{11} polarized (94) and the nonpolarized (106, 122) microwave mode is not understood.

The experimental dependences of the line-averaged plasma density on microwave power absorbed measured in ECR plasma generated by "pure" right-hand polarized waves has a region where it changes its character (Figs. 22 and 25). This occurred for two gases, argon and SF_6, at a microwave power absorbed of 500–800 W and plasma densities $N_e = (6-15) \times 10^{11} \, cm^{-3}$.

We observed similar dependencies in ECR plasma sources that did not employ mode converter. The line-averaged plasma densities measured with a microwave interferometer ($f = 35 \, GHz$) at $z = 11 \, cm$ from the source output (limiting ring aperture) are given in Fig. 26 for four gases (argon, nitrogen, oxygen, and hydrogen) at pressures of 2 mTorr. It is seen that there are a "threshold" microwave power, P_{th}, and a threshold plasma density, N_{th}, where plasma density growth with microwave power changes its character. In the oxygen plasma, P_{th} is 600 W, and $N_{th} \simeq 2 \times 10^{11} \, cm^{-3}$; in nitrogen plasmas, $P_{th} \simeq 400 \, W$, and $N_{th} \simeq 2 \times 10^{11} \, cm^{-3}$; while in argon plasma $P_{th} = 400 \, W$, but $N_{th} \simeq 3.5 \times 10^{11} \, cm^{-3}$. With the reasonable assumption of a threefold drop in plasma density from the absorption zone to the probe location, $z = 11 \, cm$, one can get a "threshold" plasma density of $10^{12} \, cm^{-3}$, which is very close to the "threshold" plasma density measured by Stevens et al. (127).

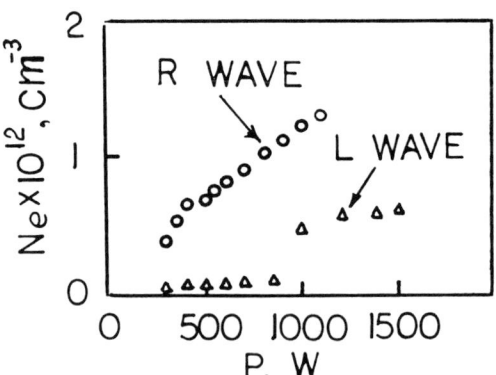

FIG. 25. Line-averaged plasma density vs. microwave power. Circles are right-hand circularly polarized TE_{11} waves with thin (nontuned) quartz window. Triangles are left-hand circularly polarized TE_{11} waves with the same window. After Ref. 127.

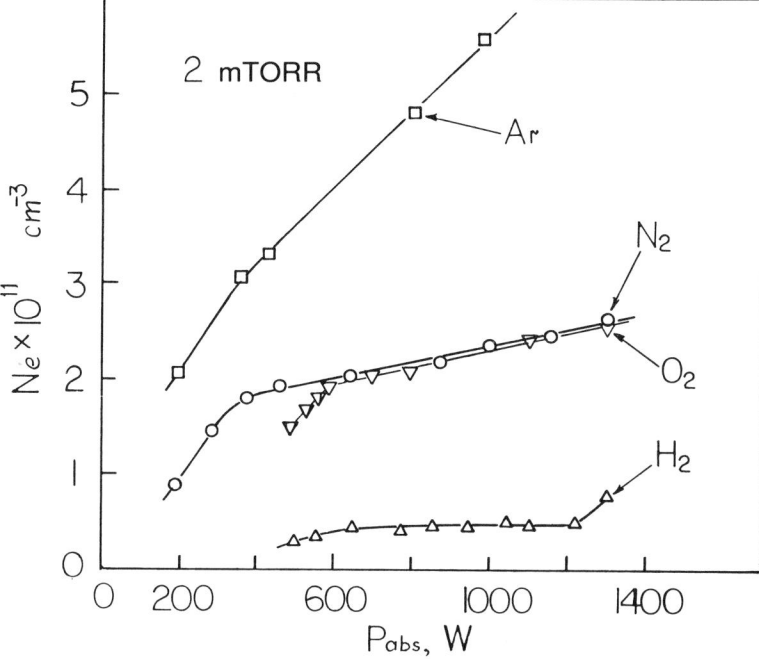

FIG. 26. Line-averaged plasma density vs. microwave power absorbed measured at $z = 11$ cm from the source output in four gases. 2 mTorr. ECR (925 G) in plasma generation chamber. After Ref. 53.

The nature of the nonlinear dependence $N_e = f(P_{abs})$ is certainly not associated with the type of microwave mode launched into the ECR plasma chamber, but with plasma ionzation efficiency, which varies with gas and pressure and is discussed later.

It is widely conceived that only the circular microwave mode can generate an azimuthally uniform plasma and hence a uniform deposited film. We observed various plasma shapes in ECR plasma sources operated with and without circular modes converters. They could be obtained by microwave tuning and by varying a magnetic field configuration. The azimuthally and radially uniform plasmas were achieved routinely when a radially uniform magnetic field was set in the microwave absorption zone, and microwaves were launched from the "high field" side, $B > 1,000$ G (*26, 86, 138, 149, 236*). (See also Fig. 27.) As a result, radially and azimuthally uniform ($\pm 2\%$) silicon nitride and silicon oxide films were deposited onto 10–15 cm diameter silicon wafers (*26, 138*).

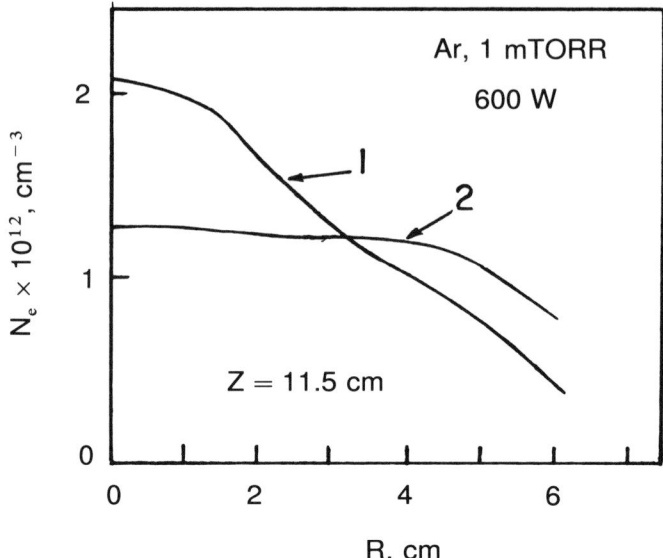

FIG. 27. Radial profiles of plasma density in the ECR chamber measured at 11.5 cm from the microwave window for two magnetic field configurations: 1—Doppler-shifted ECR field near the microwave window, $B_{vw} = 940$ G, and 2—high magnetic field, $B_{vw} = 1,010$ G. After Ref. 149.

4. ECR Plasma Chamber: Resonant Cavity, Cutoff, Scaling

An ECR plasma chamber introduced by Matsuo et al. in 1983 for plasma stream enhanced CVD had dimensions of a resonant cavity for the TE_{113} microwave mode (20). This cavity was not completely "closed" and had two apertures: in the top (microwave introduction aperture), and in the bottom (plasma extraction aperture). The authors appeared to believe that a resonant cavity was needed to provide a high level of microwave power absorption in a magnetized plasma. This approach is rooted in the "traditional" microwave technology where a resonant cavity is used as a perfectly matched plasma applicator (92) with its impedance only slightly varied as plasma impedance changes.

Another reason to use resonators as plasma chamber is the very high Q of resonators due to neglible microwave power losses in cavity walls. As a result, standing waves are formed in the cavity with a very high electric field at very low microwave power. The latter was important when operating at a low power level, $P_{inc} = 0.1–10$ W, and for achieving microwave breakdown in the absence of a magnetic field.

However, all these advantages of an "empty" resonant cavity are irrelevant for a cavity filled with an overdense plasma inserted in a magnetic field close to or higher than the ECR field. We discussed in Section II that the propagation of electromagnetic waves in a magnetized overdense plasma is controlled by plasma properties, that is, by the plasma complex dielectric constant $\varepsilon = n^2 = \alpha + j\beta$. (The exact expression for n^2 depends on the type of wave, magnetic field, and plasma density (63, 64, 143, 218).) Because of the dispersion of microwaves, the wave phase velocity, v_{ph}, varies with wave frequency, plasma parameters, and magnetic field. For instance, in an overdense plasma with $B > B_{ce}$, the phase velocity and the wavelength are much smaller than those in an empty waveguide (cavity) (63, 64). The impedance of a plasma-filled cavity is usually much smaller than the impedance of the resonant cavity.

At low plasma density, $N_e < N_{cr}$, the dominant TE_{11} microwave mode can form on azimuthally nonuniform electric field, $E^2 = (E_r^2 + E_\theta^2)^{1/2}$ (143). The plasma density follows the electric field profile and has an elliptical shape in the microwave absorption zone (143).

So far many designers and researchers use a resonant cavity as an ECR plasma chamber, mainly a TE_{113} resonator (21, 29, 48, 54, 55, 120, 143, 156, 226). In the path of ECR plasma source development, the designers followed industry needs for larger wafer diameters: from 3-inch diameter in 1977 to 6-inch diameter in 1992. Consequently, the ECR plasma chamber gradually increased its diameter: the plasma chamber used by Samukawa is 26 cm in diameter with a plasma uniformity of ±5% over a 20 cm wafer. ECR systems that use a divergent magnetic field with a collimating magnet downstream have ECR chambers smaller than the plasma chamber employed by Samukawa.

There is to my knowledge no study of chamber diameter effects on the ECR plasma. To do this, we measured plasma parameters, plasma density, electron temperature, plasma potential, and floating potential in nitrogen ECR plasmas ignited at 1 mTorr and 1 kW of microwave power in cylindrical chambers of different diameters: 2.5 cm, 5 cm, 7.6 cm, 10 cm, and 14.8 cm. We also used data of plasma density measured by Samukawa in a 26 cm diameter ECR chamber (120).

The results are given in Fig. 28, where plasma density N_e plasma potential V_{pl}, and the ratio of plasma density to the microwave power density, $P_s = P/S_{ch}$, where S_{ch} is the chamber cross-section, are plotted as functions of the chamber diameter. It can be seen that the ratio N_e/P_s grows with the chamber diameter D_{ch}, while the plasma potential decreases with D_{ch} from 50 V at $D_{ch} = 2.5$ cm to 27 V at $D_{ch} = 15$ cm. The decrease of the plasma potential with D_{ch} is a direct indication of

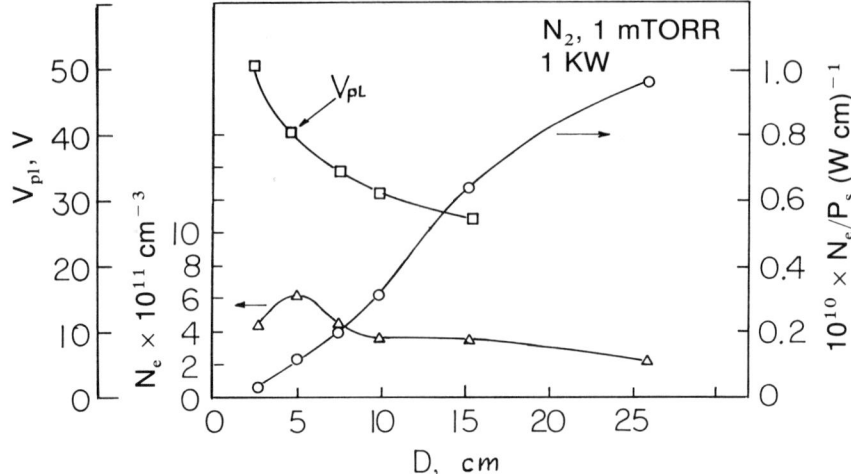

FIG. 28. Plasma density, N_e, plasma potential, V_{pl}, and the plasma density normalized to the power density, N_e/P_s, as functions of the chamber diameter. $P_s = P/S_{ch} = P/\pi R^2$. Nitrogen, 1 mTorr, $P_{abs} = 1,000$ W.

the reduction of the charged particle losses. (We did not get data on V_{pl} for $D_{ch} = 26$ cm.)

The observed dependencies are believed to be solid evidence that chamber diameter plays an important role in the ECR plasma ionization and energy balance up to diameters of 20–25 cm, which are comparable with the plasma chamber length.

Substantial plasma losses on chamber walls were also observed by Rossnagel et al. in an ASTeX ECR plasma source 15 cm in diameter and 40 cm long, which employs a mirror magnetic field (129). These losses resulted in very low source efficiency: only 14% of ions generated in the ECR plasma were extracted from the source (129). To improve the source efficiency the authors suggested (i) reducing the ratio of chamber length to its diameter, (ii) employing an axially uniform magnetic field, and (iii) placing a wafer near the uniform ECR magnetic field. All these recommendations have been independently implemented by ECR plasma sources designed and built by NEC (Japan) and Microscience, Inc. (USA).

Both types of ECR sources (NTT and Hitachi) have a limiting ring (source aperture) that terminates a chamber and determines the initial diameter of the stream expanding into the process chamber. We found that a decrease of the limiting ring aperture causes a drop in the downstream plasma density (86). To deliver a radially uniform plasma to the wafer, the diameter of the limiting ring is usually made slightly larger than the wafer holder diameter (20, 24, 25, 55).

Certainly, at 25–30 cm downstream from the limiting ring the plasma uniform area is larger than the ring diameter, but also the plasma density is 3–5 times lower than near the aperture. This is illustrated in Fig. 29, where two plasma density radial profiles measured at $z = 7.6$ cm and 38 cm from the source aperture (14 cm diameter) are shown. One can see how plasma uniformity improves with the distance from the source. In the commercial Microscience ECR 9200 CVD machine, a wafer holder

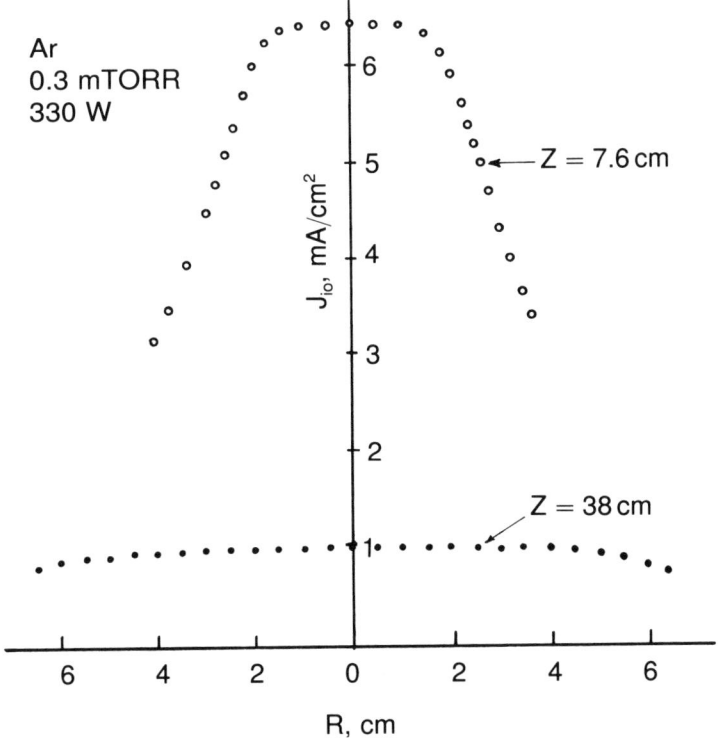

FIG. 29. Radial profiles of the ion saturation current density at two distances from the source: 7.6 cm and 38 cm. Argon 0.3 mTorr, $P_{abs} = 330$ W.

is usually placed at 12 cm from the source aperture where the plasma stream is uniform ($\pm 2\%$) over 15 cm in diameter (*42*).

The actual plasma chamber dimensions also include the wafer holder. Indeed, because of the plasma confinement by the axial magnetic field, plasma charged particle losses occur predominantly on the chamber terminating surfaces: the microwave window and the wafer holder. Therefore, plasma ionization balance is controlled primarily by the distance between the wafer holder and the window. This effect is less distinctive in ECR sources employing mirror or divergent magnetic fields.

Rossnagel *et al.* (*129*) and Caughman and Holber (*134*) studied the effects of a wafer holder biased by dc and of power supplies on stream plasma density N_e.

It was found that biasing the wafer with a positive potential decreases plasma density N_e (*129*), while negative potential increases plasma density both near the wafer and at the source aperture (*129, 134*). Dc negative biasing causes a drastic decrease (up to 40–50%) in the plasma potential near the substrate but does not affect the electron temperature (*134*). The biasing effect was found to be more pronounced for a large wafer holder and for more collimated streams. Experimental results obtained in Ref. 134 are shown in Fig. 30. It is also seen that an ECR plasma stream source employing a wafer holder with a larger diameter has a higher plasma density (at least near the wafer) than that near the smaller holder.

Our measurements of the total stream ion current, I_s, (nitrogen, 1 mTorr, 500 W) on a 25 cm diameter aluminum platen placed at 25 cm from the source output show that the increase of the platen negative potential from -30 V to -40 V caused a 30% increase in the ion current to the platen.

The observed phenomena were interpreted in Refs. 129 and 134 by the increase of the ionization by fast electrons repelled by the negative potential applied to the wafer holder. This shift in the ionization balance does not affect the bulk of slow electrons (4–6 eV), but does affect their energetic tail, "responsible" for inelastic processes, including ionization. The decrease of the plasma potential with the negative biasing can be associated with two phenomena: (i) the change in the plasma ionization balance, and (ii) the "tailoring" of the whole plasma to the wafer holder by the predominantly axial magnetic field (*134*). The latter effect should be always counted when dc or rf biasing is applied to the wafer.

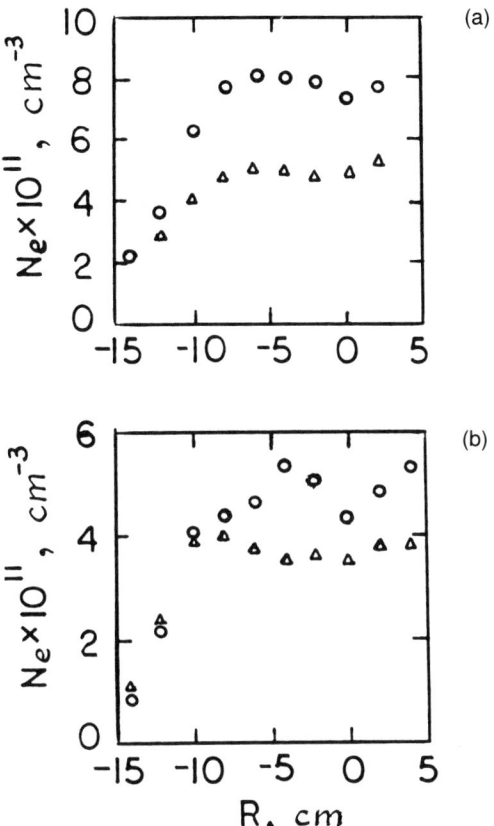

FIG. 30. Plasma density radial profiles measured in argon plasma stream at 10 cm from the wafer holder with dc bias (circles), and without dc bias (triangles). (a) 20 cm diameter wafer holder; (b) 12 cm wafer holder. 1 mTorr, microwave power 1,200 W. After Ref. 134.

C. ECR Plasma Source Characteristics and Plasma Parameters

In this subchapter we present and discuss basic characteristics and plasma parameters of the NTT/Hitachi ECR plasma source. Most data were obtained in nitrogen, argon, and oxygen plasmas, which are commonly used in ECR PACVD technologies and in many cases determine ECR plasma and stream parameters.

1. Microwave Plasma Modes

It was found that ECR plasma sources have (at least) two modes: low density and high density (*86, 93, 104, 106, 114–116, 122, 123, 127, 138, 145, 147, 149, 219*). The low-density plasma mode is characterized by relatively high reflected power (30–70%), plasma density below N_{cr}, standing wave in ECR chamber beyond the ECR zone (*123*), propagation of waves to the process chamber and wafer holder, relatively high electron temperature downstream, and unusually high floating potential $V_f = -40$–60 V.

The transition from the low mode to the high mode was studied by Popov et al. (*106, 114, 145*), Gorbatkin et al. (*104*), Neuman et al. (*122*), Carl et al. (*123*), Stevens et al. (*127, 149*), and Minomo et al. (*147*). In fact, any researcher who studies the plasma density relationship with incident microwave power, P_{inc}, has to deal with the nonlinear dependence of N_e on P_{inc}. However, in most publications the authors did not distinguish the incident microwave power P_{inc}, which is launched by the magnetron, from the absorbed microwave power, P_{abs}. And practically nobody distinguishes the absorbed microwave power from the power delivered to the ECR plasma, $P_{prop} = P_{inc} - P_{ref}$, which is always larger than the absorbed microwave power, $P_{abs} = P_{prop} - P_{trans}$. The latter is the microwave power transmitted through the ECR plasma to the wafer holder and can be detected in the process chamber and even beyond it.

To analyze different ECR plasma modes we made a series of experiments in oxygen and nitrogen ECR plasmas generated by 2.45 GHz nonpolarized microwave power of 150–1,500 W at pressures of 0.3–7 mTorr (*154*). The line-averaged plasma density, N_e, measured with a microwave interferometer (35 GHz) at 11 cm from the source output, and the reflected microwave power, P_{ref}, measured in the dummy load attached to the circulator, are plotted as functions of the incident power P_{inc} in Fig. 31.

There are three distinctive regions of microwave power with a different character of the dependence $N_e(P_{inc})$. In the first region, plasma density grows very slowly with P_{inc}, while reflected power grows or stays almost constant with P_{inc}. The second region starts at microwave power of $P_{inc} = 250$–350 W, where plasma density jumps from values of $N_{tr} = (2$–$4) \times 10^{10}$ cm^{-3} to magnitudes of $N_e = (5$–$8) \times 10^{10}$ cm^{-3}. Since at gas pressures of 0.5–5 mTorr the plasma density drops over the distance of 10–12 cm not more than three times, we could assume that the plasma density in the microwave absorption zone jumps from $N_1 = (5$–$10) \times 10^{10}$ cm^{-3} to $N_2 = (1.2$–$2.0) \times 10^{11}$ cm^{-3}. The value of N_1 is in

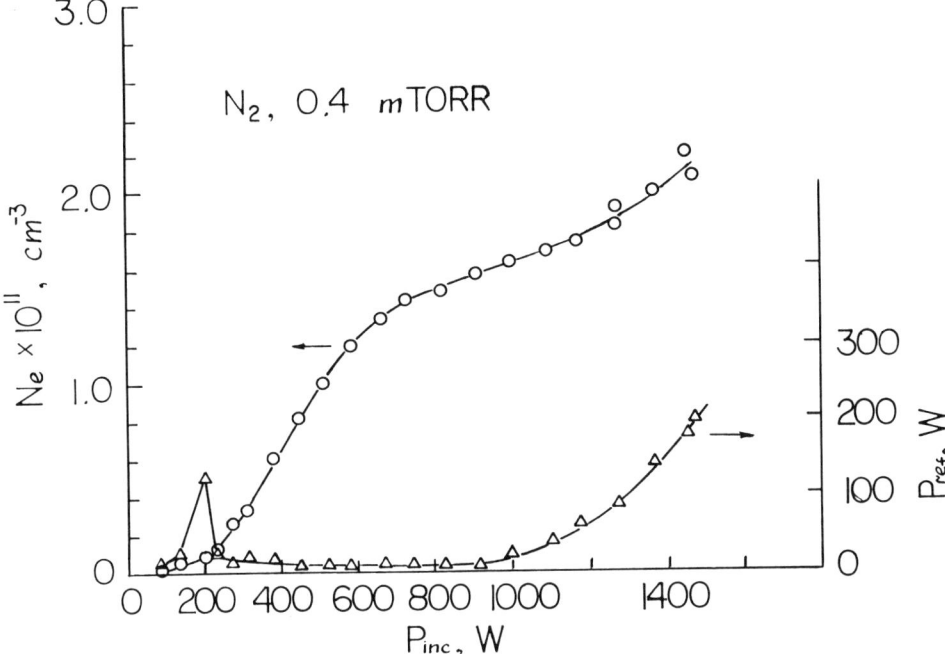

FIG. 31. Line-averaged plasma density, N_e, vs. the incident microwave power, P_{inc}, measured in nitrogen plasma stream, 0.4 mTorr, at $z = 11$ cm from the source output. Whistler uniform plasma mode ($B_{vw} = 1,100$ G).

good agreement with the critical plasma density $N_{cr} = 7.4 \times 10^{10}$ cm^{-3}. A sharp increase of plasma density at $N_e = N_{cr}$ was observed by the author in his earlier works where the plasma density was measured directly near the microwave absorption zone (*86*, *93*, *106*, *114*). The transition was accompanied by a drop in the floating potential and electron temperature and a slight growth of the plasma potential (Fig. 32). It can be achieved by other means that lead to an increase of the plasma density to a value close to N_{cr}: by an increase of gas pressure (*106*), microwave tuning (three stub tuner or EH plunger) (*93*, *147*), and by varying the magnetic field (*145*).

Similar observations were made by other authors working in different ECR plasma sources. Minomo et al. observed the transition in an Hitachi-type ECR source when the plasma density at 6 cm downstream was $N_1 = 4 \times 10^{10}$ cm^{-3} (*147*). Carl et al. studied the transition between

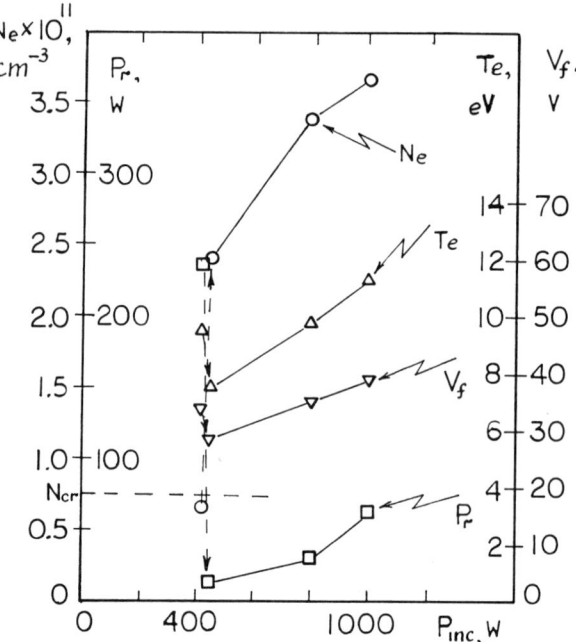

FIG. 32. Plasma parameters vs. the incident microwave power P_{inc} near the transition from the underdense plasma to the overdense plasma. Nitrogen, 1 mTorr. $z = 3$ cm from the ECR zone.

low and high plasma modes in a mirror-type ECR source and found that the transition usually occurred at a plasma density of 10^{11} cm^{-3} (123). A steep growth of ion current density, J_i, was observed by Samukawa et al. at $J_i \simeq 2.5$ mA/cm^2, which corresponds to a plasma density (nitrogen) of $N_e \simeq 5 \times 10^{10}$ cm^{-3} (120) (see Fig. 3). Stevens et al. studied microwave power absorption in ECR plasmas ignited by TE_{11} circularly polarized waves ("pure" R waves), TE_{11} left waves, and by nonpolarized circular TM_{01} waves (149). They found that both TE_{11} L waves and TM_{01} nonpolarized waves had a jump in plasma density of $\simeq 1.2 \times 10^{11}$ cm^{-3}, while TE_{11} R waves launched from the high-field side did not exhibit any jump at such a power (Figs. 22, 23, and 25).

Thus, we can conclude that the observed jump in the plasma density at $N_e = 5-10 \times 10$ cm^{-3} is caused by L wave absorption in a plasma with density near N_{cr}. Plasma density grows higher than $N_{cutoff} \simeq 1.5 \times 10^{11}$ cm^{-3}, so no L wave propagation downstream is observed in the second mode. This mode is often called the "high-density mode."

Note that since the transition from the low mode to the high mode occurs at the wide microwave power range, from 100 W (*12*) to 900 W (*127*), the observed phenomenon is believed not to be associated with the nonlinear processes caused by high microwave electric field (see Section I).

The second region is characterized by fast growth of plasma density with microwave power; sometimes N_e grows faster than linearly. This phenomenon was explained by Musil *et al.* (*158*).

The third region begins at microwave power of 500–800 W where the dependence $N_e(P_{abs})$ changes its character: it grows practically linearly but more slowly than in the second region (Figs. 22, 25, and 26). We discussed this earlier and believe that this phenomenon is attributable to the changes in plasma energy and ionization balance. Indeed, as gas pressure decreases, the heating of electrons by microwave power goes to the increase of electron temperature rather than ionization. As pressure increases, the plasma density at which the transition to the third region begins, N_{th}, shifts to larger values. This is shown in Fig. 33, where the line-averaged plasma density N_e measured at 11 cm from the source output is plotted as a function of microwave power absorbed for several oxygen pressures: 0.2–8.8 mTorr. At a pressure of 0.2 mTorr the transition begins at $N_e = 2 \times 10^{10}$ cm^{-3}, and at 8.8 mTorr it starts at $N_e = 1.7 \times 10^{11}$ cm^{-3}.

Similar trends were observed by the author in an 8 cm diameter source (*159*), by Forster and Holber (*94*), and by Rossnagel *et al.* (*115*).

The regime where plasma density grows sublinearly with microwave power absorbed and even has trends to saturation we call the "saturation" mode. From the practical point of view the preferable operation mode of ECR source is that where the plasma density is high enough to minimize the deposition time. However, high plasma density is usually accompanied by a high plasma potential (see below) $V_{pl} > 30$ V, which can increase sputtering from the walls/liner. Operational mode is chosen in each particular process, but in any case the plasma density must be higher than N_{cutoff} for the L waves to exclude its propagation downstream to wafer.

The minimum pressure at which an overdense plasma, $N_e - N_{cutoff}$, can be maintained depends on the sort of gas (ionization efficiency and ion mass), plasma chamber geometry, magnetic field structure and microwave power absorbed. For most gases used in plasma processing and at a microwave power of 400–1,300 W, the minimum pressure was found to be 0.1–0.3 mTorr, and it decreases as microwave power increases (*104, 145*).

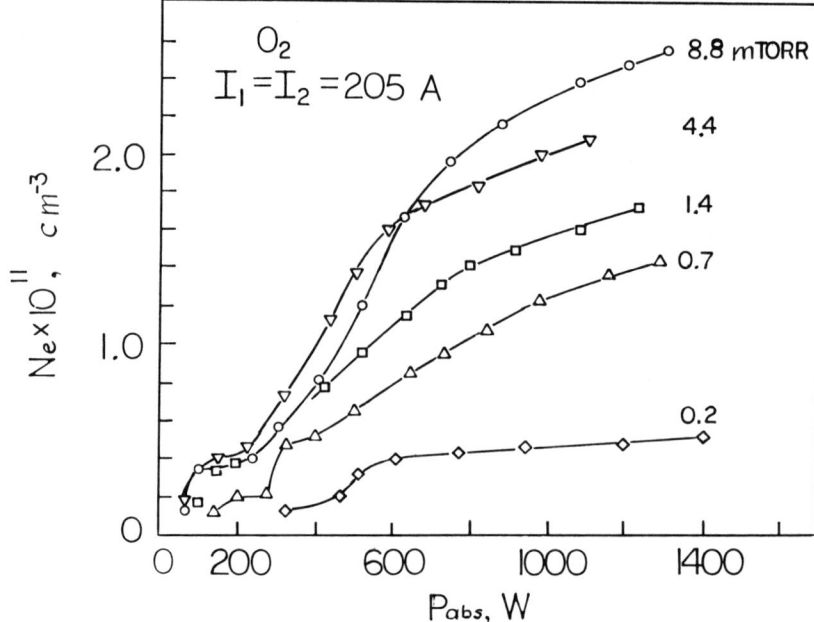

FIG. 33. Line-averaged plasma density, N_e, measured at $z = 11$ cm from the source output as a function of microwave power absorbed at five oxygen pressures. Whistler uniform plasma, $B_{vw} = 1,100$ G. After Ref. 53.

2. *Plasma Parameters as Function of Microwave Power Absorbed*

We have shown that there are three ECR plasma modes in the NTT/Hitachi type of ECR plasma sources: (i) the underdense mode, with high wave transmission through plasma and where plasma density grows very slowly with microwave power; (ii) the overdense mode, where plasma density grows faster than linearly with microwave power and without microwave propagation downstream; and (iii) overdense with slow plasma density growth, but also without microwave propagation downstream.

It is natural to expect that for each type of gas and different magnetic field structure the dependence $N_e(P_{abs})$ has its own features. We measured line-averaged plasma density at $z = 11$ cm from source output when microwave plasma was maintained in different but uniform magnetic fields. The results of experiments made in an ECR magnetic field ($I_1 = I_2 = 170$ A, $B = 925$ G), in a magnetic field of 1,100 G

(Whistler wave, $I_1 = I_2 = 205$ A), and at second harmonic resonance ($I_1 = I_2 = 85$ A, $B = 430$ G) are given in Fig. 34.

It is seen that in the first mode ($P_{abs} < 200$ W), plasma densities are essentially the same in microwave plasmas excited at all three magnetic fields. At $P_{abs} = 200$ W when plasma density at $z = 8$ cm is 2.5×10^{10} cm^{-3}, plasma density in ECR plasma begins to rise faster than linearly up to $P_{abs} = 600$ W and forms a narrow stream with the maximum plasma density on the axis. This is the second overdense plasma mode. When ECR plasma density reaches a magnitude of 2×10^{11} cm^{-3} ($P_{abs} = 600$ W), it begins to grow much more slowly with the distinctive trend to saturation. Plasma density in the Whistler mode grows faster than in a SHR plasma, probably because of the higher magnetic field and hence better plasma confinement. Plasma density in SHR increases with P_{abs} with the same slope as the ECR plasma at $P_{abs} > 600$ W.

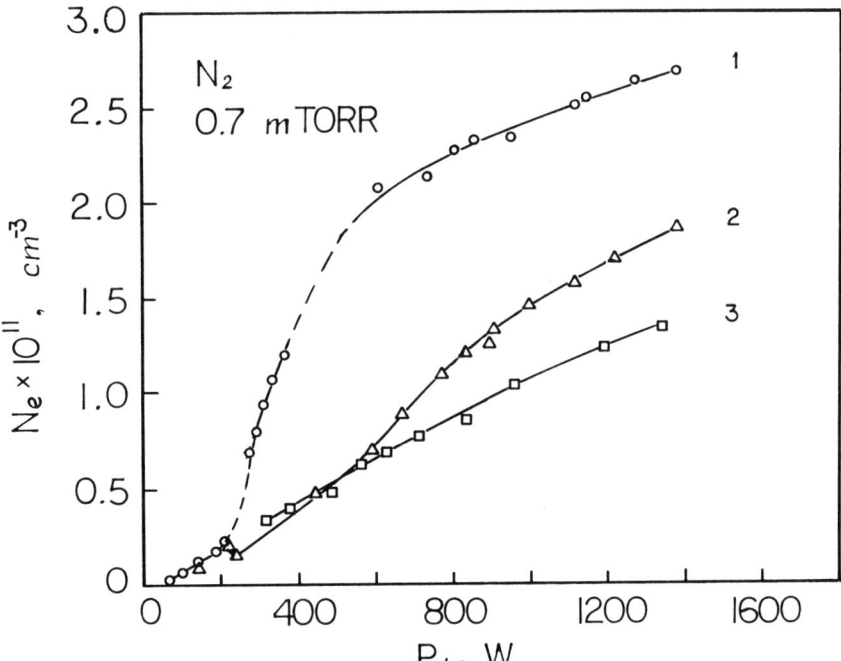

FIG. 34. Line-averaged plasma density measured at $z = 11$ cm as a function of microwave power absorbed in three microwave plasma modes: 1—ECR mode ($I_1 = I_2 = 170$ A), 2—Whistler mode ($I_1 = I_2 = 205$ A), and 3—second harmonic resonance mode ($I_1 = I_2 = 85$ A). Nitrogen, 0.7 mTorr.

The results presented clearly show that narrow-mode ECR plasma sources are more efficient than uniform Whistler and SHR plasma modes. We have also shown (Fig. 26) that the highest plasma density is in an argon ECR plasma stream. Nitrogen and oxygen plasma streams have close plasma density, and hydrogen plasma stream has the lowest plasma density.

It is widely assumed that electron temperature, T_e, grows with microwave power (*104 153, 219*). But the dependence of electron temperature on microwave power absorbed is generally rather complicated. At low plasma density, when microwaves (supposedly L waves) partially propagate through an underdense plasma, electron temperature is relatively high and grows with microwave power. It can increase with P_{inc} even downstream, probably because of the presence of the electric field of the standing waves that are formed in an ECR and a process chamber (*123, 130*), or in the narrow sites where the magnetic field starts to diverge (*113, 167*). When the plasma density reaches N_{cr}, L waves stop to propagate downstream and dissipate in one of the microwave power absorption mechanisms discussed in Section I. This results in the noticeable decrease of the electron temperature downstream. As microwave power grows further the electron temperature either continues to decrease or begins slowly to grow dependent on gas pressure and plasma density.

This is illustrated in Fig. 35, where electron temperatures measured in a 14.6 cm diameter ECR plasma chanber at $z = 10$ cm from the microwave introduction window are plotted as function of microwave power absorbed for two magnetic field profiles. Plasma density is also plotted there. It can be seen that T_e decreases at power of 200–250 W where plasma density jumps from $N_e \approx N_{cr}$ to overdense magnitudes, $N_e = (2-2.5) \times 10^{11}$ cm^{-3}. Then the electron temperature passes its minimum at $P_{abs} = 500–700$ W and begins to rise again.

At higher gas pressures ($p = 2 \times 10^{-2}$ Torr), the electron temperatures at 5 cm from the microwave window were lower, and no rise in T_e was observed up to 1,000 W. (Fig. 36). The "temperature" of fast electrons (T_{e2}) is also slightly reduced with absorbed microwave power, while the plasma density monotonically grows with power from 2×10^{11} cm^{-3} to 3.5×10^{11} cm^{-3}.

It seems to be reasonable to suggest that electron temperature behavior is controlled by two (at least) competitive processes: two-step ionization and saturation. At high pressures ($p > 10$ mTorr) and high plasma densities, the two-step ionization can prevail over the single-impact ionization that is dominant at low pressures and relatively low plasma densities (*229*).

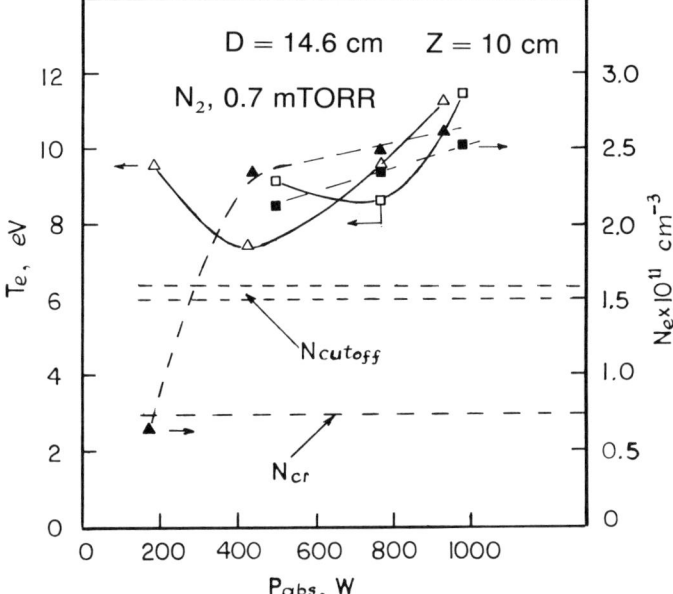

FIG. 35. Electron temperature, T_e, as a function of microwave power absorbed measured at $z = 10$ cm from the microwave window. $D_{ch} = 14.6$ cm, nitrogen, 0.7 mTorr.

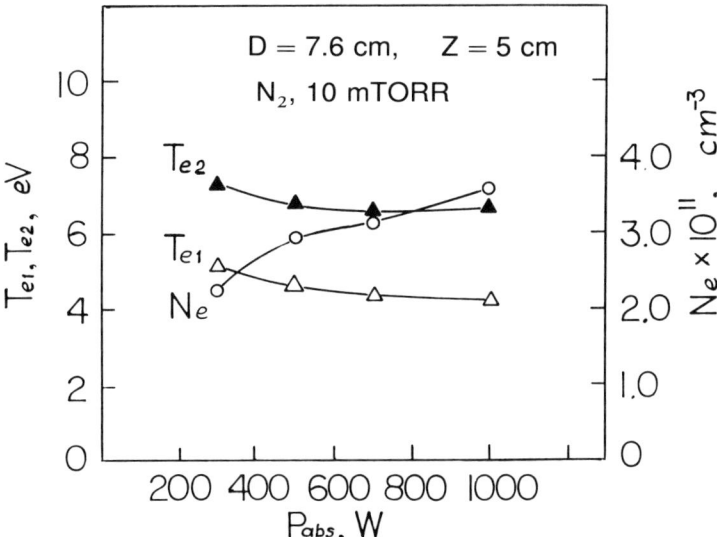

FIG. 36. Electron temperature as a function of microwave power absorbed measured at $z = 5$ cm from the microwave window. $D_{ch} = 7.6$ cm. Nitrogen, 10 mTorr.

The presence of the "second" electron temperature, T_{e2}, on Langmuir probe IVC was reported by the author in other works (*130, 159*). He attributed the appearance of fast electrons to Landau damping and to stochastic electron heating in zones where ECR heating was excluded. But certainly other processes that cause the departure of the electron energy distribution from Maxwellian can be essential or even dominant (*153*).

Plasma potential is an important parameter that determines ion energy and controls the sputtering from ECR chamber walls. It is given as a function of microwave power for various ECR plasma chamber diameters, 2.5 cm, 5 cm, 7.6 cm, and 14.6 cm, in Fig. 37. Plasma potential increases insignificantly in the large diameter chambers (7.6 cm and 14.6 cm) where it has magnitudes of 25–35 V, while V_{pl} grows substantially in small diameter chambers (2 and 5 cm) and reaches values of 50–60 V. Ions that bombard the chamber walls and microwave introduction window with such an energy can sputter walls and window materials, which causes process particulation and coating on the microwave window (*219*). This is another reason to use a larger diameter chamber, $D_{ch} > 8$ cm, even in processes where a small substrate is used.

The floating potential, V_f, in the ECR plasma chamber has values close to the ground potential. It grows insignificantly with the microwave power absorbed (Fig. 38). In the case of Maxwellian electron

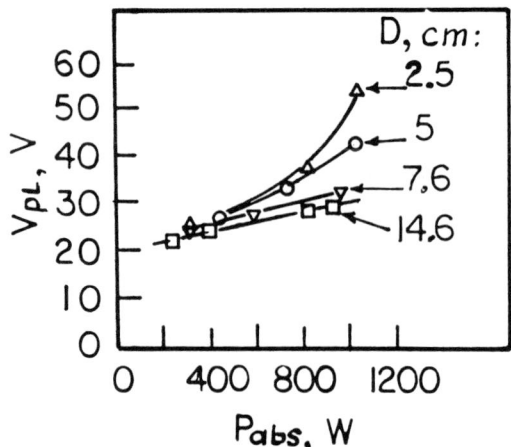

FIG. 37. Plasma potential, V_{pl}, as a function of microwave power absorbed measured in plasma chambers of different diameter at $z = 7.5$ cm from the microwave window. Nitrogen, 0.6 mTorr.

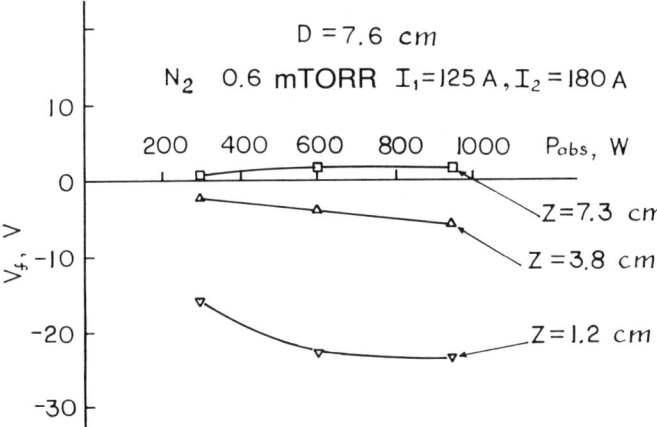

FIG. 38. Floating potential as function of microwave power absorbed measured in a 7.6 cm diameter plasma chamber at three distances from the microwave window: 1.2 cm, 3.8 cm, and 7.3 cm. Nitrogen, 0.6 mTorr. ECR magnetic field is at $z = 3$ cm.

energy distribution, it can be computed as

$$V_f = V_{pl} - 0.5\, kT_e Ln(M/m). \tag{28}$$

In the excess (or lack) of fast electrons, the experimental floating potential is larger (or smaller) than computed from Eq. (28). Some authors pointed out that the appearance of fast electrons on the probe current-voltage characteristic (IVC) was accompanied by a high level of microwave power absorption. Thus, the large negative floating potential, which departs from that computed from Eq. (28), can be an indicator of the microwave absorption zone. This is illustrated in Fig. 38, where negative floating potential exhibits a sharp growth near the microwave introduction window where the microwave absorption zone is supposed to be.

Note that very large negative floating potentials can be an incorrect reading caused by the distortion effect of the microwave electric field on the probe *IV* characteristic, which shifts V_f to negative potentials.

3. *Axial Distribution of ECR Plasma and Plasma Stream*

The microwave plasma generated in ECR (or by another microwave absorption mechanism) is transported by the static ambipolar electric field along magnetic field lines. If a substrate is located in a process chamber, the process characteristics are controlled primarily by plasma parameters in the vicinity of the substrate. The latter are determined by many factors: distance from the microwave absorption zone (surface), gas and pressure, plasma parameters in the absorption zone, magnetic field magnitude and structure, the diameter of the substrate holder, the diameter of the ECR chamber, and the process chamber diameter.

The transport of plasma along and across magnetic field lines and plasma parameters' axial distribution were studied by many authors: Matsuoka et al. (83), Gorbatkin et al. (104, 105, 219), Gottscho et al. (119, 128, 142, 167), Holber with co-workers (94, 134), Oomori et al. (133), Popov et al. (113, 130), Hopwood et al. (101, 102), and Rossnagel et al. (129). The modeling of ECR plasma transport including electron energy distribution and plasma parameters was done by Hussein et al. (112), Wang and Kushner (153), and Uhm and Choi (237). Though there is a large diversity in ECR source design and operation conditions, the physical laws that govern plasmas "extracted" or diffused from the ionization zone are essentially the same. Indeed, if we consider a plasma with zero microwave electric field, its transport is controlled by the "initial" plasma parameters in the ionization zone, and by the configuration and magnitudes of the magneic field, and the system geometry.

One of the most explored ECR plasma sources is the NTT type where plasma is transported (we do not use the word "diffused" because at pressures lower than 10 mTorr electrons are drifted in the "freefall" regime) along the divergent magnetic field lines in the process chamber. But in an ECR plasma chamber the magnetic field can be both divergent and axially uniform, dependent on the coils' currents and their positions.

a. Plasma Parameters in the ECR Chamber

We present here experimental data of plasma parameters axial distributions measured in an ECR plasma source using a Helmholtz pair of coils but with various plasma chamber diameters and lengths.

It was found that in all explored plasma chambers and magnetic field profiles, the plasma density was practically constant along the chamber axis and decreased only near the microwave introduction window and chamber end (Fig. 39). This profile was observed with both a positive and a negatiive magnetic field gradient, dB_z/dz, and when the ECR magnetic field, $B_z = 875\,G$, was placed at different positions along the plasma chamber and outside of the chamber. Even at magnetic fields substantially below ECR, the plasma density had an almost "flat" profile within most of the chamber. The distinctive decrease of the plasma density was observed at $z = 3-4\,cm$ from the microwave introduction

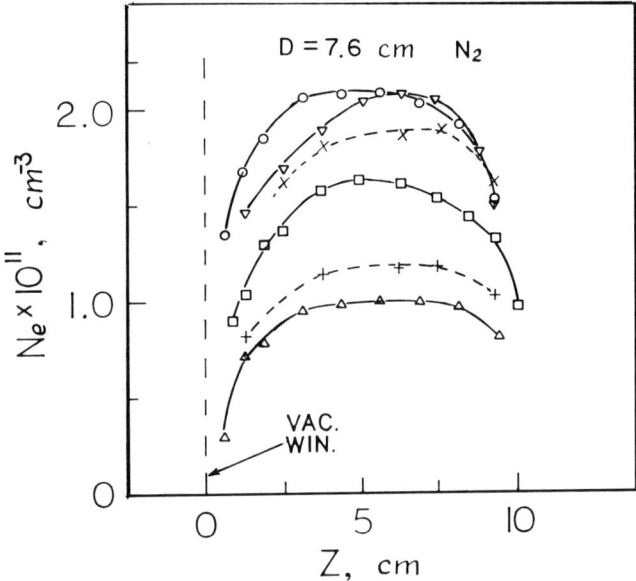

FIG. 39. Plasma density, N_e, axial distribution in microwave plasma chamber. $D_{ch} = 7.6$ cm. Nitrogen, 10 mTorr, $P_{abs} = 300$ W: △, $I_1 = I_2 = 160$ A; ▽, $I_1 = 135$ A, $I_2 = 160$ A; □, $I_1 = 190$ A, $I_2 = 85$ A; ○, $I_1 = 52$ A, $I_2 = 190$ A. Nitrogen, 1 mTorr, $I_1 = 125$ A, $I_2 = 180$ A: +, 300 W, ×, 600 W.

window. Such a profile was observed at low pressures, $p = 0.4$ mTorr, and at high pressures, $p = 10$ mTorr (see Fig. 39).

An increase of microwave power does not change the character of the plasma density axial distribution, but certainly raises plasma density (Fig. 39).

Plasma potential, V_{pl}, gradually decreases from the window with an ambipolar electric field of 0.1–0.3 V/cm. But the remarkable feature of the plasma potential axial profile is that V_{pl} still grows approaching the window in sites where the plasma density already decreases (Fig. 40). V_{pl} starts to drop only at a few millimeters from the microwave introduction window where the sheath between the window and the plasma is thought to be formed.

A measurement of the ion saturation current on the Langmuir probe, $J_i(z)$, showed that J_{i0} decreases more slowly when approaching the window, or even stays constant. Such a behavior of V_{pl} and J_{i0} together with the decrease of N_e can be if the electron temperature increases

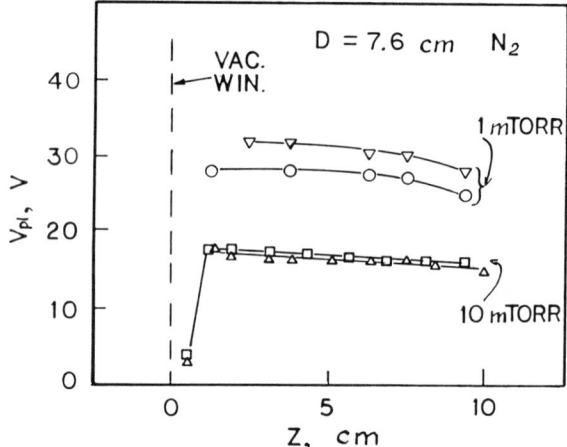

FIG. 40. Plasma potential, V_{pl}, axial distribution. Conditions and key as in Fig. 39.

toward the microwave introduction window:

$$J_i = 0.6 \, ev_{te} N_i \propto (T_e)^{1/2} N_i \qquad (29)$$

The results of measurements of electron temperatures $T_e(z)$ in a nitrogen microwave plasma (below and higher ECR) at low pressure (0.4–0.7 mTorr) and high pressure (10 mTorr) are shown in Figs. 41 and 42. The sharp surge of T_e when approaching the microwave window is seen in all chambers and pressures with their maxima at the window. Even when the magnetic field in the chamber was substantially higher than ECR, so that R waves could propagate through the overdense plasma to an ECR surface located 6–8 cm downstream, or even outside of the chamber, the greatest magnitudes of electron temperature were near the window.

This can be a clear indication that the main portion of the microwave power is observed near the window. The other remarkable phenomenon is that the maximum electron temperature is essentially the same at low and high pressures, $T_e = 10–14 \, eV$, though at higher pressures the electron temperature decreases with the distance from the window faster than at low pressures.

The presence of fast electrons on the prove IV characteristic (the second electron temperature, T_{e2}) was distinctively observed at $z = 3–8 \, cm$ from the microwave introduction window in nitrogen plasmas sustained at high pressures at all microwave power levels (150–1,000 W),

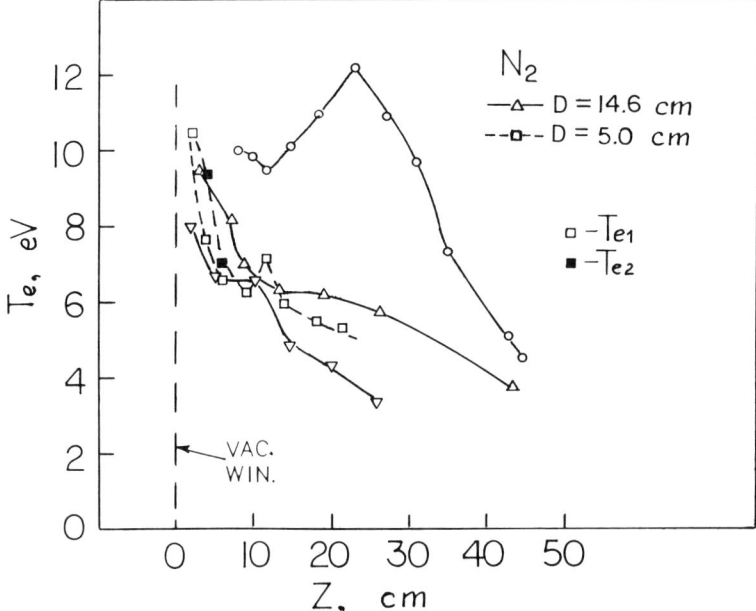

FIG. 41. Electron temperature axial distribution in two different ECR plasma chambers. Nitrogen. $D = 14.6$ cm, $P_{abs} = 500$ W, $p = 0.4$ mTorr: ▽, $I_1 = 145$ A, $I_2 = 195$ A; △, $I_1 = 155$ A, $I_2 = 195$ A. $p = 0.7$ mTorr: ○, $I_1 = 100$ A, $I_2 = 190$ A. $D = 7.6$ cm, $P_{abs} = 700$ W, $p = 0.7$ mTorr: □, $I_1 = I_2 = 165$ A.

while at low pressures the distinctive energetic tail on the probe IV characteristics was observed only at high microwave powers of 800–1,000 W.

The other interesting phenomenon was that the second electron temperature, T_{e2}, was distinctively observed in plasmas with magnetic fields higher than ECR, i.e., in plasmas where R Whistler waves can propagate. Beyond the ECR surface, $B = 875$ G, the second electron temperature disappears (*130*).

The appearance of fast electrons can be caused by various factors, some of them discussed in Refs. 121 and 153. Among these are the trapping of fast electrons by high plasma potential, or different relaxation lengths of slow and fast electrons, or heating of fast electrons by Landau damping.

At lower gas pressures the presence of fast electrons (T_{e2}) was not so distinctive, probably because of the large electron relaxation length at low pressures of 0.3–1.0 mTorr. Meanwhile, a recent study of ECR

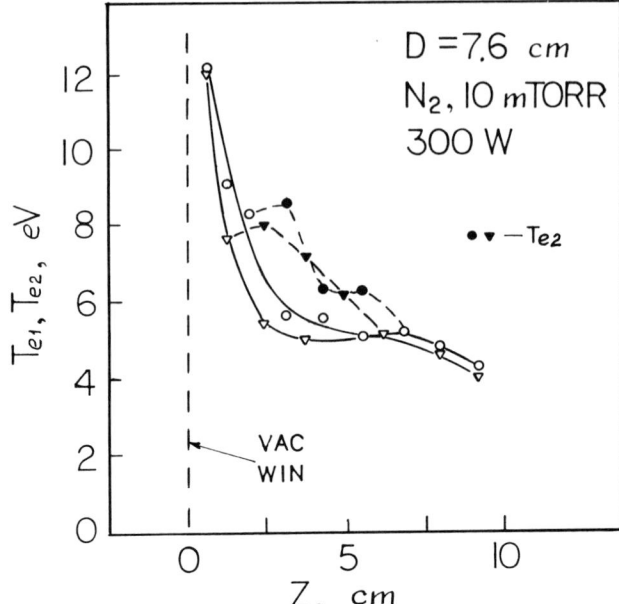

FIG. 42. Electron temperatures, T_{e1} and T_{e2}, axial distribution at high pressure. Nitrogen, 10 mTorr. $D = 7.6$ cm, $\bigcirc$, $I_1 = 52$ A, $I_2 = 190$ A; ∇, $I_1 = 135$ A, $I_2 = 160$ A.

plasma made at the University of Wisconsin showed the presence of fast ("hot") electrons far away from the microwave absorption zone, even near the substrate holder (230).

The floating potential, V_f, gradually gets more negative towards the window with a sharp jump to very large negative values of 15–20 V at $z < 2$ cm. This is illustrated in Fig. 43a, b, showing data of V_f measured in a nitrogen ECR plasma sustained at low (0.6 mTorr) and high (10 mTorr) pressures. Very large negative floating potentials in ECR plasma near the microwave introduction window raise the question of the proper potential on the chamber liner in the microwave power absorption zone. Indeed, the dielectric liner kept at the floating potential is sputtered by ion bombardment more severely than that at the grounded potential because of the larger ion energy, $V_{pl} - (-V_f)$, which is larger than V_{pl}.

b. *Plasma Stream Parameters*

Plasma parameters beyond the source output (limiting ring aperture) gradually decrease with the distance from the source. The plasma density

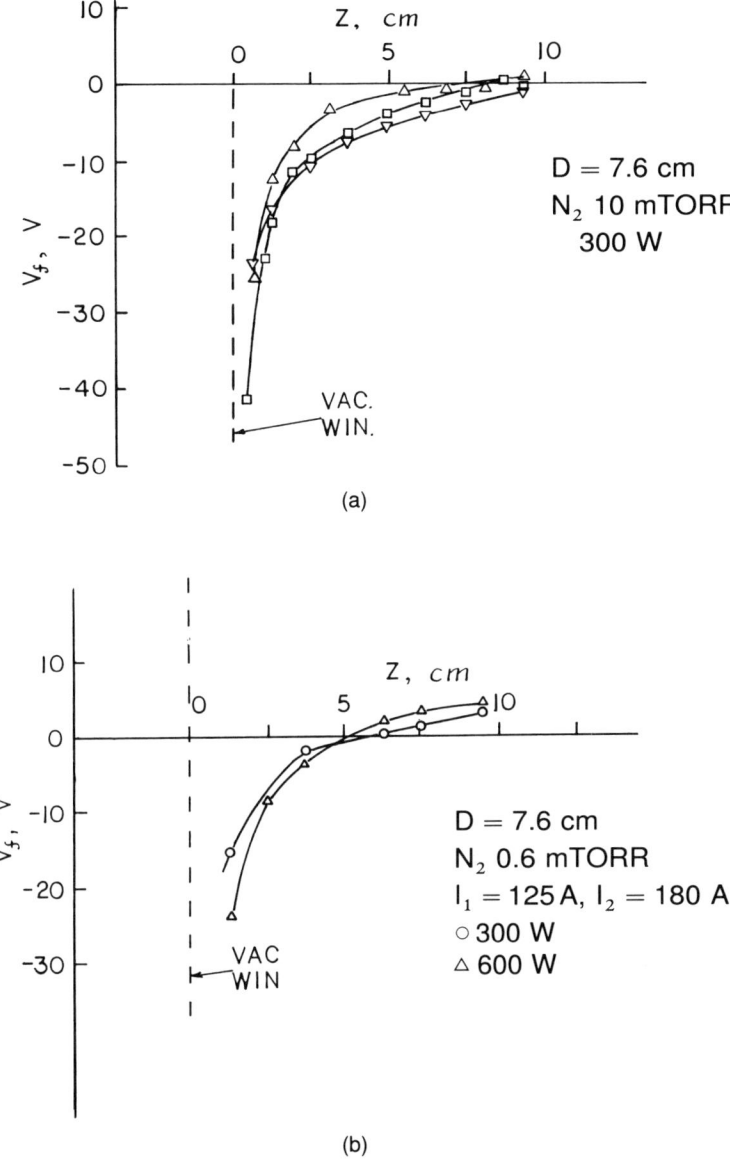

FIG. 43. Floating potential, V_f, axial distributions at low and high pressures. Nitrogen, $D = 7.6$ cm. (a) $p = 0.6$ mTorr, $I_1 = 125$ A, $I_2 = 180$ A. ○, $P_{abs} = 300$ W; △, $P_{abs} = 600$ W. (b) $p = 10$ mTorr, $P_{abs} = 300$ W: □, $I_1 = 190$ A, $I_2 = 85$ A; ▽, $I_1 = 135$ A, $I_2 = 160$ A; △, $I_1 = I_2 = 160$ A.

axial distributions, $N_e(z)$, measured in nitrogen ECR plasma streams "extracted" from 5 cm and 14.6 cm diameter ECR plasma chambers into a 32 cm diameter process chamber are shown in Fig. 44. It can be seen that plasma densities in streams extracted from the smaller diameter chamber (5 cm) decrease inside the ECR chamber ($L = 11$ cm) faster than in the 14.6 cm diameter chamber. But in the process chamber (downstream), all curves $N_e(z)$ have essentially the same axial gradient, dN_e/dz. An increase of nitrogen pressure from 0.6 mTorr to 1 mTorr did not change the plasma density axial profile.

The plasma potential, V_{pl}, decreases monotonically with the distance from the source output with the ambipolar electric field, E_{dc}, ranging from $\simeq 1$ V/cm, near the source limiting ring, to $\simeq 0.5$ V/cm, in the process chamber (Fig. 45).

The value of plasma potential depends from plasma density and

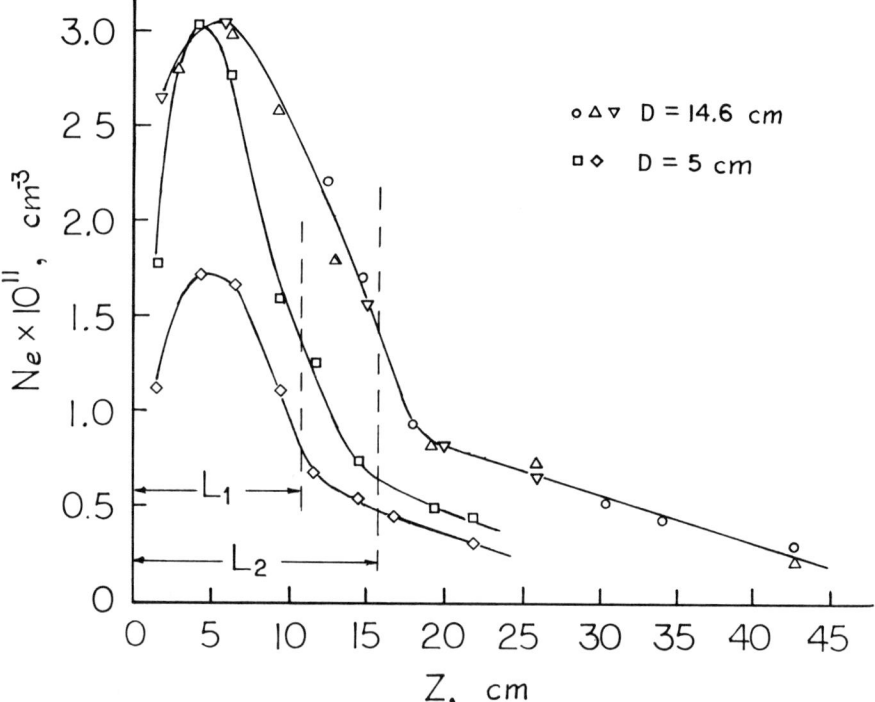

FIG. 44. Plasma density axial distribution, $N_e(z)$, along a generation and process chamber. Conditions and key as in Fig. 41.

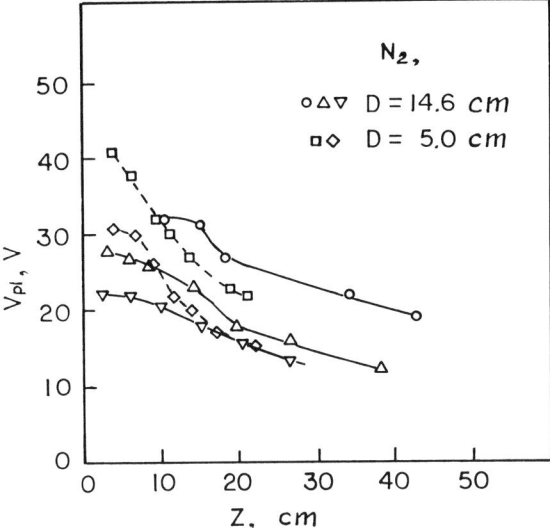

FIG. 45. Plasma potential axial distribution, $V_{pl}(z)$. Conditions and key as in Fig. 41.

electron temperature and was close to that determined by the Boltzmann law:

$$N(z) = N(0) \exp(V_{pl}(z) - V_{pl}(0))/kT_e. \tag{30}$$

The deviation of the experimental data from those computed from Eq. (30) (which is $\approx 10-30\%$) can be attributed to the non-Maxwellian character of the electron energy distribution function f_{ee}, and to the abundance or lack of fast electrons on electron f_{ee}.

It can be seen from Figs. 44 and 45 that at the same microwave power and plasma densities, plasma potentials in the lower magnetic field plasma stream (I90A/100A) are higher; also the ambipolar electric field is higher. According to Eq. (30), this can occur if the electron temperature is higher.

For these cases shown in Fig. 41, the electron temperatures are much higher in the lower magnetic field plasma stream. Moreover, at a distance of 13–15 cm, T_e begins to grow with its maximum somewhere at $z \approx 20$ cm. The cause of the elevation of electron temperature and plasma potentials at a distance far away from the microwave absorption zones is not clear. It can be attributed to the increase of charged particle losses in the area of large magnetic field gradient, dB/dz, i.e., where the

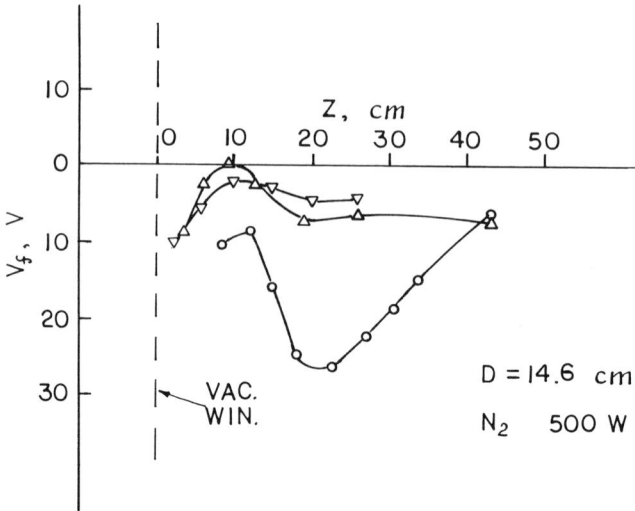

FIG. 46. Floating potential axial distribution, $V_f(z)$. Conditions and key as in Fig. 41.

radial component of the magnetic field, B_r, has substantial growth, which causes the increase of plasma radial losses (*167*).

To maintain the ionization balance, the plasma potential has to be higher and trap fast electrons. Similar electron temperature behavior was observed in an ECR plasma chamber of small diameter (5 cm) (*130*), and in a plasma stream extracted from an 8 cm diameter ECR chamber into a long (30 cm) "plasma tube" (*113*).

The negative floating potential, V_f, follows the behavior of electron temperature and plasma potential and has its "second" maximum at $z \simeq 20$ cm (Fig. 46). The same axial dependence was observed in a plasma stream propagating in a 2 cm diameter "plasma tube" (*113*).

4. ECR Plasma Parameter Pressure Dependencies

The operation pressure range is determined by the process conditions typical for ECR CVD. Higher pressure offers a higher film deposition rate, but it also raises the question of the proper wafer/substrate location because the pressure increase leads to a larger ion–neutral collision frequency and hence reduces the ion flux to the wafer located downstream. Meanwhile, the planarization of silicon oxide film was shown to be successful at low pressures of 0.1–0.3 mTorr where ions are transported from the ECR zone to the wafer stage without suffering collisions with neutral particles.

Collisions with neutral species—elastic collisions and charge exchange collisions—determine the total ion collision cross-section, and hence, its mean free path, λ_i. The collision cross-section is different for each gas; therefore, the pressure dependencies of plasma parameters are expected to be different in each gas.

a. Electron Temperature at Different Pressures

Pressure dependencies of plasma parameters in the microwave power absorption zone are certainly not the same as those downstream near the substrate holder. Most of the researchers used Langmuir diagnostic techniques to measure N_e, T_e, V_{pl}, V_f, and f_{ee}, mainly in the wafer zone primarily because of the distorting effects of the microwave electric field (*231*) and high magnetic field on the probe IVC (*232*). Lee et al. measured plasma parameters both in the ECR zone and downstream within a wide pressure range from 0.5 mTorr to 70 mTorr (*95*). They found that in argon ECR plasma at a low pressure of 1 mTorr, the electron energy distribution function in the ECR zone was between Maxwellian and Druyvesteyn distributions with an average energy of 7.14 eV. This means that the ECR plasma has the larger fraction of slow electrons and a smaller population of fast electrons compared with those in a Maxwellian electron energy distribution function.

We measured plasma parameters in argon plasma excited in a 14.6 cm ECR chamber at pressures $p = 0.1$–5 mTorr, in the ECR zone and downstream. We did not observe a "tail" in either the "narrow" (ECR plasma) mode or in the uniform ($B_{wv} > 1,000$ G) mode. However, in the smaller diameter plasma chamber ($D = 8$ cm), we observed the second electron temperature, $T_{e2} > T_{e1}$, at a pressure of 0.5 mTorr even at $z = 27$–30 cm from the microwave absorption zone (*159*).

Probe diagnostics made in nitrogen plasmas (8 cm diameter chamber) near the ECR (microwave absorption zone) showed a clear presence of two electron temperatures, $T_{e2} > T_{e1}$, at all pressures used, from 0.6 to 20 mTorr. But in oxygen ECR plasmas sustained in a small plasma chamber of 5 cm diameter, no distinctive fast electrons with the second electron temperature, T_{e2}, were observed at pressures of 1–5 mTorr and microwave power from 100 to 1,000 W.

Electron temperatures T_{e1} and T_{e2}, and floating and plasma potentials V_f, and V_{pl}, decrease with gas pressure in the ECR zone and downstream. This was observed by many authors working in different gases: Gorbatkin et al. (*104, 219*) and Oomori et al. (*133*) measured T_{e1} in argon and chlorine ECR plasmas and streams. Lee *et al.* in argon and oxygen (*95*) and Popov in nitrogen, oxygen, and argon (*26, 86, 130*) diagnosed T_e along the chamber axis within a large range of pressures (0.2–50 mTorr).

Electron temperature measured near the microwave absorption zone vs. gas (nitrogen) pressure is shown in Fig. 47. It is seen that both T_{e1} and T_{e2} gradually decrease with pressure. The electron temperature vs. argon pressure measured in a narrow ECR plasma stream and a uniform Whistler stream at 46.5 cm from the microwave introduction window (31 cm from the source aperture) is plotted in Fig. 48. It is seen that even far away from the microwave absorption zone, the electron temperature in the ECR stream is higher than in the Whistler wave uniform stream. Note that the second electron temperature (presence of fast electrons) was not observed in the probe *IV* characteristics in either the ECR or the Whistler plasma stream.

b. *Plasma and Floating Potentials*

Plasma and floating potentials, V_{pl} and V_f, as functions of argon pressure were measured by Gorbatkin *et al.* in the ECR zone and downstream (*104*). They calculated V_{pl} as the sum of the measured

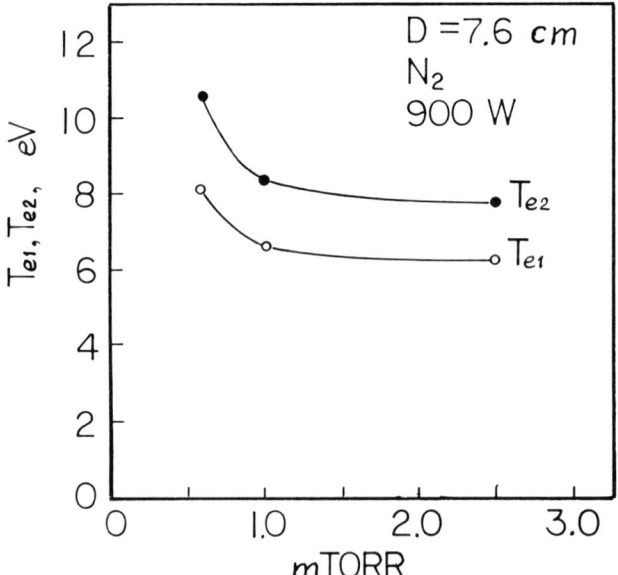

FIG. 47. Electron temperatures, T_{e1} and T_{e2}, vs. gas pressure. $D = 7.6$ cm, nitrogen, $I_1 = 125$ A, $I_2 = 180$ A, $P_{abs} = 900$ W.

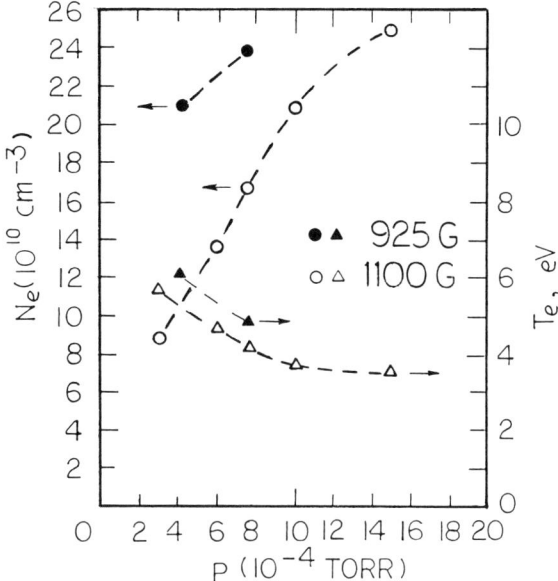

FIG. 48. Electron temperature, T_e, and plasma density, N_e, as functions of gas pressure for ECR narrow mode ($B = 925$ G), and for uniform mode ($B = 1,100$ G). Argon, $z = 31$ cm from the source output.

floating potential V_f and the voltage in the probe sheath:

$$V_{pl} = V_f + (kT_e/2e)\ln(M/2.3m). \tag{31}$$

This formula is correct with the assumption that electrons have a Maxwellian energy distribution function, but it was shown earlier that this is not the case in an argon ECR plasma (95).

We determined a plasma potential as the departure of the dependence $\ln(I_e)$ vs. V_{pr} from the linear one. The results of measuring V_{pl} as a function of pressure at $z = 9$ cm from the ECR zone are shown in Fig. 49. It can be seen that plasma potential decreases with pressure approximately as $V_{pl} \propto (p)^{-1/2}$, which is in good agreement with data obtained by Gorbatkin (104).

Similar dependencies were observed by other authors (95, 104, 133, 219). In fact, in all gases used in plasma processing, and in plasma chambers of 10–20 cm diameter, plasma potentials downstream decrease slowly from 20 to 25 V at 0.2–0.4 mTorr to 10–15 V at 10 mTorr. In plasma chambers of smaller diameter, 5–8 cm, plasma potentials are higher at the same pressures, $V_{pl} = 35$–40 V.

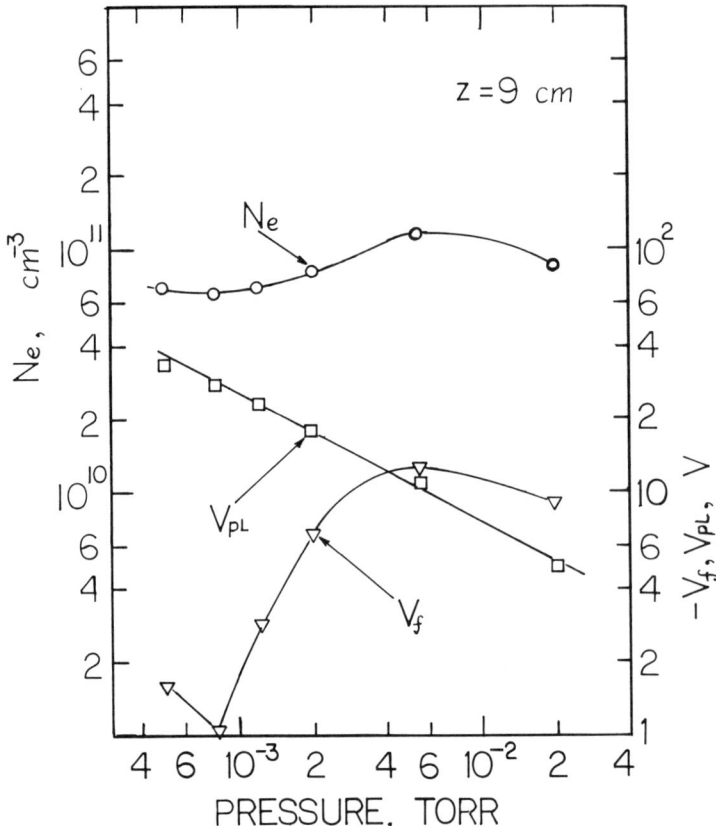

FIG. 49. Plasma potential, V_{pl}, floating potential, V_f, and plasma density, N_e, as functions of gas pressure. $D = 7.6$ cm, nitrogen, $z = 9$ cm. Note that floating potentials are negative.

c. *Plasma Density vs. Gas Pressure*

If electron temperature and plasma potentials gradually reduce their values with pressure both in an ECR zone and downstream, the plasma density relationships with pressure are different in the ionization zone (ECR zone) and downstream. With the possible exception of the microwave absorption zone, the dependence of N_e vs. pressure always has a maximum at a particular pressure (or within the range of pressures). At low pressure (typically below 1 mTorr), plasma density grows with pressure; at higher pressure, plasma density slowly but gradually decreases. The pressure where plasma density has its maximum magnitude varies with gas, distance from the microwave absorption zone, magnetic field structure, ECR plasma mode, and microwave power absorbed.

The author of this paper is not aware of a systematic study of the plasma density–pressure relationship when each aforementioned factor was varied separately. Therefore, he discusses here the effect of some factors provided the necessary data are available.

Line-averaged plasma densities measured in an oxygen uniform ($B > 1{,}100\,\text{G}$) plasma stream at $z = 10$ cm from the ECR zone are shown for three microwave powers (500 W, 600 W, and 1,000 W) in Fig. 50. It is seen that plasma density grows with pressure up to 5–8 mTorr and then slowly decreases. The greater the microwave power absorbed, the higher the pressure at which the plasma density saturates. This phenomenon can be explained by the higher electron temperature at large microwave power, which by the ionization downstream compensates the ion flux losses in charge exchange collisions.

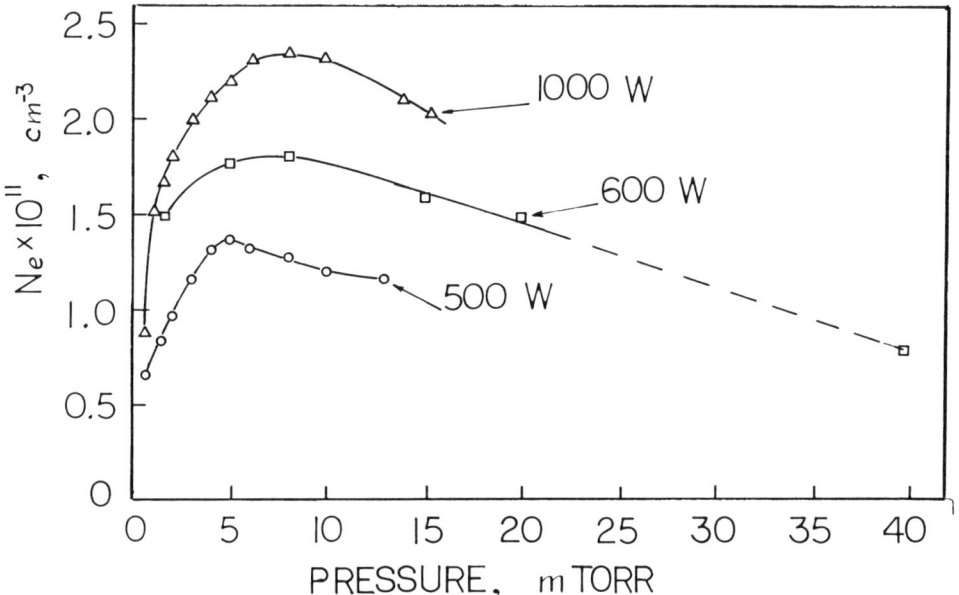

FIG. 50. Line-averaged plasma density, N_e, vs. gas pressure in a uniform plasma mode ($B > 1100\,\text{G}$) at different microwave powers absorbed. Oxygen, $z = 10$ cm from the source output. After Ref. 53.

Plasma density measured by Oomori et al. in an argon plasma stream at 35 cm from the ECR zone did not show any saturation with the pressure up to 10 mTorr (133). This observation contradicted results obtained by Gorbatkin et al., who observed plasma density saturation downstream at 2–5 mTorr and did not reach plasma density saturation in the ECR zone even at pressures of 18 mTorr (104).

We observed plasma density saturation in argon and nitrogen ECR plasmas at pressure of 20–25 mTorr, but the Langmuir probe was at $\simeq 2$ cm from the ECR zone. When the probe was at $z = 9$ cm, the saturation was observed at 5–8 mTorr (see Fig. 49), which was the same pressure as in oxygen plasma. We also did not find a significant effect of type of microwave mode on the plasma density–pressure dependence, though plasma density in the ECR narrow mode was always higher on the axis than density in the uniform Whistler mode (Fig. 49) that was reported in our previous papers (138, 159).

It was noticed that in a plasma chamber of smaller diameter, the "saturation pressure" is lower than in a large-diameter plasma chamber (113).

From a practical point of view, a pressure range of 2–6 mTorr is optimum to achieve high plasma density even at large distances from the source where the stream diameter is large enough to provide a uniform plasma for wafers of 12–15 cm diameter. With this optimistic statement we finish the overview of ECR plasma parameters and come to the applications of ECR plasma source for PE CVD of thin films.

IV. ECR Plasma Sources in PA CVD of Thin Films

We discuss in this section some applications of ECR plasma sources in low-pressure (0.1–50 mTorr) plasma-assisted (enhanced) chemical vapor deposition of thin films (100–10,000 Å thick). This particular technology covers only a small segment of all ECR CVD applications (some of them are listed in the Introduction). The author has chosen it as an appropriate example and illustration of ECR CVD technology, which already has found its industrial applications. The author's experience in this technology gives him a chance to include in this book some results of his work that might be found interesting and useful. Of course, the analyses of all ECR CVD technologies that are now under development could be the subject of a special book and cannot be properly described in one chapter.

A. Silicon Nitride Film ECR CVD

1. Review of Previous Work

A pioneer work in the application of ECR plasma stream for PA CVD was done by Matsuo and Kiuchi (*20, 156*). They inserted a 10 cm diameter wafer in a low-pressure (0.1–0.2 mTorr) plasma stream "extracted'. into the process chamber from a 20 cm diameter ECR plasma chamber by a divergent magnetic field. The plasma "forming" gas (nitrogen) was introduced into the ECR chamber, and silane was fed into the process chamber above the substrate holder. Without additional heating and at a microwave power of 100–300 W, the deposition rate was 200–350 Å/min. By increasing the silane flow rate twice and the nitrogen flow rate three times, Matsuo increased the deposition rate twice (*55*).

In the following years many researchers used Matsuo-type ECR plasma stream sources (NTT-type) for depositing low- (even room-) temperature silicon nitride films. Hirao *et al.* studied extensively how deposition process conditions—microwave power, gas pressure and composition, flow rate ratio—affect film characteristics: stoichiometry, refractive index, transmittance, and resistivity (*21*). They operated in a larger pressure range (0.2–5.0 mTorr) and microwave power range ($P = 50$–500 W). They found that the deposition rate gradually (though not linearly) grew with microwave power and silane flow rate. The same authors studied hydrogen incorporation in silicon nitride film and hydrogen effect on film properties (*31*).

Barbour *et al.*, using a Microscience (uniform Whistler and ECR plasmas), source studied effects of the deposition conditions (pressure, substrate location, silane/nitrogen flow rate ratio) on hydrogen content in silicon nitride films (*232*). Shapoval *et al.* deposited room-temperature silicon nitride films in an ECR plasma stream at pressures of 1–10 mTorr using nitrogen and a mixture if 25% $SiH_4 + Ar$ (*23*). The authors introduced silane via a gas distribution ring placed a few centimeters above the substrate, which was thought to improve film azimuthal uniformity.

The author of this paper has done considerable work on the deposition of silicon nitride film onto silicon wafers of different diameters—from 1 to 8 inches. He used various ECR plasma modes: narrow (ECR/Doppler shifted ECR) and uniform (Whistler mode), overdense and underdense ECR plasmas, high magnetic field (1,100 G) and low magnetic field (SHR) microwave plasmas (*138, 234*). Here he presents and discusses some results of this work, together with a description of the ECR CVD equipment.

2. Silicon Nitride Films Deposited in ECR and SHR Microwave Plasmas

The experimental setup is shown in Fig. 51. It was based on a Microscience ECR machine 9200. Microwave power $P = 100-1,000$ W (at a frequency of 2.45 GHz) was introduced into a stainless steel plasma chamber ($D = 14.6$ cm, $L = 16$ cm) via a rectangular waveguide, through a matched microwave introduction quartz window. The diameter of the window was slightly larger than the chamber diameter, so such an arrangement was thought to provide a more uniform microwave introduction into the chamber (120).

The reflected power was measured with an rf diode installed in the dummy load attached to the three-port circulator. A three-stub tuner was used to improve the matching of the impedance of the microwave transmission line and that of microwave magnetoactive plasma. Two adjustable magnetic coils surrounded the plasma chamber, forming a

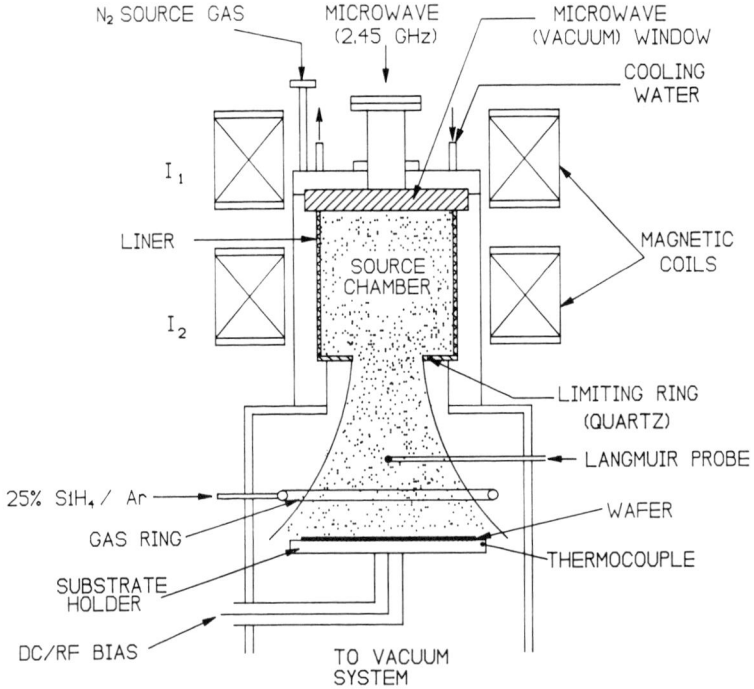

FIG. 51. Schematic of the ECR plasma source and experimental setup. I_1 and I_2 are magnetic coil currents. After Ref. 234.

Helmholtz pair ($z = 4$ cm). Each coil was independently fed by a dc power supply with a maximum coil current of 90 A, which generated a magnetic field strength of 910 G in the coil center. The magnetic field profile, which was typical for uniform plasma stream mode, was given in Fig. 13 (Section II). One can see that in this mode the magnetic field of $B \gtrsim 1,000$ G is radially and axially uniform within 5–7 cm, beginning from the microwave introduction window, and then decreases toward the source output. The ECR zone is located inside the source chamber at a distance of 3 cm from the limiting ring aperture (8 cm diameter), which terminates the source chamber and determines the stream initial diameter.

The magnetic field radial uniformity at the ECR zone is better than $\pm 2\%$. Such a magnetic field structure provides conditions for a very uniform 10–12 cm diameter plasma in the source chamber, and a 15–20 cm diameter uniform plasma downstream near the wafer (*26, 114, 120*).

In the narrow plasma mode, the magnetic field profile can be also radially uniform along the chamber, but the magnetic field strength was 875 G or 925 G (Fig. 11). The narrow plasma mode can also be maintained in decreasing or increasing magnetic fields, $dB/dz < 0$ and $dB/dz > 0$, dependent on the coil currents, I_1 and I_2. But the strength of magnetic field near the microwave introduction window in the narrow plasma mode was always 875 G or 925 G (see Fig. 11).

Nitrogen was introduced into the upper part of the ECR chamber. (In some cases we used a special gas distributed ring placed near the introduction window.) Silane diluted in argon (25% SiH_4) was introduced into the process chamber via a gas distribution ring of diameter larger than the waver holder. This ring had many tiny apertures of ≈ 1.0 mm diameter, with the whole cross-section slightly larger than that of the gas tubing.

The process chamber was 35 cm in diameter and 40 cm in height and was equipped with a few flanges used for load lock, reactive gas introduction, and plasma diagnostics: Langmuir probes and optical emission spectroscopy. A wafer holder was placed at 25 cm from the limiting ring aperture on the heater, which was capable of being heated up to 800°C. A thermocouple attached to the holder was used to monitor the holder temperature. The actual wafer temperature was unknown but was assumed to differ from the thermocouple reading by not more than 100°C.

The pumping system consisted of mechanical and turbomolecular pumps that provided a vacuum of 10^{-6} Torr. Silane and nitrogen mass flow rates were varied with mass flow controllers (MFC) from 6 to 40 sccm. The ratio of SiH_4/N_2 was varied from 0.1 to 2 at operating pressures of 0.4–4 mTorr using MFCs and a throttle valve.

We deposited silicon nitride films on 10 cm, 15 cm, and 20 cm silicon wafers typically for 10 min. The substrate holder diameter was always 30% larger than the wafer diameter in order to eliminate the edge effect on film radial uniformity.

Silicon nitride film radial distribution across wafers of different diameters is shown in Fig. 52. It can be seen that 10 and 15 cm diameter wafers have a very uniform films, while the 20 cm wafer has a clear decrease of film thickness toward the wafer edge. The 20 cm diameter had a 25 cm diameter holder, which seemed not to be large enough.

Films deposited onto the 10 cm diameter waver in the second harmonic resonance (SHR) plasmas had also good radial and azimuthal uniformity (Fig. 53). The latter was of special interest because the wavelengths of waves absorbed near the microwave introduction window are close to those in free space, $\lambda_0 = 12.24$ cm. Therefore, one can

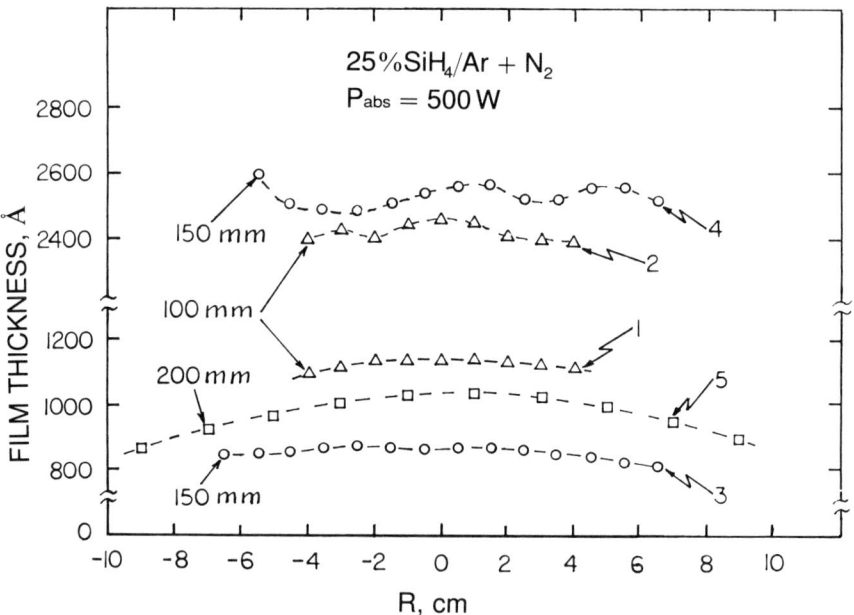

FIG. 52. Silicon nitride film thickness across wafers of different diameter: 100 mm, 150 mm, and 200 mm. Variation of thickness between each run were due to different deposition time: from 5 to 13 min. $z = 25$ cm. After Ref. 145.

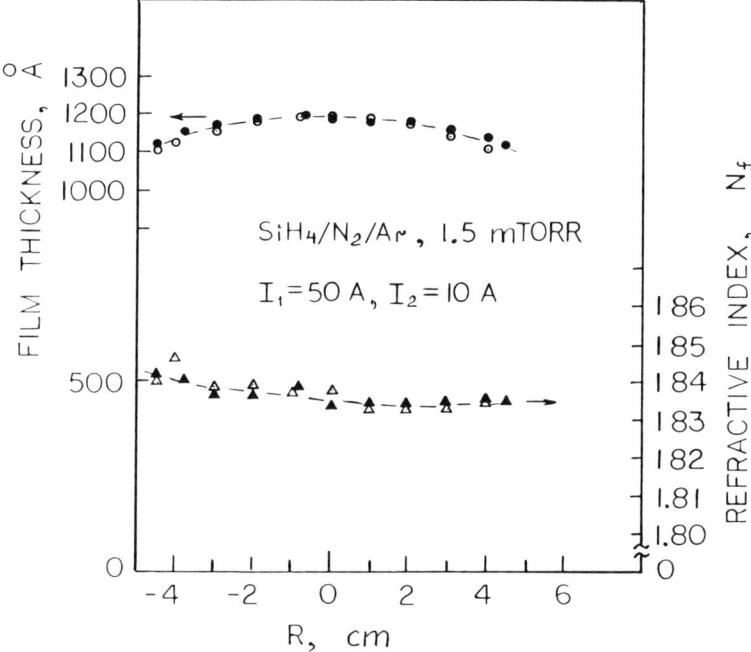

FIG. 53. Silicon nitride film thickness and refractive index, N_f across 10 cm diameter wafer. Second harmonic resonance: $I_1 = 50$ A, $I_2 = 10$ A ($B = 430$ G). 1.5 mTorr, $z = 25$ cm. $\bigcirc, \triangle$, x direction; $\bullet, \blacktriangle$, y direction. After Ref. 234.

expect a standing wave pattern in the plasma density radial profile (231). However, the film radial profile is typically more uniform than the plasma density.

The film deposition rate, W_g, was found to grow with the total gas pressure at a fixed ratio of SiH_4/N_2 flow rates. This is illustrated in Fig. 54, where W_g is plotted as a function of gas pressure for narrow and uniform plasma modes (138). It can be seen that film thickness grows as $W_g \propto p^\alpha$ ($\alpha = 1.2–1.5$), which is faster than nitrogen and argon plasma densities increase with pressure (see Figs. 48–50). This result, which was discussed in Ref. 138, contradicts the sublinear film growth dependence with pressure that was observed by Hirao et al. (21). It can be also noticed that the dependence $W_g(p)$ is the same in both plasma modes, ECR (narrow) and Whistler (uniform).

Film thickness was found to grow linearly with silane flow rate (Fig. 55), which is typical for conventional plasma-enhanced CVD processes and was observed by others (21).

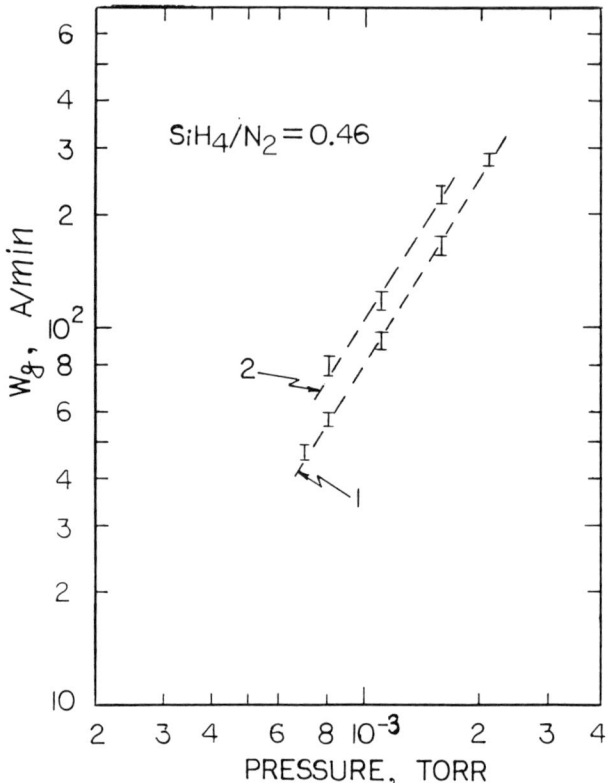

FIG. 54. Silicon nitride film growth rate, W_g, as a function of the total gas pressure. 1—ECR narrow plasma mode ($B = 925$ G): $P_{abs} = 300$ W, $I_1 = 80$ A, $I_2 = 75$ A; 2—uniform plasma mode ($B = 1,100$ G): $P_{abs} = 500$ W, $I_1 = I_2 = 84$ A. After Ref. 138.

It is interesting to compare the silicon nitrode film growth rate, W_g, in ECR and second harmonic resonance plasmas. The results of depositions made at a pressure of 0.7 mTorr and a microwave power of 200 W are shown in Fig. 56 as functions of silane partial pressure. One can see that deposition in the ECR plasma stream is $\approx 25\%$ faster than in the SHR plasma stream. This can be explained by the higher ECR plasma density and electron temperature (234).

The variation of nitrogen flow rate on film growth rate is not so dramatic. Film grew as $W_g \propto N_2^\gamma$, where $0.2 < \gamma < 0.5$ (Fig. 57). The role

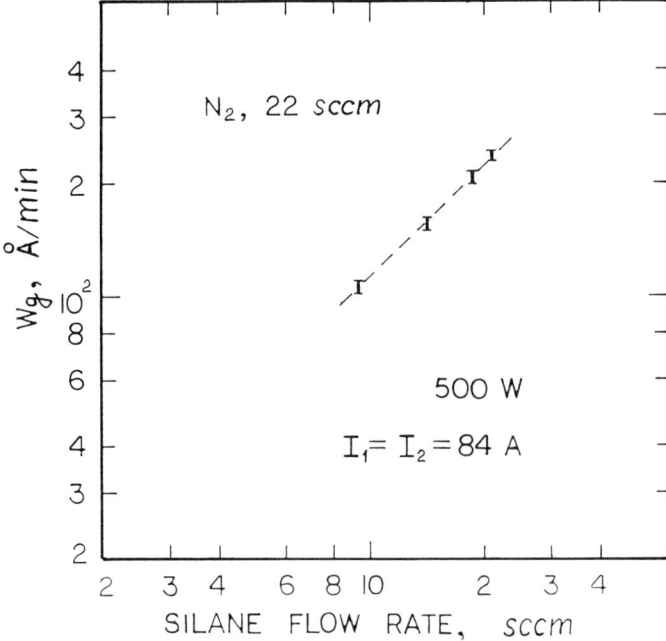

FIG. 55. Silicon nitride film growth rate, W_g, as a function of silane flow rate. Nitrogen flow rate is 22 sccm. Uniform plasma mode, $I_1 = I_2 = 84$ A. $P_{abs} = 500$ W. After Ref. 138.

of nitrogen ions in silicon nitride film formation was studied by Hirao et al. in silane/nitrogen ECR plasma where plasma density was controlled by the nitrogen molecules and atom ionization (*31*). In our ECR plasma the electron density "forming" gas is argon. Nevertheless, the increase of nitrogen flow rate (molecules concentration) causes the sublinear growth of plasma density (see Fig. 49), and hence the sublinear film growth rate.

The film refractive index, N_f, decreases as nitrogen flow rate grows (Fig. 57), and it rises with silane flow rate. The stoichiometric refractive index, 1.95, was observed when the silane/nitrogen flow rate ratio was 0.4–0.6. The content of hydrogen (Si–H bonds) could be decreased with an increase in nitrogen concentration and by an increase of the process temperature: the Si–H bond practically disappeared from film infrared absorption spectra when the film was deposited at $T = 300°C$ (*138*).

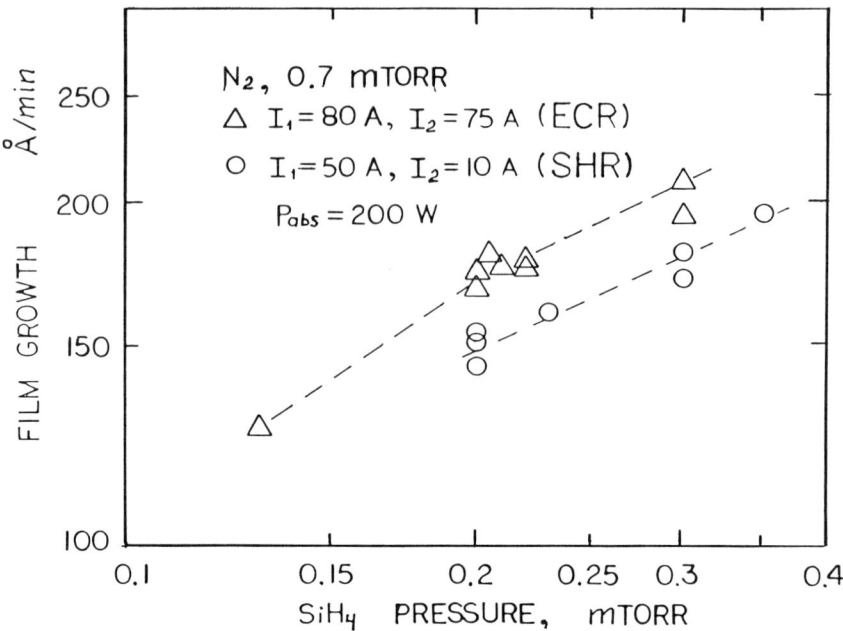

FIG. 56. Silicon nitride film growth rate in ECR and SHR plasma streams as functions of silane partial pressure. △, ECR mode; ○, SHR mode. After Ref. 234.

B. Silicon Oxide Films

1. Review of Previous Work

The first efforts to use ECR plasmas for the formation of silicon oxide films were done by oxidation of a silicon wafers in an oxygen ECR plasma by Bardos *et al.* (*16, 17, 212*). The authors used cw and pulsed microwave plasmas in uniform magnetic fields, $B = 1.3-1.5 B_{ce}$, at an oxygen pressure of 3 mTorr and a microwave power of 1 kW. Bardos *et al.* found that an oxide film could be formed only if the floating potential at the wafer location was close to zero or positive, which could be achieved when plasma density was higher than $N_{min} = 10^{12} cm^{-3}$. This might indicate the detrimental effect of ion bombardment on oxide film formation.

In the early 1980s, Miyake *et al.* applied a Hitachi ECR plasma stream source for 4 cm diameter wafer oxidation at a pressure of 0.1–0.2 mTorr and a microwave power of 150 W (*24*). A wafer was placed on a 7.5 cm

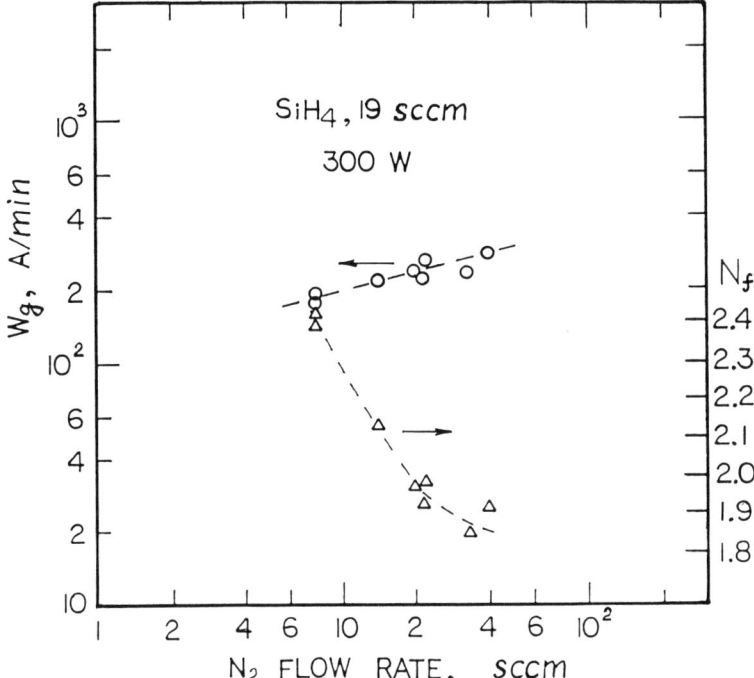

FIG. 57. Silicon nitride film growth rate, W_g, and film refractive index, N_f, as functions of nitrogen flow rate. ECR plasma mode. After Ref. 138.

diameter quartz support heated up to 800°C. However, the oxidation process was too slow, with oxide growth of $W_g = 1-2$ Å/min.

Recently Carl et al. used an ECR plasma for the relatively low-temperature oxidation (400–500°C) of a silicon wafer placed downstream at 37 cm from the microwave introduction window (58, 107). At microwave powers of 400–700 W and an oxygen pressure of 1 mTorr, the plasma density near the wafer was $3-6 \times 10^{11}$ cm^{-3}, and the oxide layer growth rate was as high as 30–50 Å/min (107, 232).

A breakthrough in silicon dioxide film technology was made by Matsuo in the 1980s, when he introduced the ECR plasma CVD of SiO_2 film using a mixture of silane and oxygen (20). At a pressure of 0.2 mTorr and silane and oxygen flow rates of 10 sccm, Matsuo deposited room-temperature dioxide films with deposition rates from 200 Å/min to 400 Å/min as microwave power rose from 50 W to 250 W. By increasing silane and oxygen gas flow rates up to 30 sccm each (and total gas pressure to 0.6 mTorr) he increased the oxide film growth rate, W_g, three times, up to 1,200 Å/min at 300 W of microwave power.

In the following 10 years this type of ECR CVD technology was accepted by most researchers as the mainstream in ECR CVD of silicon dioxide films technology, primarily for IC final passivation, and interlevel isolation. For many research groups the ECR silicon dioxide deposition process became a "rich" target of extensive research aimed at understanding the plasma chemistry kinetics and its effect on film properties.

Herak et al. studied silicon dioxide films deposited in ECR plasma in a rectangular WR 284 waveguide at low power levels (1–50 W) (27). Bulat et al. studied SiO_2 film deposition in an ECR plasma stream ignited in a mixture of silane diluted in argon with a mix of N_2O/O_2 (221). The authors believed that a high-density plasma generates in such a mixture the sufficient amount of oxygen atoms needed for better refractive index uniformity. They studied effects of rf bias on film deposition rate and film properties and found that film wet etch rate drastically dropped as the substrate negative bias exceeded -50 V, while the dry etching rate remained practically constant with rf bias (33). The film growth rate decreased almost by half as the negative bias grew from -30 V to -200 V. This was explained by ion physical sputtering of growing film materials, which can be substantial in an argon-rich plasma at negative bias of $V_b > -30$ V. However, at higher negative bias the film stress was lower and better than that in the conventional PE CVD system.

Andosca et al. deposited silicon dioxide films onto a silicon wafer placed at 22 cm from the 9 cm diameter aperture in ECR plasma source limiting ring (28). The authors ignited the ECR plasma at 300 W of microwave power in a mixture of oxygen and silane diluted in helium at a total pressure of 3 mTorr (28). Because of the small amount of silane (2%), the film growth rate was relatively low, $W_g = 30$–80 Å/min. The authors studied effects of rf bias and substrate temperature on film deposition rate, index of refraction, film etching rate, and film FTIR spectra (28).

2. *ECR CVD of Silicon Dioxide Film in the Microscience 9200 System*

We began to explore ECR plasmas for SiO_2 film deposition back in 1988. Popov and Waldron deposited a uniform silicon oxide film on 7.6 and 10 cm diameter silicon wafers placed at 19 cm from the ECR plasma source output (26). Microwave plasma was excited in an ECR 9200 machine at uniform plasma mode with a magnetic field of 1,100 G near the microwave introduction window, and a magnetic field of 875 G near the source output. At such a magnetic field configuration, a radially uniform plasma was sustained at microwave powers from 400 W to 1,000 W.

Silicon dioxide film deposition was performed in a mixture of oxygen and silane diluted in argon (15% SiH_4) at a wide range of pressure, $p = 0.1–10$ mTorr, and a microwave power of 400–700 W. We found that silicon dioxide film grew linearly with silane flow rate, by was practically constant with gas pressure up to 5–6 mTorr (26). At $p > 5$ mTorr, the deposition rate began to grow very fast with gas pressure, but the quality of the film was poor. Films deposited at pressures below 3 mTorr and temperatures higher than 200°C had properties better than (or comparable to) those of thermal oxide films.

Recently we deposited silicon dioxide onto a 10 cm diameter silicon wafer and studied film growth rate W_g and film refractive index N_f at various silane flow rates, gas pressures, and microwave power levels. The dependencies of W_g and N_f on silane partial pressure with fixed oxygen partial pressure, 0.6 mTorr, are given in Fig. 58 for microwave power of

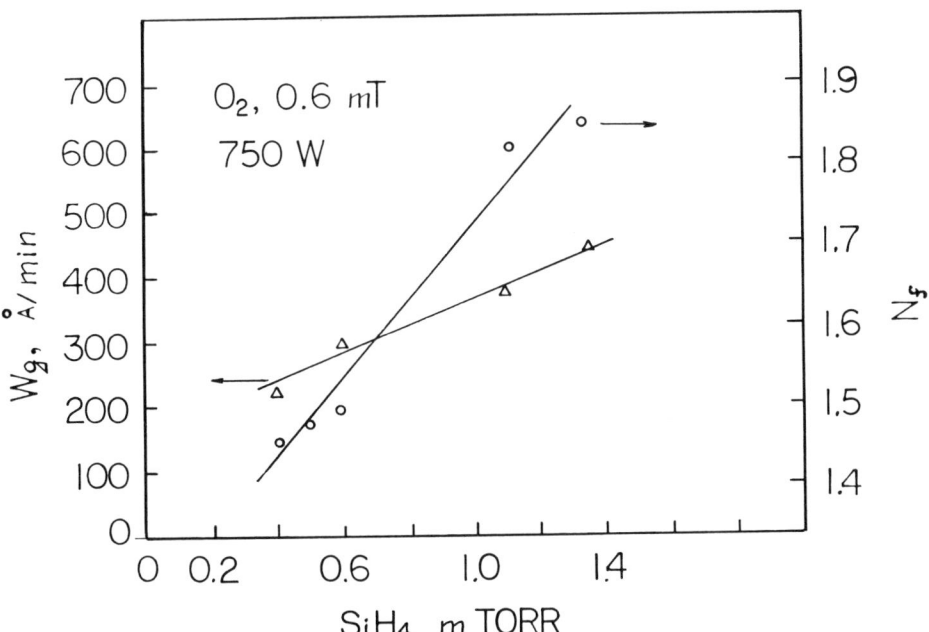

FIG. 58. Silicon dioxide film deposition rate, W_g, and film refractive index, N_f, as functions of silane partial pressure. ECR plasma mode. $P_{abs} = 750$ W.

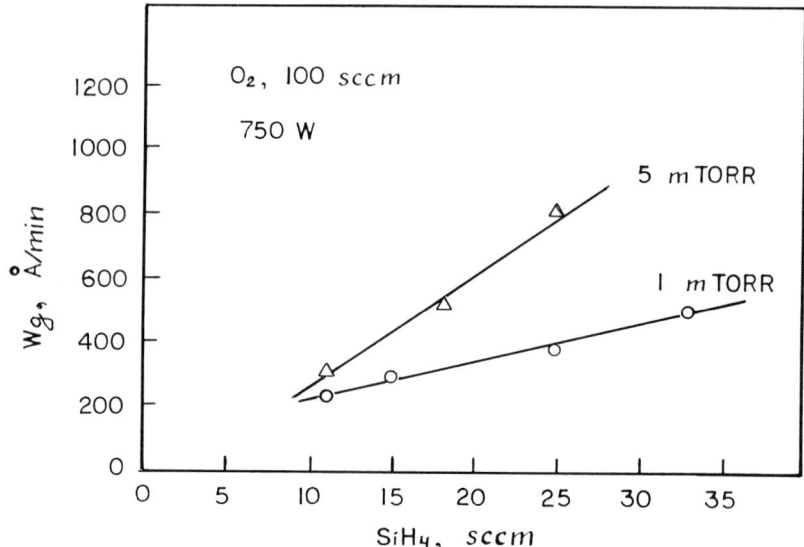

FIG. 59. Silicon dioxide film growth rate, W_g, as a function of silane flow rate obtained at two gas pressures: 1 mTorr and 5 mTorr. Oxygen flow rate 100 sccm. ECR plasma mode, $P_{abs} = 750$ W.

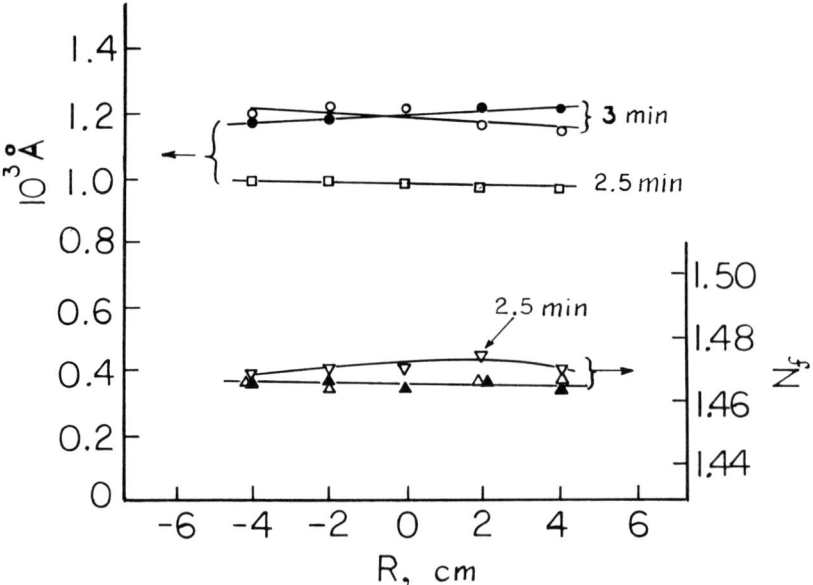

FIG. 60. Silicon dioxide film thickness and refractive index uniformity. 10 cm diameter wafer. $\bigcirc, \triangle$, in x direction; $\bullet, \blacktriangle$, in y direction. 1 mTorr, uniform plasma mode ($B = 1,100$ G), $P_{abs} = 500$ W.

750 W. It is seen that both W_g and N_f increase approximately linearly with silane pressure. The required refractive index of 1.46 was achieved at the ratio $SiH_4/O_2 = 0.8-1$, which was in good agreement with results obtained by other authors (20, 55).

Film growth rate increased linearly with silane flow rate at both chosen total gas pressures, 1 and 5 mTorr (Fig. 59). At a higher pressure of 5 mTorr, the silicon dioxide film grew faster than at 1 mTorr, which was expected, and reached values of 700–800 Å/min. The dioxide film thickness and refractive index radial profiles measured in both directions (x and y) on a 10 cm diameter silicon wafer deposited at 1 mTorr are presented in Fig. 60. A good uniformity ($\pm 2\%$) is seen for both film thickness and refractive index.

C. ECR PLASMA PLANARIZATION WITH SILICON DIOXIDE FILMS

1. Silane/Oxygen Planarization

Planarization of submicron multilevel interconnections became a critical issue in the development of VLSI technology. The conventional rf CVD that operates at medium pressures of 10–100 mTorr seems to be unsuitable for this job because its ion mean free path is comparable with the thickness of the rf electrode sheath. This makes problematic submicron gap filling with a high aspect ratio.

In 1986 Machida and Oikawa (29) explored the principles of planarization technology that was suggested in 1976 by Kennedy (233), in ECR plasma CVD of silicon dioxide. The planarization utilized the higher ion sputtering yield of material on a surface that is at 30–40° to the ion trajectory, than on a surface perpendicular to the ion velocity (233). Using this approach the formation of the reentrant angle is prevented and the gap (or trench) between metal lines is filled without voids and "soft spots" (51).

To accomplish a controllable planarization, the trajectory of ions oncoming to the growing film surface must be highly controllable, that is, the radial component of ion velocity must be much smaller than its vertical component (with the respect to waver surface). To accelerate plasma ions to the surface of the growing silicon dioxide film, Machida and Oikawa applied an rf (13.56 MHz) bias to the wafer holder placed downstream of the NTT ECR source (29). ECR plasma was ignited in a mixture of silane, oxygen, and argon; each flow rate was varied from 10 to 30 sccm. The total gas pressure was not presented in this paper, but it is believed to be between 0.1 and 1 mTorr. Microwave power was 100–400 W, and rf power was 50–500 W.

By using the ECR/rf arrangement, the metal (Al) interconnections of 0.5 micron width and height were perfectly planarized without visible aluminum pattern degradation. The latter is of special interest because the planarization was performed for 34 min using only chilling water for substrate cooling.

In the late 1980s ECR planarization technology was developed by specialists from Lam Research, Sumitomo Metal, and Intel. The ECR source they used differed from the NTT type by its horizontal orientation, which is thought to reduce the contamination on a wafer from microwave window and chamber walls, and by the incorporation of an auxiliary downstream (shaping) magnet behind the wafer holder, which improved plasma radial uniformity, and hence film radial uniformity (*25*).

This ECR source was used for very low-pressure, 0.1 mTorr, silicon dioxide film planarization onto 10 cm and 15 cm diameter wafers with aluminum pattern by Chiang and Fraser (*30*). The planarization process had three stages: (i) SiO_2 deposition without rf bias that provided a thin protective coating; (ii) rf bias application for the simultaneous deposition and sputtering; and (iii) argon addition to increase planarization.

The exact dc voltage generated between the wafer holder and plasma potential (which is close to ground) by rf power was not given in Ref. 30, but the authors showed that the application of rf bias not only created resputtering for good step coverage, but produced a more dense film. The hydrogen content was found to be controlled by rf bias and by the silane/oxygen flow rate ratio. The good planarization was achieved on structures with a 0.8 micron wide space and a 1.3 micron deep trench (*30*).

The deposition characteristics of the Sumitomo/Lam ECR system such as the film deposition rate as a function of microwave power, or the effects of rf biasing on plasma parameters are not available. It is hoped that such data will someday be available for open publication.

2. *Metalorganic ECR CVD and Planarization of SiO_2 Films*

In the last few years, electron cyclotron resonance plasmas were explored for metal organic chemical vapor deposition of silicon oxide films. Pai *et al.* used the divergent plasma stream extracted from an ECR source for planarization of silicon dioxide films using a mix of oxygen with tetraethylorthosilicate (TEOS) and tetramethylcyclotetrasiloxane (TMCTS) (*51,52*). The use of TEOS (Si(OC$_2$H$_5$) or TMCTS (C$_4$H$_{16}$Si$_4$O$_4$) allows replacement of a flammable silane and also offers a PA CVD process with a higher deposition rate.

Pai *et al.* deposited silicon dioxide film onto a 12.5 cm diameter silicon wafer with an aluminum patterned structure of 0.5–1.0 micron height and 0.7–2 micron width, which was placed on an aluminum holder downstream. TEOS (or TMCTS) vapor was introduced into the process chamber via a special quartz ring placed a few centimeters above the wafer holder, while oxygen was introduced into the ECR chamber (*52*).

A good-quality silicon dioxide film was reported in Ref. 52 when film was deposited at pressures of 30–40 mTorr and a flow rate ratio O$_2$/TEOS = 3:1. The deposition time was typically 10–15 min, which allowed deposition of 2 microns of SiO$_2$ film without aluminum pattern degradation by ion bombardment, and without wafer backside gas cooling. The actual wafer temperature was not known, while the wafer holder temperature was about 300°C.

The deposition rate, W_g, grew linearly with TEOS flow rate within the total pressure range of 10–30 mTorr, while the flow rate ratio O$_2$/TEOS was 2 and 3. W_g grew linearly with microwave power but decreased with the flow rate ratio O$_2$/TEOS (Fig. 61). The application of rf bias also

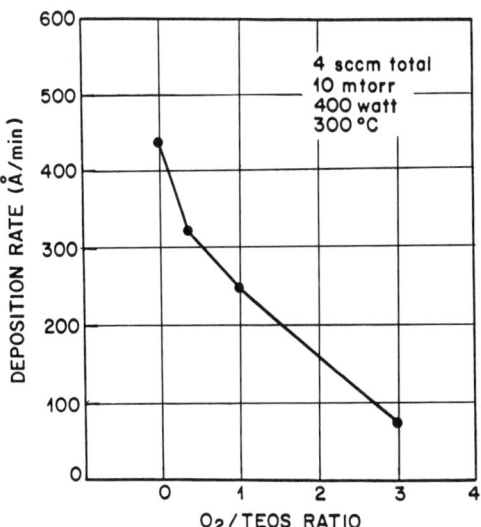

FIG. 61. Silicon dioxide film growth rate as a function of O$_2$/TEOS flow rate ratio. 10 mTorr, P_{abs} = 400 W, T_w = 300°C. After Ref. 52.

caused a decrease in the deposition rate: The increase of negative bias from 0 to -100 V resulted in a film growth drop from 1,600 Å/min to 1,200 Å/min (52). However, it was shown by the authors (51, 52) that to planarize SiO_2 film with a good step coverage a negative bias voltage higher than 100 V is needed.

Other metalorganic substances are considered as promising candidates for ECR CVD of boron and phosphorus doped silicon oxide films. This might be important because of the toxic and flammable nature of the B_2H_4 and PH_3 gases that are typically used in silicon IC technology.

We performed a series of experiments aimed at evaluating ECR plasma techniques for the deposition of boron-doped SiO_2 film (53). As a precursor we have chosen a liquid metalorganic substance, tris(trimethylsiloxy)boron (TTMSB), which has a low evaporation temperature. Other candidates for P- and B-doped silicon oxide films are trimethylphosphite (TMP) and trimethylborate (TMB). Chemical structures of these liquid metalorganic substances are given in Fig. 62.

The plasma forming gas, oxygen, was introduced in the ECR plasma chamber, while vapor of TTMSB was introduced into the process chamber via a distribution ring located a few centimeters above the wafer holder. TTMSB was stored in a stainless steel container (Schumacher). Vapor flow rate and pressure in the process chamber were controlled by container temperature and a mass flow controller. Because of the low operation pressure, $p = 1-60$ mTorr, no carrier gas was used. The typical operation liquid temperature in the container was 35–40°C, while the temperature of vapor line tubing and process chamber walls was kept 10–20°C higher than in the container in order to avoid vapor condensation on cold spots on tubing and chamber walls.

A 10 cm diameter silicon wafer was introduced into the process chamber via a load lock and was fixtured on the heating stage. Plasma density was monitored with a Langmuir probe and a microwave interferometer (35 GHz). The following typical operation conditions were required for the deposition of stoichiometric SiO_2 film with a refractive index of 1.45–1.46: The O_2/TTMSB flow rate ratio was 1.5–2.0 with an oxygen flow rate of 60 and 100 sccm, and the TTMSB flow rate was 30–50 sccm. The total gas pressures varied from 10 to 60 mTorr, and the microwave power was 500–800 W (53).

We found that the film deposition rate, W_g, grew with TTMSB vapor pressure and with the oxygen flow rate (Fig. 63). It was 200 Å/min at 20 sccm of O_2, and 800 Å/min at 80 sccm of O_2. For comparison, the film growth rate in a mixture of TES and oxygen is also given in Fig. 63. One can see that it decreases with oxygen flow rate, which indicates an excess of oxygen atoms required for film formation.

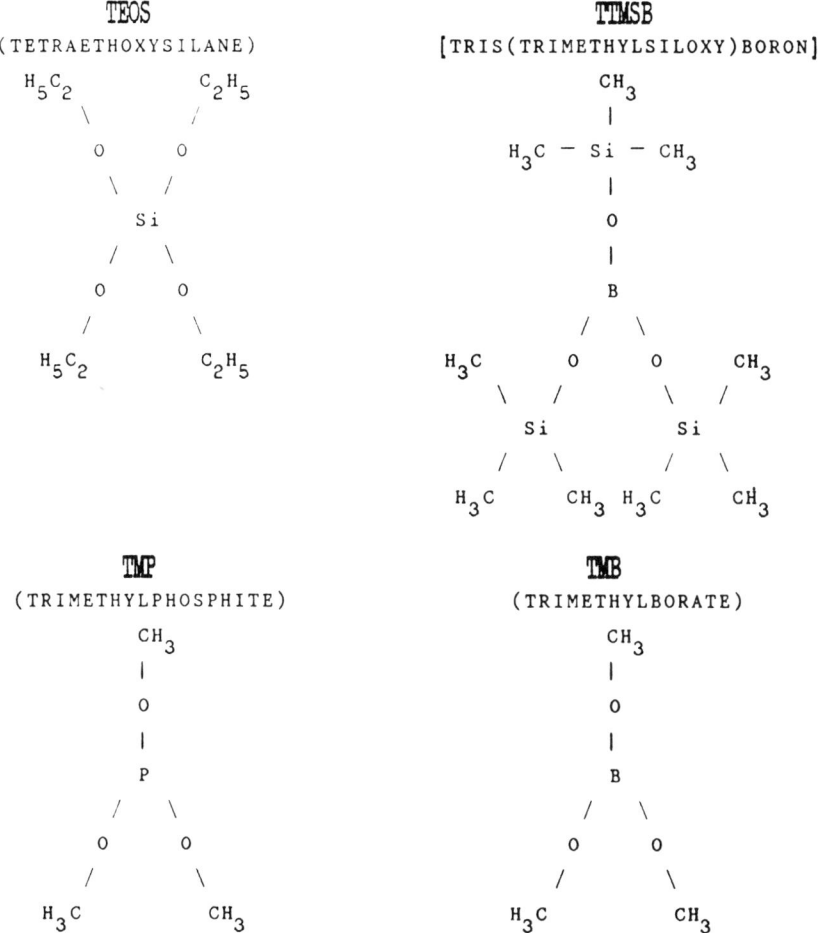

FIG. 62. Chemical structures of liquid precursors used in ECR MOCVD of doped and undoped silicon dioxide films.

The film refractive index was found to grow with TTMSB vapor flow rate and decrease with oxygen flow rate. Film uniformity improved as gas pressure grew from 10 mTorr ($\pm 15\%$) to 60 mTorr ($\pm 2\%$). However, factors that control film uniformity are not yet understood.

FTIR of doped silicon dioxide film showed that boron concentration varied from 3% to 10% dependent on TTMSB flow rate and the total gas pressure.

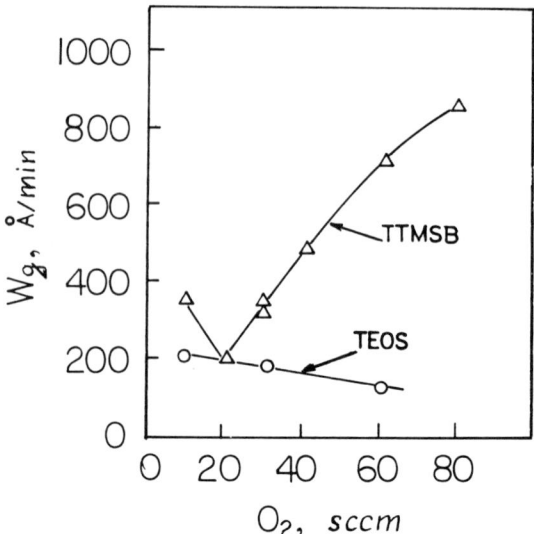

FIG. 63. Silicon dioxide film growth rate as function of oxygen flow rate. TTMSB flow rate was 50 sccm, TEOS flow rate was 37 sccm. ECR plasma mode, $I_1 = 135$ A, $I_2 = 215$ A ($B = 925$ G), $P_{abs} = 600$ W. After Ref. 53.

V. Conclusion

We presented here some aspects of the design, physics and characteristics of electron cyclotron resonance plasma sources that are currently used (or might be used) in plasma assisted chemical vapor deposition of thin films. It was practically impossible to discuss all plasma phenomena as well as give a detailed description of plasma deposition technologies. In fact, all processes that occur in ECR CVD of thin films could be broken down into four major categories:

(1) microwave power absorption and plasma ionization;
(2) plasma transport;
(3) plasma-chemical reactions in volume; and
(4) growth of a thin film on the substrate surface.

The author of this paper placed an emphasis on what he thought is a "specific" and less studied aspect of ECR plasma sources: microwave power absorption. On the other hand, he dealt here with issues that are more familiar to him and where he contributed somehow. As to plasma chemistry and plasma–surface interactions that are "directly" involved in film formation, he left them to the appropriate specialists.

He also hopes that this paper might be of help both to source designers and to process technologists.

Acknowledgments

The author wishes to thank Doctors A. J. Lichtenberg, J. Musil, and J. E. Stevens for their valuable comments and recommendations.

References

1. B. Lax, W. P. Allis, and S. C. Brown, *J. Appl. Phys.* **21**, 1297 (1950).
2. D. C. Kelly, H. Margenau, and S. C. Brown, *Phys. Rev.* **4**, 1367 (1957).
3. S. C. Brown, *IRE Trans. on Microwave Theory and Techniques*, January, 69 (1959).
4. M. Ericson, C. S. Ward, S. C. Brown, and S. J. Buchsbaum, *J. Appl. Phys.* **33**(8), 2429 (1962).
5. R. J. Kerr, R. A. Dandl, H. O. Eason, A. C. England, M. C. Becker, and W. B. Ord, *Bull. Am. Phys. Soc.* **7**, 290 (1962).
6. D. B. Miller, P. Gloersen, E. F. Gibbons, and D. J. BenDaniel, Third Annual Symposium on the Engineering Aspects of MHD, Univ. of Rochester (March 1962).
7. D. B. Miller, D. J. BenDaniel, E. F. Gibbons, and P. Gloersen, *AIAA Preprint 63002* (1963).
8. T. Consoli and R. B. Hall, *Fusion Nucleaire* **3**, 237 (1963).
9. R. Bardet, T. Consoli, and R. Geller, *Physics Letters* **1**, 67 (1964).
10. A. I. Anisimov, V. N. Budnikov, V. L. Vinogradov, V. E. Golant, S. L. Nanobashvili, A. A. Obukhov, A. P. Pakhomov, A. D. Pilya, and V. I. Fedorov, The Third Conference on Plasma Physics and Controlled Fusion, Novosibirsk, Vol. 11, p. 399 (1968).
11. J. Musil and F. Zacek, *Plasma Phys.* **13**, 471 (1971).
12. H. Dreicer, D. B. Henderson, and J. C. Ingraham, *Phys. Rev. Letters* **26**, 1616 (1971).
13. B. D. McVey and J. E. Scharer, *Phys. Fluids* **17**, 142 (1974).
14. R. Geller, *Appl. Phys. Lett.* **16**, 401 (1970).
15. H. Tamagawa, Y. Okamoto, and C. Arutagawa, *Jpn. J. Appl. Phys.* **11**, 1226 (1972).
16. L. Bardos, G. Longar, I. Stoll, J. Musil, and F. Zacek, *J. Phys. D: Appl. Phys.* **8**, L195 (1975).
17. G. Longar, J. Musil, and L. Bardos, *Czech. J. Phys.* **B30**, 688 (1980).
18. K. Suzuki, S. Okudaira, N. Sakudo, and I. Kanomata, *Jpn. J. Appl. Phys.* **16**, 1979 (1977).
19. N. Sakudo, K. Tokiguchi, H. Koike, and I. Kanomata, *Rev. Sci. Instrum.* **48**, 762 (1977).
20. S. Matsuo and M. Kiuchi, *Jpn. J. Appl. Phys.* **22**, L210 (1983).

21. T. Hirao, K. Setsune, M. Kitagawa, T. Kamada, K. Wasa, and T. Izumi, *Jpn. J. Appl. Phys.* **26**, 2015 (1987).
22. C. Keqiang, Z. Erli, W. Jinfa, Z. Hausheng, G. Zuoyao, and Z. Bangwei, *J. Vac. Sci. Technol.* **A4**, 828 (1986).
23. S. Y. Shapoval, C. Y. Kopecky, and O. E. Balvinsky, *Sov. J. Surf. Sci.* **3**, 92 (1989).
24. K. Miyake, S. Kimura, T. Warabisako, H. Sunami, and T. Tokuyama, *J. Vac. Sci. Technol.* **A2**, 496 (1984).
25. D. R. Denison, C. Chiang, and D. B. Fraser, Abstract 180, p. 261, "The Electrochemical Society Extended Abstracts," Vol. 89-1, Los Angeles (1989).
26. O. A. Popov and H. Waldron, *J. Vac. Sci. Technol.* **A7**, 914 (1989).
27. T. V. Herak, T. T. Chau, D. J. Thomson, S. R. Mejia, D. A. Buchanan, and K. C. Kao, *J. Appl. Phys.* **65**, 2457 (1989).
28. R. G. Andosca, W. J. Varhue, and E. Adams, *J. Appl. Phys.* **72**, 1126 (1992).
29. K. Machida and H. Oikawa, *J. Vac. Sci. Technol.* **B4**, 818 (1986).
30. C. Chiang and D. B. Fraser, Abstract 179, p. 260, "The Electrochemical Society Extended Abstracts," Vol. 89-1, Los Angeles (1989).
31. T. Hirao, T. Kawada, M. Kitagawa, K. Setsune, K. Wasa, A. Matsuda, and K. Tanaka, *Jpn. J. Appl. Phys.* **27**(4), 528 (1988).
32. Y. Hattori, D. Kruangam, T. Toyama, H. Okamoto, and Y. Hamakawa, *J. Non-cryst. Solids* **97/98**, 1079 (1987).
33. Y. Sugiyama, S. Fujimori, H. Yamazaki, and I. Hatakayama, *Appl. Phys. Lett.* **56**(15), 1403 (1990).
34. A. Chayahara, H. Yokoyama, T. Imura, and Y. Osaka, *Jpn. J. Appl. Phys.* **26**, L1435 (1987).
35. S. Y. Shapoval, V. T. Petrashov, O. A. Popov, A. O. Westner, M. D. Yoder, Jr., and C. K. C. Lok, *Appl. Phys. Letters* **57**, 1885 (1990).
36. H. Kawarada, K. S. Mar, and A. Hiraki, *Jpn. Appl. Phys.* **26**, L1032 (1987).
37. H. Kawarada, K. Nishimura, T. Ito, J. Suzuki, K. S. Mar, Y. Yokota, and A. Hiraki, *Jpn. J. Appl. Phys.* **27**, L683 (1988).
38. C. R. Eddy, Jr., B. D. Sartwell, and D. L. Youchison, *Surf. Coatings Technol.* **48**, 69 (1991).
39. F. S. Pool and Y. H. Shing, *J. Appl. Phys.* **68**(1), 62 (1990).
40. I. Nagai, A. Ishitani, H. Kuroda, M. Yoshikawa, and N. Nagai, *J. Appl. Phys.* **67**(6), 3890 (1990).
41. P. W. Pastel and W. J. Varhue, *J. Vac. Sci. Technol.* **A9**(3), 1129 (1991).
42. S. Y. Shapoval, O. A. Popov, M. D. Yoder, Jr., C. K. C.L ok, and A. Z. Drybanski, submitted for publication.
43. K. Kobayashi, M. Hayama, S. Kawamoto, H. Miki, *Jpn. J. Appl. Phys.* **26**, 202 (1987).
44. S. R. Mejia, R. D. McLeod, K. C. Kao, and H. C. Card, *Rev. Sci. Instrum.* **57**(3), 493 (1986).
45. T. Watanabe, M. Tanaka, K. Azuma, M. Nakatami, T. Sonobe, and T. Shimada, *Jpn. J. Appl. Phys.* **26**, L288 (1987).
46. R. D. Knox, V. L. Dalal, and O. A. Popov, *J. Vac. Sci. Technol.* **A9**, 474 (1991).
47. K. Shirai and S. Gonda, *J. Appl. Phys.* **67**(10), 6281 (1990).
48. H. Yamada and Y. Torii, "Extended Abstracts of 17th Conf. Solid State Device and Materials, Tokyo," p. 305 (1985).
49. I. Nagai, T. Takahagi, A. Ishitani, H. Kuroda, and M. Yoshikawa, *J. Appl. Phys.* **64**(10), 5183 (1988).

50. Y. Tanaka, Y. Kunitsugu, I. Suemune, Y. Honda, Y. Kan, and M. Yamanishi, *J. Appl. Phys.* **64**(5), 2778 (1988).
51. C. S. Pai, J. F. Miner, and P. D. Foo, Electrochem. Soc. Extended Abstracts (1991).
52. C. S. Pai, J. F. Miner, and P. D. Foo, *J. Electrochem. Soc.* **139**(3), 850 (1992).
53. O. A. Popov, S. Y. Shapoval, M. D. Yoder, and A. A. Chumakov, *J. Vac. Sci. Technol.* **A12**, 300, March/April (1994).
54. T. Ono, C. Takahashi, and S. Matsuo, *Jpn. J. Appl. Phys.* **23**, L534 (1984).
55. S. Matsuo, "Extended Abstracts of 16th Conf. Solid State Devices," p. 459, Kobe (1984).
56. Y. Manabe and T. Mitsuyu, *Jpn. J. Appl. Phys.* **30**, 334 (1990).
57. G. Salbert, D. K. Reinhard, and J. Asmussen, *J. Vac. Sci. Technol.* **A8**(3), 2919 (1990).
58. D. A. Carl, D. W. Hess, and M. A. Lieberman, *J. Vac. Sci. Technol.* **A8**(3), 2924 (1990).
59. S. Y. Shapoval, O. E. Balvinskii, I. V. Malikov, A. A. Chumakov, and L. A. Niselson, *Appl. Surf. Sci.* **45**, 257 (1990).
60. A. A. Balmashnov, K. S. Golovanivsky, E. Camps, E. M. Omeljanovsky, A. V. Pakhomov and A. Y. Polyakov, *Sov. J. Tech. Phys. Lett.* **12**, 1486 (1986).
61. E. M. Omeljanovsky, A. V. Pakhomov, and A. Y. Poplyakov, *J. Electronic Mater.* **18**(6), 659 (1989).
62. P. Singer, *Semiconductor International*, July (1992).
63. T. Stix, "Theory of Plasma Waves." McGraw-Hill, New York, 1962.
64. V. L. Ginzburg, "Propagation of Electromagnetic Waves in Plasma." Gordon and Breach, New York, 1961.
65. W. P. Allis, S. J. Buchsbaum, and A. Bers, "Waves in Anisotropic Plasma." M.I.T., Cambridge, Massachusetts, 1963.
66. A. F. Kuckes, *Plasma Physics* **10**, 397 (1968).
67. E. Canobbio, *Nucl. Fusion* **9**, 27 (1969).
68. V. N. Budnikov, V. E. Golant, and A. A. Obukhov, *Phys. Lett.* **31A**, 76 (1970).
69. M. Porkolab, V. Arunasalam, and R. A. Ellis, *Phys. Rev. Lett.* **28**, 340 (1972).
70. L. A. Ferrari, A. W. McQuade, R. J. LaHaye, *Phys. Fluids* **17**, 171 (1974).
71. J. Musil and F. Zacek, "Experimental Study of Absorption of Intense Electromagnetic Waves in Magnetoactive Plasmas." Academia, Prague, 1975.
72. R. W. Motley, S. Bernabei, W. M. Hooke, and D. L. Jassby, *J. Appl. Phys.* **46**, 3286 (1975).
73. R. Bardet, P. Briand, L. Dupas, C. Gormezano, and G. Melin, *Nucl. Fusion* **15**, 865 (1975).
74. Y. Sakamoto, *Jpn. J. Appl. Phys.* **16**, 1993 (1977).
75. D. J. BenDaniel, H. Hurwitz, Jr., and G. W. Sutton, *Phys. Fluids* **6**, 884 (1963).
76. M. Dahimene and J. Asmussen, *J. Vac. Sci. Technol.* **B4**, 126 (1985).
77. J. Musil, *Vacuum* **36**, 161 (1986).
78. Shimada, I. Watanabe, and Y. Torii, *Nucl. Instrum. Methods* **B21**, 178 (1987).
79. K. Tokiguchi, N. Sakudo, and H. Koike, *Rev. Sci. Instrum.* **57**, 1526 (1986).
80. M. Pichot and A. Durandet, *Rev. Sci. Instrum.* **59**, 1072 (1987).
81. J. Asmussen and M. Dahimene, *J. Vac. Sci. Technol.* **B5**, 328 (1987).
82. H. Amemiya, H. Oyama, and Y. Sakamoto, *J. Phys. Soc. Jpn.* **56**, 2401 (1987).
83. M. Matsuoka and K. Ono, *Appl. Phys. Lett.* **50**, 1864 (1987).
84. H. Amemiya, K. Shimizu, S. Kato, and Y. Sakamoto, *Jpn. J. Appl. Phys.* **27**, L927 (1988).
85. B. Petit and J. Pelletier, *Jpn. J. Appl. Phys.* **26**, 825 (1987).

86. O. A. Popov, *Surf. Coatings Technol.* **36,** 917 (1988).
87. L. Mahoney, M. Dahimene, and J. Asmussen, *Rev. Sci. Instrum.* **59,** 448 (1988).
88. A. Yonesu, Y. Takeuti, A. Komori, and Y. Kawai, *Jpn. J. Appl. Phys.* **27,** L1746 (1988).
89. E. G. Bustamante, E. Anabirate, M. A. G. Calderon, and J. M. Senties, *J. Appl. Phys.* **64,** 1, 68 (1988).
90. P. K. Shufflebotham, D. J. Thomson, and H. D. Card, *J. Appl. Phys.* **64**(9), 4398 (1988).
91. K. Shirai, T. Iizuka, and G. Gonda, *Jpn. J. Appl. Phys.* **28,** 897 (1989).
92. J. Asmussen, *J. Vac. Sci. Technol.* **A7,** 883 (1989).
93. O. A. Popov, *J. Vac. Sci. Technol.* **A7,** 894 (1989).
94. J. Forster and W. Holber, *J. Vac. Sci. Technol.* **A7,** 899 (1989).
95. Y. H. Lee, J. E. Heidenreich, and G. Fortuno, *J. Vac. Sci. Technol.* **A7,** 903 (1989).
96. J. S. McKillop, J. C. Foster, and W. M. Holber, *J. Vac. Sci. Technol.* **A7,** 908 (1989).
97. E. Ghanbari, I. Trigor, and T. Nguyen, *J. Vac. Sci. Technol.* **A7,** 918 (1989).
98. J. Asmussen, J. Hopwood, and F. C. Sze, *Rev. Sci. Instrum.* **61,** 250 (1990).
99. H. Amemiya and S. Ishii, *Jpn. J. Appl. Phys.* **28,** 2289 (1989).
100. B. H. Quon and R. A. Dandl, *Phys. Fluids* **B1,** 2010 (1989).
101. J. Hopwood, R. Wagner, D. K. Reinhard, and J. Asmussen, *J. Vac. Sci. Technol.* **A8,** 2904 (1990).
102. J. Hopwood, D. K. Reinhard, and J. Asmussen, *J. Vac. Sci. Technol.* **A8,** 3103 (1990).
103. R. R. Burke, J. Pelletier, C. Pomot, and L. Vallier, *J. Vac. Sci. Technol.* **A8,** 2931 (1990).
104. S. M. Gorbatkin, L. A. Berry, and J. B. Roberto, *J. Vac. Sci. Technol.* **A8,** 2893 (1990).
105. C. C. Tsai, L. A. Berry, S. M. Gorbatkin, H. H. Haselton, and W. L. Stirling, *J. Vac. Sci. Technol.* **A8,** 2900 (1990).
106. O. A. Popov, *J. Vac. Sci. Technol.* **A8,** 2909 (1990).
107. D. A. Carl, D. W. Hess, and M. A. Lieberman, *J. Appl. Phys.* **68,** 1863 (1990).
108. M. Geisler, J. Kieser, E. Rauchle, and R. Wilhelm, *J. Vac. Sci. Technol.* **A8,** 908 (1990).
109. H. Nihei, J. Morikawa, and N. Inoue, *Jpn. J. Appl. Phys.* **29,** L822 (1990).
110. G. E. Guest, M. E. Fetzer, and R. A. Dandl, *Phys. Fluids* **B2,** 1210 (1990).
111. E. A. den Hartog, H. Persin, and R. C. Woods, *Appl. Phys. Lett.* **57,** 661 (1990).
112. M. A. Hussein and G. A. Emmert, *J. Vac. Sci. Technol.* **A8,** 2913 (1990).
113. O. A. Popov, W. Hale, and A. O. Westner, *Rev. Sci. Instrum.* **61,** 300 (1990).
114. O. A. Popov and A. O. Westner, *Rev. Sci. Instrum.* **61,** 303 (1990).
115. S. M. Rossnagel, S. J. Whitehair, C. R. Guarneri, and J. J. Cuomo, *J. Vac. Sci. Technology* **A8,** 3113 (1990).
116. P. K. Shufflebotham and D. J. Thomson, *J. Vac. Sci. Technol.* **A8,** 3713 (1990).
117. W. M. Holber and J. C. Foster, *J. Vac. Sci. Technol.* **A8,** 3720 (1990).
118. R. K. Porteous and D. B. Graves, *IEEE Trans. Plasma Sci.* **19,** 204 (1991).
119. T. Nakano, N. Sadeghi, and R. A. Gottscho, *Appl. Phys. Lett.* **58,** 459 (1991).
120. S. Samukawa, S. Mori, and M. Sasaki, *J. Vac. Sci. Technol.* **A9,** 85 (1991).
121. T. D. Mantei and T. E. Ryle, *J. Vac. Sci. Technol.* **B9,** 29 (1991).
122. G. Neuman and K.-H. Kretschmer, *J. Vac. Sci. Technol.* **B9,** 334 (1991).
123. D. A. Carl, M. C. Williamson, M. A. Lieberman, and A. J. Lichtenberg, *J. Vac. Sci. Technol.* **B9,** 339 (1991).
125. T. D. Mantei and S. Dhole, *J. Vac. Sci. Technol.* **B9,** 26 (1991).
126. M. Matsuoka and K. Ono, *J. Vac. Sci. Technol.* **A9,** 691 (1991).

127. J. E. Stevens, J. L. Cecchi, Y. C. Huang, and R. L. Jarecki, Jr., *J. Vac. Sci. Technol.* **A9,** 696 (1991).
128. S. J. Pearton, T. Nakano, and R. A. Gottscho, *J. Appl. Phys.* **69,** 4206 (1991).
129. S. M. Rossnagel, K. Schatz, S. J. Whitehair, C. R. Guarneri, D. N. Ruzic, and J. J. Cuomo, *J. Vac. Sci. Technol.* **A9,** 702 (1991).
130. O. A. Popov, *J. Vac. Sci. Technol.* **A9,** 711 (1991).
131. M. Tanaka, R. Nishimoto, S. Higashi, N. Harada, T. Ohi, A. Komori, and Y. Kawai, *J. Phys. Soc. Jpn.* **60,** 1600 (1991).
132. D. A. Buchanan and G. Fortuno-Wiltshire, *J. Vac. Sci. Technol.* **A9,** 804 (1991).
133. T. Oomori, M. Tuda, H. Ootera, and K. Ono, *J. Vac. Sci. Technol.* **A9,** 722 (1991).
134. J. B. O. Caughman II and W. M. Holber, *J. Vac. Sci. Technol.* **A9,** 3113 (1991).
135. S. Samukawa and T. Nakamura, *Jpn. J. Appl. Phys.* **30,** L1330 (1991).
136. M. A. Lieberman and A. J. Lichtenberg, *Phys. Rev.* **5A,** 1852 (1972).
137. S. Pongratz, R. Gesche, K.-H. Kretschmer, G. Lorenz, M. Hafner, and J. Zink, *J. Vac. Sci. Technol.* **B9,** 3493 (1991).
138. S. Y. Shapoval, V. T. Petrashov, O. A. Popov, M. D. Yoder, P. Maciel, and C. K. C. Lok, *J. Vac. Sci. Technol.* **A9,** 3071 (1991).
139. R. A. Dandl and G. F. Guest, *J. Vac. Sci. Technol.* **A9,** 3119 (1991).
140. H. Nishimura and S. Matsuo, *Jpn. J. Appl. Phys.* **30,** L1767 (1991).
141. J. Pelletier and M. J. Cooke, *J. Vac. Sci. Technol.* **B7**(1), 59 (1989).
142. D. J. Trevor, N. Sadeghi, T. Nakano, J. Derouard, R. A. Gottscho, P. D. Foo, and J. M. Cook, *Appl. Phys. Lett.* **57,** 1188 (1990).
143. S. Iizuka and N. Sato, *J. Appl. Phys.* **70,** 4165 (1991).
144. V. E. Golant and A. D. Pilya, *Sov. Phys. Usp.* **14,** 413 (1972).
145. O. A. Popov, S. Y. Shapoval, and M. D. Yoder, Jr., *Plasma Sources Sci. Technol.* **1,** 7 (1992).
146. H. Nihei, J. Morikawa, D. Nagahara, H. Enomoto, and N. Inoue, *Rev. Sci. Instrum.* **63,** 1932 (1992).
147. S. Minomo, K. Kondo, Y. Yoshizako, Y. Ishida, and M. Tanaguchi, *Rev. Sci. Instrum.* **63,** 2547 (1992).
148. G. King, F. C. Sze, P. Mak, T. A. Grotjohn, and J. Asmussen, *J. Vac. Sci. Technol.* **A10,** 1265 (1992).
149. J. E. Stevens, Y. C. Huang, R. L. Jarecki, and J. L. Cecchi, *J. Vac. Sci. Technol.* **A10,** 1270 (1992).
150. A. Ghanbari, M. S. Ameen, and R. S. Heinrich, *J. Vac. Sci. Technol.* **A10,** 1276 (1992).
151. P. Mak, G. King, T. A. Grotjohn, and J. Asmussen, *J. Vac. Sci. Technol.* **A10,** 1281 (1992).
152. T. J. Castagna, J. L. Shohet, K. A. Ashitani, and N. Hershkowitz, *J. Vac. Sci. Technol.* **A10,** 1325 (1992).
153. Y. Wang and M. J. Kushner, *J. Appl. Phys.* **72,** 33 (1992).
154. O. A. Popov (unpublished).
155. J. Datlov, J. Teichmann, and F. Zacek, *Phys. Lett.* **17**(1), 30 (1965).
156. S. Matsuo, in "Thin-Film Deposition Processes and Techniques," p. 147 (1984).
157. V. Dusek and J. Musil, *Czech. J. Phys.* **40,** 1185 (1990).
158. J. Musil and F. Zacek, *Plasma Phys.* **12,** 17 (1970).
159. O. A. Popov, A. O. Westner, and A. Z. Drybanski, *Rev. Sci. Instrum.* **61,** 4432 (1992).
160. G. M. Batanov and K. A. Sarksyan, *JETP Lett.* **13,** 384 (1971).
161. M. Porkolab, V. Arunasalam, and R. A. Ellis, Jr., *Phys. Rev. Lett.* **29,** 1438 (1972).
162. R. P. H. Chang, M. Porkolab, and B. Grek, *Phys. Rev. Lett.* **28,** 206 (1972).

163. J. Musil, F. Zacek, and P. Schmiedberger, *Plasma Phys.* **16,** 735 (1974).
164. G. M. Batanov and K. A. Sarksyan, *Proc. Leveden Phys. Inst.* **73,** 81 (1975).
165. M. A. G. Calderon and J. M. Perez, *Plasma Phys.* **23,** 203 (1981).
166. I. R. Gekker, "The Interaction of Strong Electromagnetic Fields with Plasmas." Clarendon Press, Oxford, 1982.
167. R. A. Gottscho and M. A. Lieberman, *in* "Physics of Thin Films" (M. H. Francombe and J. L. Vossen, eds.). Academic Press, New York, 1994.
168. V. Kopecki, J. Musil, and F. Zacek, *Plasma Phys.* **17,** 1147 (1975).
169. S. N. Golovato, R. A. Breun, J. R. Ferron, R. H. Goulding, N. Hershkowitz, S. F. Horne, and L. Yujiri, *Phys. Fluids,* **28,** 734 (1985).
170. V. A. Arkhipenko and V. N. Budnikov, *Sov. Phys.-Tech. Phys.* **16,** 1858 (1972).
171. J. Musil and F. Zacek, *Czech J. Phys.* **B20,** 337 (1970).
172. A. J. Lichtenberg, M. J. Schwartz, and D. T. Tuma, *Plasma Physics* **11,** 101 (1969).
173. T. Kawamura, H. Momota, C. Namba, and Y. Terashima, *Nuclear Fusion* **11,** 339 (1971).
174. V. A. Godyak, *Sov. Phys.-Tech. Phys.* **16,** 1073 (1972).
175. M. V. Krivosheev, *Sov. Phys.-Tech. Phys.* **14,** 612 (1969).
176. B. Grek and M. Porkolab, *Phys. Rev. Lett.* **30,** 836 (1973).
177. V. N. Budnikov, K. M. Novik, and V. I. Fedorov, *Sov. Phys.-Tech. Phys.* **17,** 1108 (1973).
178. F. F. Chen, "Introduction to Plasma Physics." Plenum Press, New York, 1974.
179. J. Musil, F. Zacek, V. Dyatlov, and J. Teichmann, *Plasma Phys.* **11,** 961 (1969).
180. P. A. Tulle, *Plasma Phys.* **15,** 368 (1973).
181. K. Hothker, H. Amemiya, H. J. Belitz, W. Bieger, H. Oyama, Y. Shigueoka, and K. Yano, *Nucl. Instrum. Methods Phys. Res.* **B63,** 484 (1992).
182. A. M. Messiaen and P. E. Vandenplas, *Phys. Lett.* **2,** 193 (1962).
183. H. J. Schmitt, G. Meltz, and P. J. Freyheit, *Phys. Rev.* **139A,** 1432 (1965).
184. A. M. Messiaen and P. E. Vandenplas, *Phys. Fluids* **12,** 2406 (1969).
185. P. E. Vandenplas, "Electron Waves and Resonances in Bounded Plasmas." J. Wiley and Sons, New York, 1968.
186. F. Jaeger, A. J. Lichtenberg, and M. A. Lieberman, *Plasma Phys.* **14,** 1073 (1972).
187. V. M. Glagolev, N. A. Krivov, and Yu. V. Skosyrev, *Nucl. Fusion, Suppl. 329* (1972).
188. R. W. Motley, S. Bernabei, T. K. Chu, W. M. Hooke, and D. L. Jassby, *Bull. Am. Phys. Soc.* **18,** 1314 (1973).
189. P. A. Raimbault and J. L. Shohet, *Plasma Phys.* **17,** 327 (1975).
190. S. L. Musher, A. M. Rubenchik, and B. I. Sturman, *Plasma Phys.* **20,** 1131 (1978).
191. T. K. Chu and H. W. Hendel, *Phys. Rev. Lett.* **29,** 634 (1972).
192. M. Ono, M. Porkolab, and R. P. H. Chang, *Phys. Rev. Lett.* **38,** 962 (1977).
193. E. Valeo, C. Oberman, and F. W. Perlins, *Phys. Rev. Lett.* **28,** 340 (1972).
194. L. D. Landau, *J. Phys. USSR* **10,** 25 (1946).
195. J. E. Willett, *J. Appl. Phys.* **33**(3), 898 (1962).
196. V. V. Zheleznyakov, *Izv. Vuz. Radiophyz.* **2,** 858 (1959).
197. N. E. Andreev, A. M. Sergeev, and G. L. Stenchikov, *Sov. J. Plasma Phys.* **10**(1), 10 (1984).
198. W. P. Allis and S. J. Buchsbaum, *Nucl. Fusion* **2,** 49 (1962).
199. T. Ohkubo, K. Kawahara, and K. Matsuoka, *Nucl. Fusion* **21,** 1320 (1981).
200. R. M. Gilgenbach, M. E. Read, and K. E. Hacket, *Nucl. Fusion* **21,** 319 (1981).
201. J. Taillet, *Am. J. Phys.* **37,** 42 (1969).
202. G. M. Zaslavsky and B. V. Chirikov, *Sov. Phys. Dokl.* **9,** 989 (1965).
203. M. Seidl, *Plasma Phys.* **6,** 597 (1963).

204. A. I. Akhiezer and A. S. Bakai, *Sov. Phys.-Dokl.* **16**, 1065 (1972).
205. V. A. Godyak, *Sov. J. Plasma Phys.* **2**(1), 78 (1976).
206. M. A. Lieberman and A. J. Lichtenberg, *Plasma Phys.* **15**, 125 (1973).
207. V. A. Godyak, O. A. Popov, and A. H. Khanna, *Sov. J. Plasma Phys.* **2**, 560 (1976).
208. O. A. Popov and V. A. Godyak, *J. Appl. Phys.* **57**, 59 (1985).
209. M. A. Lieberman, *IEEE Trans. Plasma Sci.* **16**(6), 638 (1988).
210. V. A. Godyak and R. B. Pejak, *Phys. Rev. Lett.* **65**(8), 996 (1990).
211. V. A. Godyak, R. B. Piejak, and B. M. Alexandrovich, *Phys. Rev. Lett.* **68**(1), 40 (1992).
212. L. Bardos, *Vacuum* **38**(8), 642 (1988).
213. B. D. Musson, F. C. Sze, D. K. Reinhard, and J. Asmussen, *J. Vac. Sci. Technol.* **B9**, 3521 (1991).
214. K. A. Buckle, K. Pastor, C. Constantine, and D. Johnson, *J. Vac. Sci. Technol.* **B10**(3), 1133 (1992).
215. S. Y. Shapoval, I. A. Maximov, and A. A. Chumakov, *Vacuum* **42**, 45 (1992).
216. O. A. Popov and A. O. Westner, *Plasma Sources Sci. Technol.* (to be published).
217. T. Namura, I. Arikata, O. Fukumasa, M. Kubo, and R. Itatani, *Rev. Sci. Instrum.* **63**, 21 (1992).
218. Y. Yasaka, A. Fukuyama, A. Hatta, and R. Itatani, *J. Appl. Phys.* **72**(7), 2652 (1992).
219. S. M. Gorbatkin and L. A. Berry, *J. Vac. Sci. Technol.* **A10**, 3104 (1992).
220. S. Samukawa, M. Sasaki, and Y. Suzuki, *J. Vac. Sci. Technol.* **B8**, 1062 (1990).
221. E. S. Bulat, G. Ditmer, C. Herrick, and S. Hankin, *J. Vac. Sci. Technol.* **A10**, 1402 (1992).
222. Y. H. Shing, *Solar Cells* **27**, 331 (1989).
223. K. Shirai and S. Gonda, *J. Appl. Phys.* **67**(10), 6286 (1990).
224. N. Sakudo, T. Tokiguchi, H. Koike, and I. Kanomata, *Rev. Sci. Instrum.* **54**, 681 (1983).
225. Y. Torii, M. Shimada, and I. Watanabe, *Rev. Sci. Instrum.* **63**(4), 2559 (1992).
226. S. Matsuo, H. Yoshihara, and S. Yamazaki, U.S. Patent 4,401,054, Aug. 30, 1983.
227. J. Musil, F. Zacek, and V. M. Budnikov, *Czech J. Phys.* **B23**, 736 (1973).
228. A. F. Harvey, "Microwave Engineering." Academic Press, London, 1963.
229. G. Bekefi, "Radiation Processes in Plasmas." Wiley, New York, 1966.
230. N. Hershkowitz, private communication.
231. B. Lipschultz, I. Hutchinson, B. LaBombard, and A. Wan, *J. Vac. Sci. Technol.* **A4**, 1810 (1986).
232. J. C. Barbour, H. J. Stein, O. A. Popov, M. Yoder, and C. A. Outten, *J. Vac. Sci. Technol.* **A9**, 480 (1991).
233. T. N. Kennedy, *J. Vac. Sci. Technol.* **13**, 1135 (1976).
234. O. A. Popov, S. Y. Shapoval, and M. D. Yoder, *J. Vac. Sci. Technol.* **A10**, 3055 (1992).
235. M. C. Williamson, A. J. Lichtenberg, and M. A. Lieberman, *J. Appl. Phys.* **72**, (9), 3924 (1992).
236. J. E. Stevens and J. L. Ceechi, *Jpn. J. Appl. Phys.* **32**, (1993).
237. Han S. Uhm and Eun M. Choi, *Phys. Fluids.* **B5**(6), 1902 (1993).
238. N. Sadeghi, T. Nakano, D. Trevor, and R. Gottscho, *J. Appl. Phys.* **70**, 2552 (1991).
239. S. Samukawa, *J. Vac. Sci. Technol.* **A11**(5), 2572 (1993).

Unbalanced Magnetron Sputtering

SUZANNE L. ROHDE

Department of Mechanical Engineering
University of Nebraska—Lincoln
Lincoln, Nebraska

I. Introduction 235
II. Motivation for Unbalanced Magnetron Sputtering 236
   A. Sputtering in General 236
   B. Electron Motion in a Magnetic Field 240
   C. Magnetron Sputtering 241
   D. Ion-Assisted Deposition 243
III. Development of Unbalanced Magnetron Based Techniques 249
   A. Precursors to the Unbalanced Magnetron 249
   B. Unbalanced Magnetron Sputtering 255
IV. Principles of Unbalanced Magnetron Sputtering 259
V. Applications of UBM Sputtering 272
   A. Elemental Thin Films 272
   B. Films for Electronic and Optical Applications 274
   C. Films for Corrosion Protection 275
   D. Films for Wear and Abrasion Resistance 277
VI. Commercial Applications of UBM Sputtering 281
VII. Potential Future of UBM Technology 283
   Acknowledgments 285
   References 285

I. Introduction

Unbalanced magnetron sputtering is a relatively new technique for thin film deposition that has evolved from previous glow discharge processes and, more specifically, from magnetron sputtering. Developed in the mid-1970s, magnetron sputtering has been the technique of choice for many research and industrial thin film applications, as it is particularly well suited to applications that demand low deposition temperatures and a fairly high degree of uniformity over a large substrate area. When combined with the process-control techniques that were developed in the early 1980s, such as high-rate reactive sputtering, magnetron sputtering became more cost effective, competing in many applications with evaporated and chemical vapor deposited coatings. However, like other sputtering techniques, the use of magnetron sputtering has generally been limited to the deposition of thin films onto near-planar substrates, small parts that can be rotated under the cathode, or components that can be passed between two closely spaced cathodes.

Like most sputtered and evaporated films, magnetron-sputtered coatings typically exhibit columnar microstructures. Since sputtering is a line-of-sight process, "shadowing" can occur as the grains grow outward from the substrate/film interface, producing voided regions both within the grains and at the grain boundaries. The presence of these pockets and open grain boundaries reduces film density and resistance to oxidation and corrosion. Investigations into ion-assisted processes throughout the 1980s led to the understanding that ion bombardment, particularly low-energy bombardment, can improve adatom mobility, permitting atoms to move about on the surface and fill some of these voids. Thus, if ion bombardment could be incorporated into the deposition process, the effects of many of these defects could be reduced or eliminated. The natural place to start was to extract ions from the discharge in the near-substrate region so as to achieve increased ion bombardment during film deposition. The concept of the unbalanced magnetron (UBM) emerged from this idea. The benefits and limitations of this new technique will be examined in this chapter, along with the motivation for its development, the basic principles of its operation, research and commercial applications, and some speculation as to the future of UBM technology.

II. Motivation for Unbalanced Magnetron Sputtering

A. SPUTTERING IN GENERAL

If a solid or a liquid, but most commonly a solid, of any material is subjected to a stream of impinging particles of sufficient energy, some of the surface atoms will acquire enough energy that they can be liberated from the surface. The process of ejecting atoms from the surface is known as "sputtering," and the exact mechanism by which it occurs is still under investigation. As they scatter throughout the vacuum environment, some of the atoms come to rest on the substrate surface and coalesce to form a "sputter-deposited" thin film. Typically, the high-energy particles used in sputtering are ions since their mass is much larger than electrons and they can easily be accelerated toward the cathode by means of an applied electric potential. One of the simplest sources of a sizable flux of ions is a dc-diode plasma. As a starting point for this discussion, consider a low-pressure—e.g., 100 mTorr—inert gas between two closely spaced electrodes, the cathode being a metallic species that is to be sputtered and the anode being a grounded electrode, often the walls of the discharge chamber (1). If a sufficiently large electric potential—e.g., 1,500 V—is applied between these two electrodes, a few electrons will be emitted from the cathode. Once free from the surface, these electrons interact with the electric field, gaining energy as they move through the vacuum toward the anode. There exists, however, a finite probability for collision between the electrons and other particles in the plasma. This probability is expressed as the *mean free path* between collisions:

$$\lambda_e = 4 \times 2^{1/2} \lambda, \tag{1}$$

$$\lambda = 2.330 \times 10^{-20} T/(p\delta_m^2), \tag{2}$$

where λ_e (cm) is the electron mean free path and λ (cm) is the mean free path of a gas molecule. The mean free path of the molecule is a function of the molecule diameter, δ_m (cm); temperature, T (K); and pressure, p (Torr).

If the energy of the accelerated electron is sufficiently large, ionization of a gas atom can occur, liberating yet another electron. While this process occurs continuously, without another source of electrons the plasma current would quickly build to a maximum and then decay to zero. Since this decay seldom occurs, there must be an additional source of electrons that maintain the discharge; this source is, in fact, the secondary electrons emitted from the cathode as a result of ion bombardment by the positively ionized gas atoms.

Once ignited, it can be assumed that the typical dc-diode sputtering plasma is quasi-neutral; that is, the discharge is composed of nearly the same number of ions and electrons. In this case, electrons are the dominant charged species within the plasma since their low mass allows them to move more quickly than the ions. Thus, this quasi-neutral state is maintained by electron movement in response to changes in the electric field. If the frequency of the variations in the electric field is less than the plasma frequency, the electrons will be able to move quickly enough to offset localized changes in the applied potential or slight variations in the charge density. This plasma frequency is defined as

$$\omega_p = \sqrt{\frac{ne^2}{\varepsilon_0 m}}, \qquad (3)$$

where n is the density of carriers in the plasma. Thus, the plasma frequency is solely dependent on the plasma density. Like most plasmas, dc-diode plasmas can be characterized by the electron temperature, T_e; the carrier density, n; and the plasma potential, V_p. For a typical dc-processing plasma, the electron temperature is a few electron volts, the carrier density is on the order of 10^{12} to 10^{15} carriers/m^3, and the plasma potential is very near the anode potential, usually ground.

While a dc-diode process is relatively straightforward, the technique has several drawbacks as a method for thin film deposition. One of the most significant of these is the requirement that the cathode, and thus the sputtered species, must be conductive; however, this can be overcome by using an ac potential with a high frequency. Since above about 50 kHz the electrodes can be coupled through an impedance, only the directly coupled electrode must be conductive. Additionally, the electrons in the plasma will now have sufficient energy to directly ionize gas atoms, reducing the dependence on the secondary generation of electrons at the cathode. At frequencies above a few megahertz, the ions can no longer move quickly enough to offset the changing field and little accumulation of ions occurs during the portion of the cycle in which the electrode is acting as a cathode. This technique is known as rf sputtering. Typically, frequencies of 13.56 or 27 MHz are used, since these are the FCC-approved wavelengths for medical and industrial use. The greatest advantage of rf-sputtering over dc-diode sputtering is that nonconductive target materials can be sputtered.

For both rf- and dc-sputtering, the gas pressure required to sustain the discharge is in the 0.1 to 10^{-3} Torr range, leading to contamination problems in some applications. Triode sputtering allows the discharge pressure/voltage to be reduced to some degree by introducing an independent source of electrons to sustain the discharge rather than relying on the generation of secondary electrons at the cathode. The most common source is a thermionic emitter or hot filament. If care is used to prevent contamination due to filament evaporation, the cleanliness of the resulting films can be further enhanced. Unfortunately, triode systems are difficult to implement on an industrial scale, and the deposition rates that are achieved are little better than those of diode systems.

In fact, all of these techniques are quite limited from a commercial point of view because of their low deposition rates. In each case, the discharge expands to fill the vacuum chamber, leading to poor ionization efficiency. Additionally, substrate heating is often a problem for heat-sensitive substrates such as semiconductors or polymers. This heating is primarily due to plasma expansion away from the sputtering cathode; as the plasma expands, the substrates are engulfed and a relatively high degree of electron heating results. Researchers realized that both of these shortcomings could be addressed if the discharge could be confined in the near-target region, and this was the primary motivation for development of magnetron sputtering in the mid-1970s.

Magnetron sputtering differs from other sputtering techniques in that magnetic fields are used to confine the processing plasma near the cathode (2). The electrons of interest here are primarily the secondary electrons that are ejected from the cathode surface following an ion collision with the cathode. As in dc-diode sputtering, the presence of these secondary electrons is crucial for maintaining the discharge in magnetron sputtering. Curved magnetic fields can be set up above the

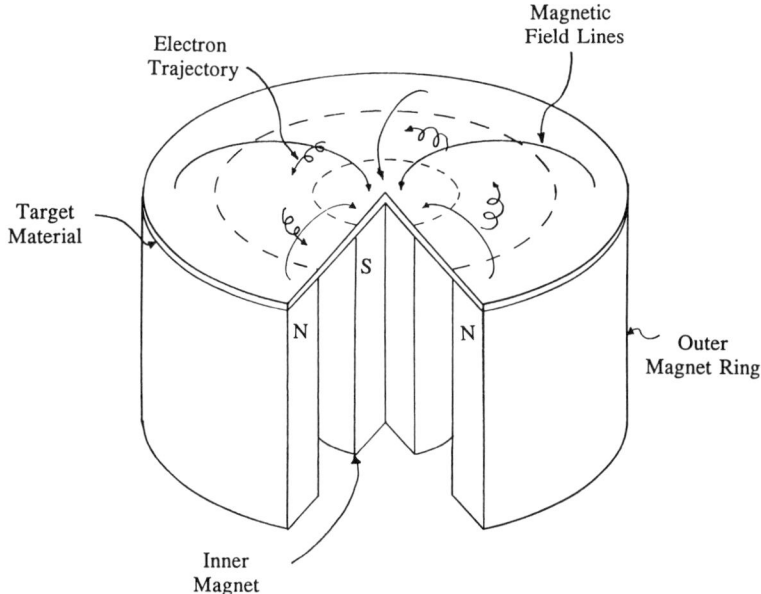

FIG. 1. Circular, planar magnetron cathode schematic, illustrating the magnetic confinement and electron trajectories. After Rohde and Münz (3).

cathode surface by arranging magnets behind the cathode as illustrated in Fig. 1 (3). These fields provide the mechanism for confining the electrons. To understand how this can lead to increased deposition rates, it is helpful to first consider the motion of an electron in a magnetic field.

B. Electron Motion in a Magnetic Field

As previously discussed, the carriers of interest are the electrons rather than the ions since their low mass gives them higher mobilities in the plasma. If it is assumed that the secondary electrons near the cathode surface are at rest, the equation defining electron movement within a combined electric/magnetic field is simply

$$m\frac{d\vec{v}}{dt} = q\vec{E} + q(\vec{v} + \vec{B}), \tag{4}$$

where q = electronic charge, m = electron mass, $\vec{E}$ = electric field, $\vec{v}$ = particle velocity, and $\vec{B}$ = magnetic field.

In the simplest case, i.e., when the electric field is zero and the magnetic field is uniform, the force on the electron can be expressed as

$$F_{\text{magnetic}} = q(\vec{v} \times \vec{B}). \tag{5}$$

Thus, when the electron movement is perpendicular to the applied magnetic field, the electrons will drift along the $\vec{B}$-field lines on an orbital path, as shown in Fig. 2a. The radius of this orbit can be expressed as

$$r_g = \frac{mv_\perp}{qB} \quad \text{or} \quad \frac{\sqrt{2m(\text{KE})}}{qB}, \tag{6}$$

which is the radius of gyration or the Lamor radius. Since $\text{KE} = \frac{1}{2}mv^2$, when the kinetic energy of this electron is large, the corresponding Lamor radius is also large. The frequency of this gyration is given by

$$\omega_c = q\frac{\vec{B}}{m}. \tag{7}$$

The parallel component of the velocity, $\vec{v}_\parallel$, is unaffected by the presence of the magnetic field, of course. Thus, the resulting electron movement is a helical trajectory.

In the case of a uniform nonzero electric field acting in concert with a uniform magnetic field, two different scenarios can be defined for parallel and perpendicular fields. In the case of parallel $\vec{E}$ and $\vec{B}$ fields,

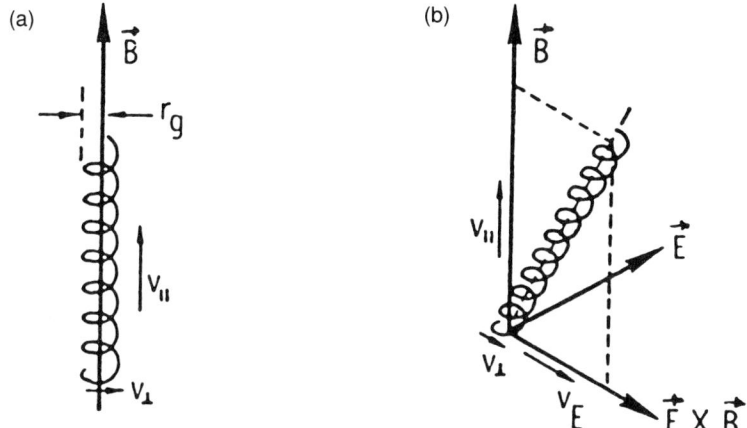

FIG. 2. Motion of an electron (a) under a uniform magnetic field and (b) under uniform perpendicular magnetic and electric fields. After Thornton (*18*).

the electron follows the same type of helical path previously described: It is accelerated by the $\vec{E}$ field, and thus the pitch of the helix is increased with distance. If, instead, the $\vec{E}$ field is perpendicular to the $\vec{B}$ field, the electron is alternately accelerated and decelerated as it travels with or against the constant force of the electric field, resulting in a net drift that is perpendicular to both $\vec{E}$ and $\vec{B}$. The $\vec{E} \times \vec{B}$ drift, as it is known, is shown schematically in Fig. 2b, where $\vec{B}$ is into the page, $\vec{E}$ is oriented vertically, and the net drift is to the right. The velocity of the drift term is then simply $v_{E \times B} = E/B$. It is this drift that is used to increase ionization efficiency in magnetron sputtering.

C. Magnetron Sputtering

The magnetron cathode offers significant improvements in deposition rates with the added benefit of reduced substrate heating. Until very recently, the goal in magnetron sputtering was to achieve increased ionization at or very near the target (cathode) surface by using strong, carefully balanced magnetic fields to bend the trajectories of the secondary electrons ejected from the target. The convoluted spiral- or hop-like paths of the electron across the target surface, illustrated in Fig. 1, are a result of the $\vec{E} \times \vec{B}$ drift. Using such a magnetic arrangement, a dense well-defined plasma can be maintained very near to the surface of the target. Ejected electrons that are unable to leave the near-cathode region expend their energies locally, increasing the probability of ionization near the target and, thereby, the sputtering rate.

Confining the plasma, as such, is beneficial in several ways: (1) higher deposition rates (i.e., higher sputtering rates) are achieved; (2) sputtering of atoms from the substrate and walls of the chamber is avoided; (3) substrate heating during deposition is reduced; and (4) the "working" gas pressure required to sustain the discharge is reduced. This method has been used quite successfully to produce high-quality, low-impurity films at reasonable deposition rates.

One of the most significant drawbacks to the use of magnetron sputtering is that the targets typically erode in a "racetrack" fashion. For a solid circular target, this type of erosion produces a large amount of waste, as high as 75%, and an annular distribution of the sputtered atom density across the target (4). Many other target geometries have been evaluated in an effort to improve deposition uniformity and target utilization efficiency (5,6). Conventional magnetrons have been used to deposit a wide variety of materials on conductive and nonconductive substrates and have been utilized commercially in many applications. Perhaps the most notable application of magnetron sputtering is in the semiconductor industry, where the high rates and low substrate heating are essential.

Because magnetron sputter deposition can be accomplished at much lower substrate temperatures than chemical vapor deposition (CVD), researchers were hopeful that magnetron sputtering would be competitive in tool coating applications. However, the success of magnetron sputtering in this area has, for several reasons, been lower than originally anticipated. One problem is "shadowing" of the tool or machine component. Since most tools and machine components are three-dimensional shapes, unlike planer semiconductor substrates, and sputtering is in general a line-of-sight process, the primary flux falls only on the portion of the component that is facing the cathode surface; hence, the remainder of the component lies in its "shadow" and is not coated. Rotation of the substrates can be used to even out the thickness of the coating, but the coating will still be applied in layers that may have varying chemical compositions. In the early 1980s a number of multicathode systems were developed that facilitated more uniform deposition by sandwiching the components between two tightly spaced magnetron cathodes, as illustrated in Fig. 3 (7). This method has been used successfully to coat small components such as eyeglass frames, since the substrate-to-target distances can be small, i.e., within a few centimeters. For larger components, the required substrate-to-target distances in-

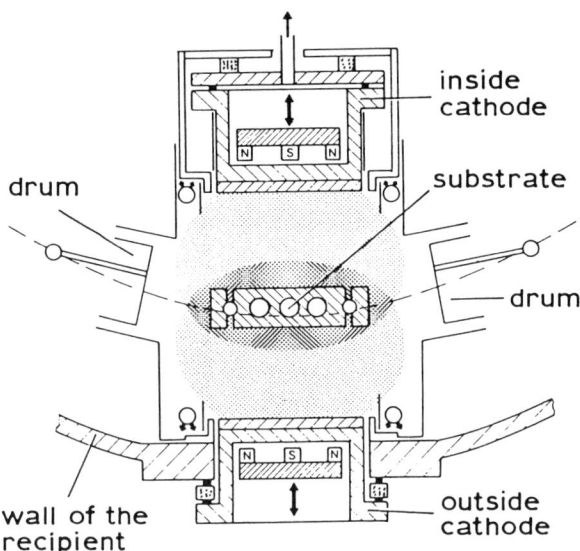

FIG. 3. Double magnetron cathode arrangement for coating contoured parts. After Münz and Göbel (7).

crease and thus the deposition rates decrease, since the deposition rate varies inversely with the square of the substrate to target distance. Additionally, as the component moves outside the near cathode region, the intensity of plasma decreases and the microstructure of the coating changes from one of dense columnar grains to one of more open, porous columnar grains. This open, voided structure has poor electrical conductivity, low mechanical strength and corrosion resistance, and a matte surface appearance. This microstructure is clearly undesirable. There are two means of obtaining the desired microstructure: (1) increase the substrate temperature, increasing surface diffusion; (2) use "ion assisted deposition," a technique in which a stream of ions impinge on the surface, increasing adatom mobility. Both of these techniques result in a more dense microstructure. In many applications, increasing the temperature is deleterious to the substrate, leaving ion-assisted deposition as the only acceptable alternative.

D. Ion-Assisted Deposition

The effectiveness of ion-assisted deposition in altering film microstructure is perhaps best illustrated through the molecular dynamics (MD) simulations and three-dimensional cascade calculations (TRIM) of Müller (8, 9). Presented in Figs. 4a–c are the results of two-dimensional molecular dynamics simulations of vapor phase growth of Ni thin films. The Ni vapor arrives perpendicular to the substrate surface with an energy of 0.1 eV. In Fig. 4a the results of the simulation in the absence

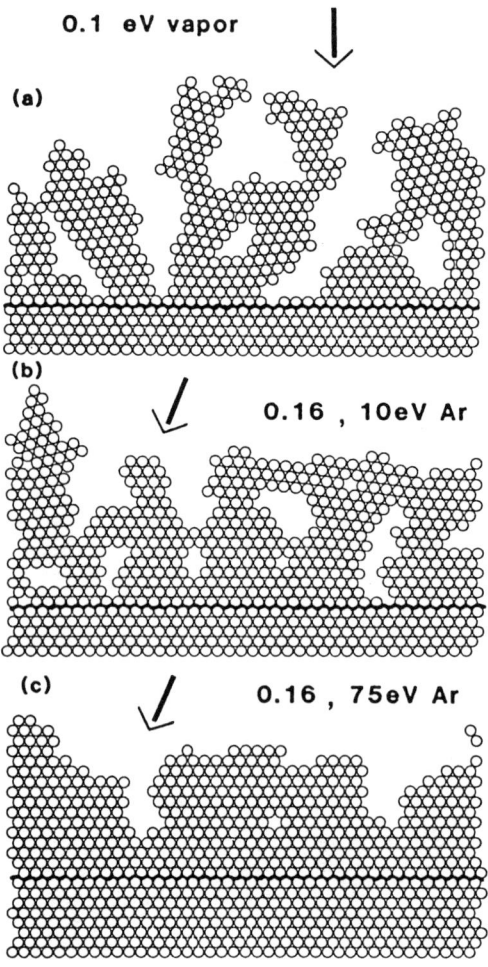

FIG. 4. Simulated microstructure for condensation of Ni vapor with a kinetic energy of 0.1 eV, arriving normal to the substrate surface (a) in the absence of ion bombardment, (b) with 10 eV Ar ions bombarding the surface at a J_i/J_v ratio of 0.16, and (c) with 75 eV Ar ions bombarding the surface at a J_i/J_v ratio of 0.16. After Müller (8).

of ion bombardment are presented. In this case, the film is quite voided and the packing density is only approximately 0.73 of the theoretical value. When ion bombardment from a beam of Ar ions is simulated at an angle 30° off the film normal, the microstructure changes. The result obtained for 10 eV Ar ions arriving at an ion-to-vapor-flux ratio (J_i/J_v) of 0.16 is shown in Fig. 4b; the packing density of this film is near 0.84. When the ion energy is increased to 75 eV, as shown in Fig. 4c, the packing density is approximately equal to the theoretical density. This simulation clearly illustrates the effect of ion bombardment, particularly the effect of increasing ion energy on film density (8).

Very similar results were obtained by Müller (8) in a related series of simulations in which the ion energy was held constant while the ion-to-vapor-flux ratio was varied. Illustrated in Fig. 5a is the simulated Ni film growth without ion bombardment and, in Fig. 5b, the ion-bombardment-induced microstructure generated by 50 eV ions at a J_i/J_v ratio of 0.04. In Fig. 5c, the J_i/J_v ratio has been increased to 0.16, while the energy of the impinging Ar ions is fixed. The packing density increases with the increasing J_i/J_v ratio, and from an initial value of 0.73 approaches the theoretical density of 1.0 at a J_i/J_v ratio of 0.16. While this study simulated epitaxial evaporative film growth, similar trends have been observed in sputter-deposited single and polycrystalline films.

Since the early 1980s, the effects of ion-assisted deposition (IAD) and the mechanisms controlling ion/surface interactions have been studied extensively. These studies have shown that ion-assisted deposition can bring about changes in the morphology, composition, nucleation behavior, film stress, and orientation of many types of thin films deposited on a great variety of substrate materials ranging from pure metal (10) to carbon (11, 12) and optical films (13). Ion bombardment can also be used to manipulate the kinetics of film growth so as to permit deposition of a variety of chemically and crystallographically metastable materials (14). A very comprehensive review is available from Harper et al. (15), and a review of ion/surface interactions in vapor phase epitaxy has been compiled by Greene et al. (16).

In the area of vapor-deposited films, ion-assisted deposition is typically performed either by (1) aiming a beam of inert gas ions of a predetermined energy at the substrate or (2) extracting ions from the discharge by using anegative bias applied to the substrate. Ion-beam techniques, while effective in low deposition rate situations such as MBE, typically provide a rather small flux of higher-energy ions. In magnetron sputtering, ions are almost always extracted from the discharge by applying a negative potential of 50 to 200 V to the substrate. However,

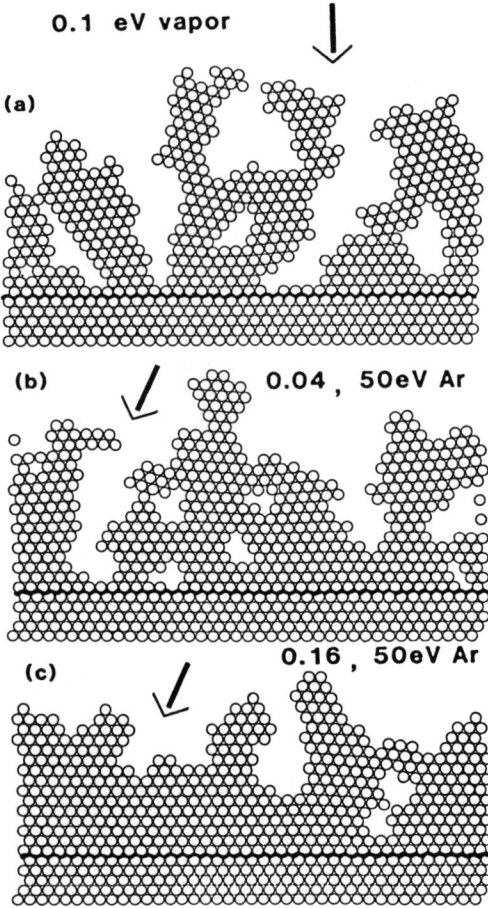

FIG. 5. Simulated microstructure for condensation of Ni vapor with a kinetic energy of 0.1 eV, arriving normal to the substrate surface (a) in the absence of ion bombardment, (b) with 50 eV Ar ions bombarding the surface at a flux ratio $J_i/J_v = 0.04$, and (c) with 50 eV Ar ions bombarding the surface at a flux ratio $J_i/J_v = 0.16$. After Müller (8).

the ion-to-neutral ratio is fairly low; that is, approximately 0.05 to 0.10 ions per deposited atom for most conventional dc magnetrons (17).

In general, the effects of ion bombardment have been thought to mimic an increase in substrate temperature. Illustrated in Fig. 6a is Thornton's structure zone model (18) relating microstructural formation to changes in total pressure and substrate temperature. A very similar relationship was observed by Messier et al. (19) in correlating changes

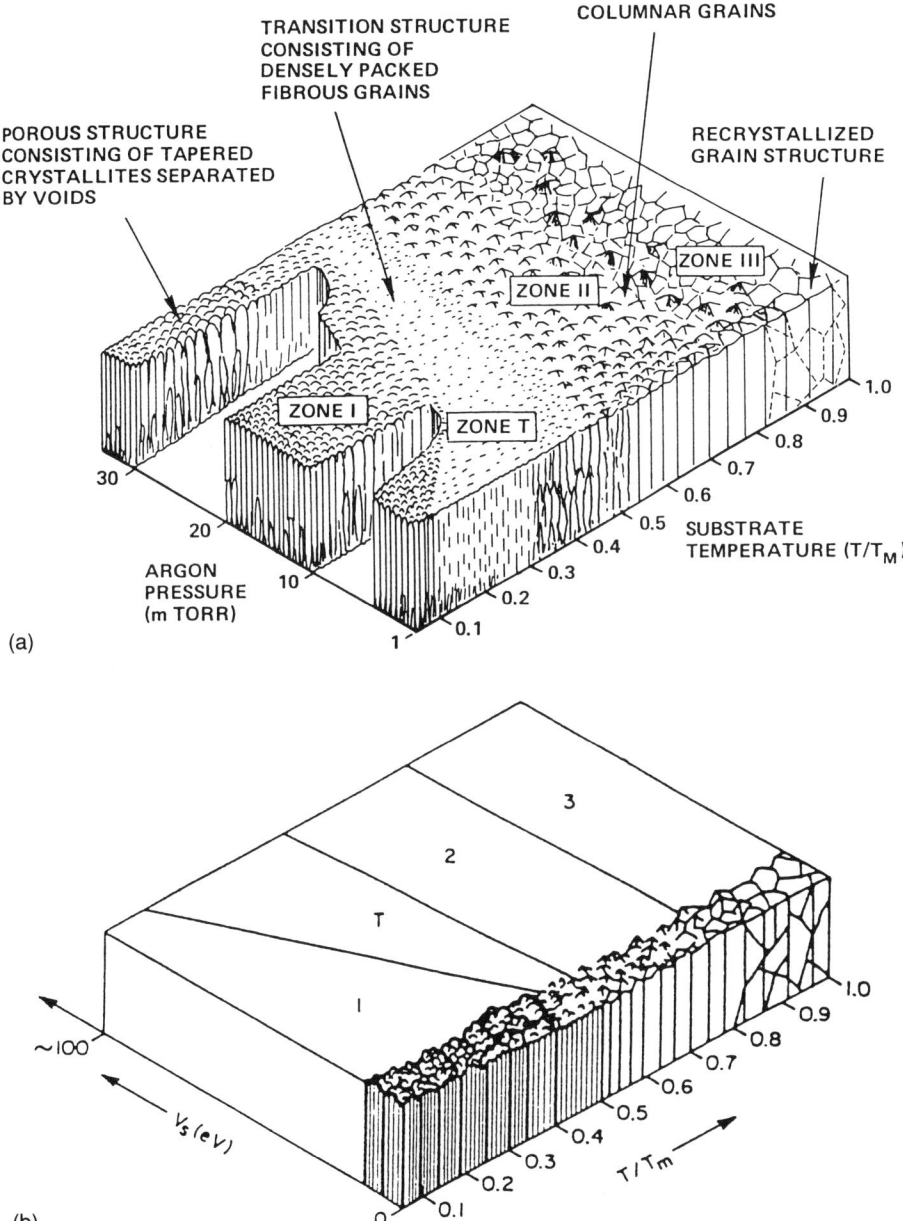

FIG. 6. Structure zone models (a) illustrating the influences of discharge pressure and substrate temperature and (b) illustrating the combined effects of substrate bias voltage and substrate temperature on the grain structure of sputtered metal film. After Thornton (*18*) and Messier *et al.* (*19*), respectively.

in film microstructures with changes in substrate temperature and applied substrate bias, as illustrated in Fig. 6b. However, a specific relationship between the energy input during ion bombardment and effective temperature increase has not been defined. The morphological changes associated with ion-assisted deposition vary markedly with the energy of the impinging particles. For instance, Petrov et al. (20) examined the influence of both the bias potential (V_s) and the substrate temperature ($T_s = 300$ to $900°C$) on the growth of polycrystalline TiN films on high-speed steel, and showed that the observed dislocation density, N_d, increases in a stepwise fashion with increasing bias potential. The ion-to-vapor-flux ratio used in this work was approximately 0.3 to 0.4 ions per deposited Ti atom, depending on the applied substrate bias potential, as shown in Fig. 7. It is useful to note that the ion-to-deposited-atom ratio in this work is higher than the values that are estimated for most conventional magnetrons (17).

These studies illustrate three important concepts. First, both the energy level of the incoming ions and the ion-to-deposited-atom-flux ratio are needed to quantify the level of ion bombardment. Second, the number of ions impinging on the substrate is directly proportional to the applied substrate potential, until saturation is reached, thus fixing the

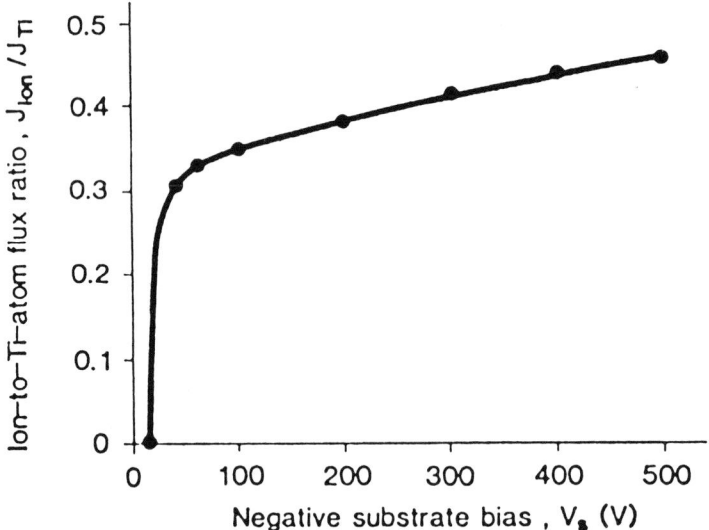

FIG. 7. Variation in the ion-to-Ti-atom ratio with increasing negative applied substrate bias potential. After Petrov et al. (20).

number of ions that can be extracted from the discharge under a particular set of discharge conditions. Third, while the ion current density can be augmented by increasing the discharge charge current, this generates a corresponding increase in the deposition rate. This means that the ion-to-deposited-atom ratio is nearly constant for a particular system geometry, gas species, and cathode material. Thus, to achieve a greater range of ion-bombardment levels, some means of independently controlling the ion current density in the substrate region is required.

III. Development of Unbalanced Magnetron Based Techniques

The primary motivation for development of the "unbalanced magnetron" was to combine the advantages of high magnetron sputtering rates with high-flux, low-energy ion bombardment. However, the development of the unbalanced magnetron has not been quite so straightforward.

A. Precursors to the Unbalanced Magnetron

As early as 1980, Hata *et al.* (*21*) published an article about a "new" magnetron configuration, illustrated in Fig. 8. This new configuration used an electromagnetic coil surrounding the chamber to increase the parallel component of the leakage magnetic field lines, in order to increase the deposition rates of ZnO films. In this work, thin films of ZnO were deposited on glass substrates and successful piezoelectric thin films were produced.

Reactive sputtering of the Zn cathode was performed both in a mixed $Ar-O_2$ and a pure O_2 discharge, at a total pressure of 0.03 to 1 Torr. The substrates in thus work were mounted on the anode and so presumably were at ground potential Thus, the aim of this research was not to evaluate ion-assisted deposition, but rather to produce an increase in the deposition rate. At a substrate-to-target distance, d_{s-t}, of 17 mm, the reported rates were, in fact, fairly high, ranging from 15 μm/hr in a (50% O_2:50% Ar) atmosphere to 9 μm/hr in a pure O_2 discharge. However, the d_{s-t} was very small, and deposition rates typically decrease as the inverse of the d_{s-t} squared. The average power density delivered to the target was 6 W/cm^2; this is appreciably smaller than in most industrial magnetron systems, which range from 10 to 20 W/cm^2. The authors suggest that this technique could also be used with rf sputtering.

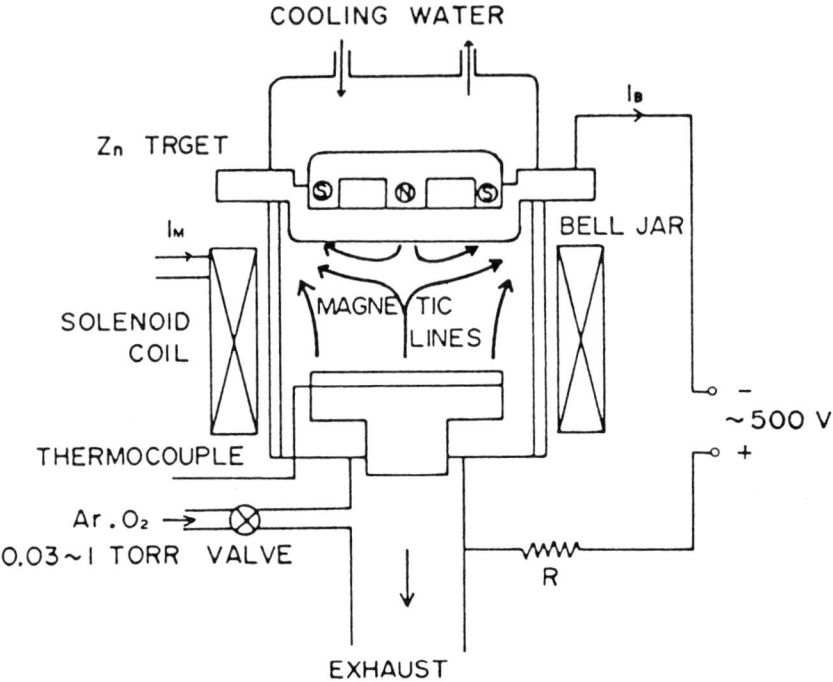

FIG. 8. Schematic illustration of the new magnetron design developed by Hata et al., to obtain enhanced deposition rates of reactively sputtered ZrO thin films (21).

Little explanation is provided as to how the presence of the additional magnetic coil is related to the increase in the deposition rate, and it appears that only one level of magnetic field strength was investigated; that is, the case of approximately 700 G due to the magnetron and 100 G due to the coil. However, this appears to be the first published use of an additional magnetic coil with a conventional magnetron coil.

A second precursor system developed by Naoe et al. (22), like the previous system, was motivated by a need for high deposition rates rather than an interest in ion-assisted deposition. The goal was to produce permalloy films at high deposition rates, while subjecting the films to the intense heating and high-energy γ electron bombardment typical of dc-diode sputtering at small d_{s-t}.

Deposition of permalloy films requires sputtering of magnetic materials such as Ni and Fe. Deposition of magnetic materials is, at best, difficult to perform using magnetron sputtering. A magnetic target

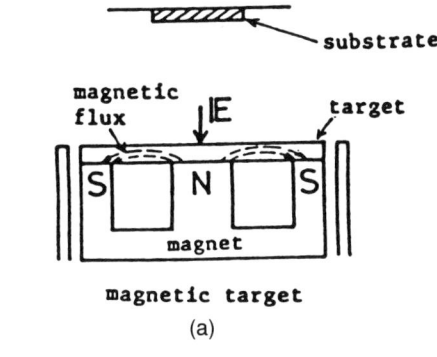

magnetic target

(a)

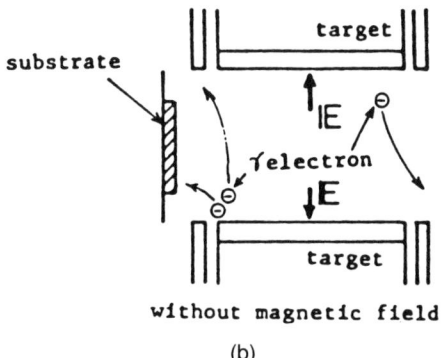

without magnetic field

(b)

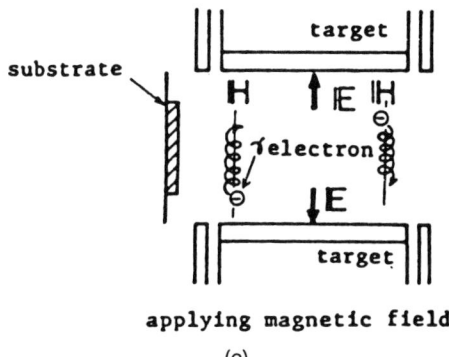

applying magnetic field

(c)

FIG. 9. (a) "Short-circuiting" in magnetron sputtering due to magnetic target materials. (b) An alternative high-rate sputtering source. (c) An improved high-rate source, which uses a parallel magnetic field to confine the electrons, thus eliminating excessive heating of the substrates. After Naoe et al. (22).

material, if sufficiently thick, will essentially short-circuit the magnetic field, as shown in Fig. 9a, eliminating the plasma confinement and enhanced deposition rates typically achieved in magnetron-sputtering systems (22).

Alternatively, an opposing pair of planar dc-diode sputtering cathodes can be used to generate an intense discharge capable of high-rate deposition so long as d_{s-t} is kept small, but this configuration often results in intense heating of the substrates by high-energy γ electrons emitted from the cathodes, as illustrated in Fig. 9b. Naoe et al. (22,23) were able to reduce this substrate heating by introducing a parallel magnetic field, as illustrated in Fig. 9c. This parallel field serves to direct the γ electrons toward one of the two cathodes, confining the discharge. By confining the discharge in this fashion, researchers were able to maintain adequate deposition rates under conditions of reduced substrate temperatures. The overall system geometry is presented in Fig. 10. The strength of the "focusing magnetic field" necessary to confine the discharge for this work varied from approximately 100 G for 8-cm

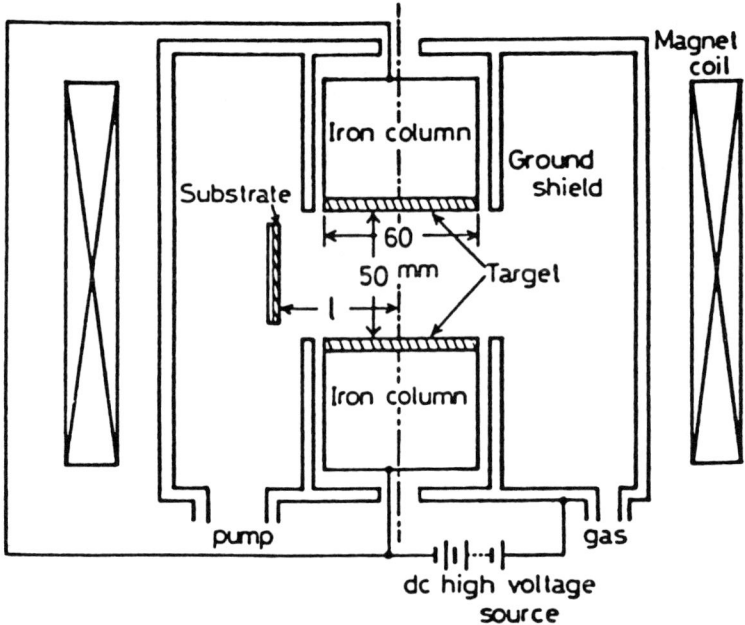

FIG. 10. Schematic illustration of the magnetically enhanced sputtering system developed by Naoe et al. (22).

diameter iron targets to approximately 400 G for 6-cm diameter iron targets (22).

Since space was not a problem and the field strengths required were relatively modest, magnetic coils could be used very effectively in this case. In later versions of this system. Naoe et al. showed that similar deposition-rate enhancements could be obtained—on the order of about 50 times those of typical dc-diode systems for Co–Cr. However, better target utilization was obtained if the electromagnet was replaced by permanent magnets.

The Naoe system provides valuable insight into the utility of magnetic confinement of the discharge. The scale-up potential for this successful system is unfortunately quite low, however, since the d_{s-t} distances must be quite small in order to achieve high deposition rates. Additionally, the substrates must be placed outside of the discharge to eliminate high-energy electron bombardment. None the less, the work of Naoe and colleagues (22–25) did spark the experimentation of Window and Savvides.

A 1985 article, by Window et al., described the use of a magnetically confined Penning source, Fig. 11, to deposit copper and TiN (26). Although this system was influenced by Naoe et al. (22–25), it differs from the previous systems in that it was designed specifically to produce enhanced ion bombardment. The source was designed to achieve high rates of deposition, similar to the dc magnetron, but with a much higher ion flux than could be achieved by merely biasing the substrates. Specifically, a source that could produce "an intense plasma which extends well away from the cathode surface and close to the substrate surface" was required (26). Thus, enough ions would be present in the near substrate region to be accelerated across the sheath at the substrate, enhancing the level of ion bombardment of the growing film. The modified Penning source, shown in Fig. 11, used two conical cathodes, spaced 16 mm apart, with a perpendicular magnetic field. The maximum magnetic field produced between the two cathodes was 70 mT (700 G) when the coil current was 30 A. However, a more moderate magnetic field of 14 mT (140 G) was used for film deposition.

The physical properties of the TiN films that were produced with this source were generally good, although at bias voltages exceeding -50 V the films tended to spall, indicating a fairly high degree of stress within the film and/or poor adhesion. The hardness of 2 μm TiN films deposited on glass substrates at 500°C using a -100 V bias was reported to be $H_v = 2{,}500\,\text{kgf/mm}^2$. The films showed a strong (111) orientation for bias voltages exceeding -50 V, and at higher biases (near -300 V) the

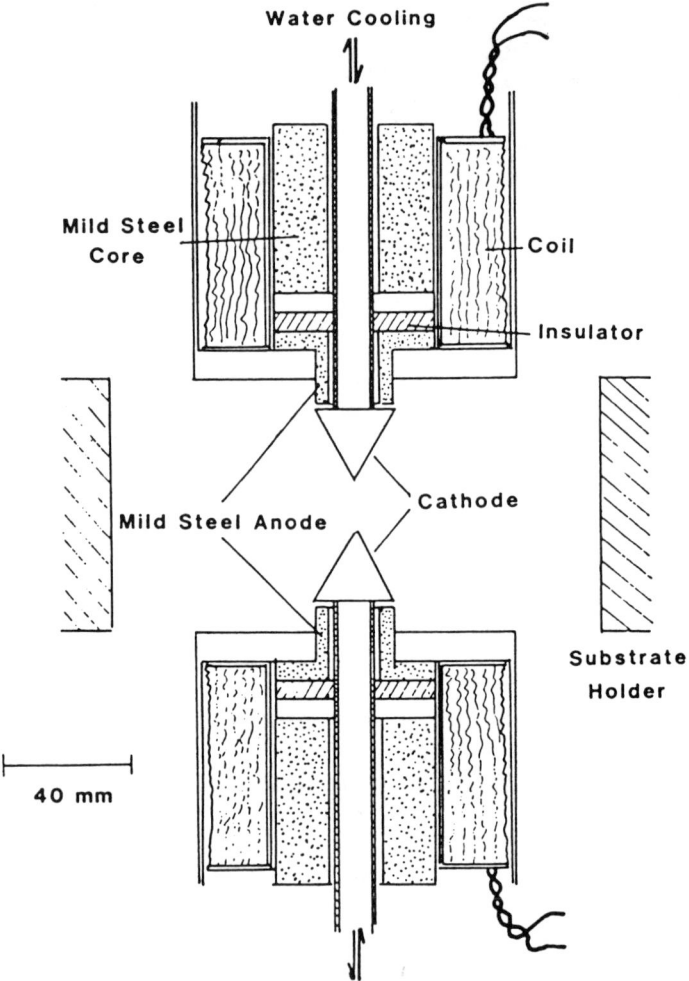

FIG. 11. Schematic of a magnetically confined Penning type source for thin film deposition with enhanced ion bombardment. After Window et al. (26).

films began to regain their random orientation. As would be expected, film density increased with increasing bombardment energy. The most significant outcome of this work was the development of a magnetron-like source that was capable of providing 0.4 ions per deposited atoms at a distance of 72 mm and $V_s = -100$ V. This ratio is larger than for most conventional magnetrons, which are on the order of 0.1 ion per deposited atom under the same conditions. However, using this configuration, the ion flux at the substrate decreases as the magnetic field increases (i.e., as the discharge confinement near the target increases), and thus this technique has limitations similar to those of a conventional magnetron.

B. Unbalanced Magnetron Sputtering

The term "unbalanced magnetron" was first used by Window and Savvides (17) in a study of seven planar magnetrons with differing magnetic field configurations. These configurations were grouped into three cases, as shown in Fig. 12. The intermediate case was that of the 'ideal' conventional magnetron in which the inner and outer magnets are perfectly balanced so as to maximize the ionization near the target and minimize electron/ion heating of the substrates. In reality, nearly all magnetrons exhibit some degree of type I or type II behavior. The study examined the effect of the seven magnetic field configurations on the deposition rate, self bias voltage, and substrate ion current. The configurations included both permanent magnet and electromagnet based systems, as illustrated in Fig. 13. All of the type I configurations (i.e., 3, 4, and 5) had positive self-bias potentials of less than $+3$ V. Thus, when a 100-mm diameter grounded probe was immersed in the plasma, a small ion current ranging from 0.5 to 2 mA was collected. Conversely, for type II configurations (i.e., 1, 2, 6, and 7) the self-bias potentials ranged from -16 V to -23 V, for configurations 2 and 7, respectively, measured at a diatance of 100 mm from the target. A summary of these results is shown in Table I.

One of the most significant findings of this study was that the type II magnetrons yielded a five- to nearly 100-fold increase in the ion current collected by the -100 V probe compared to the type I magnetrons. The ion-to-deposited-atom ratio for configuration 2 was 2:1 (type II), while for configuration 5 it was 0.00025:1 (type I). Thus, type I configurations provided very little bombardment of the substrates, while type II configurations provided considerable ion bombardment at low ion energies. Hence, type II configurations could be used in place of conventional magnetron geometries to provide increased ion bombardment and thus further alter the growth kinetics and morphology of sputter-deposited thin films.

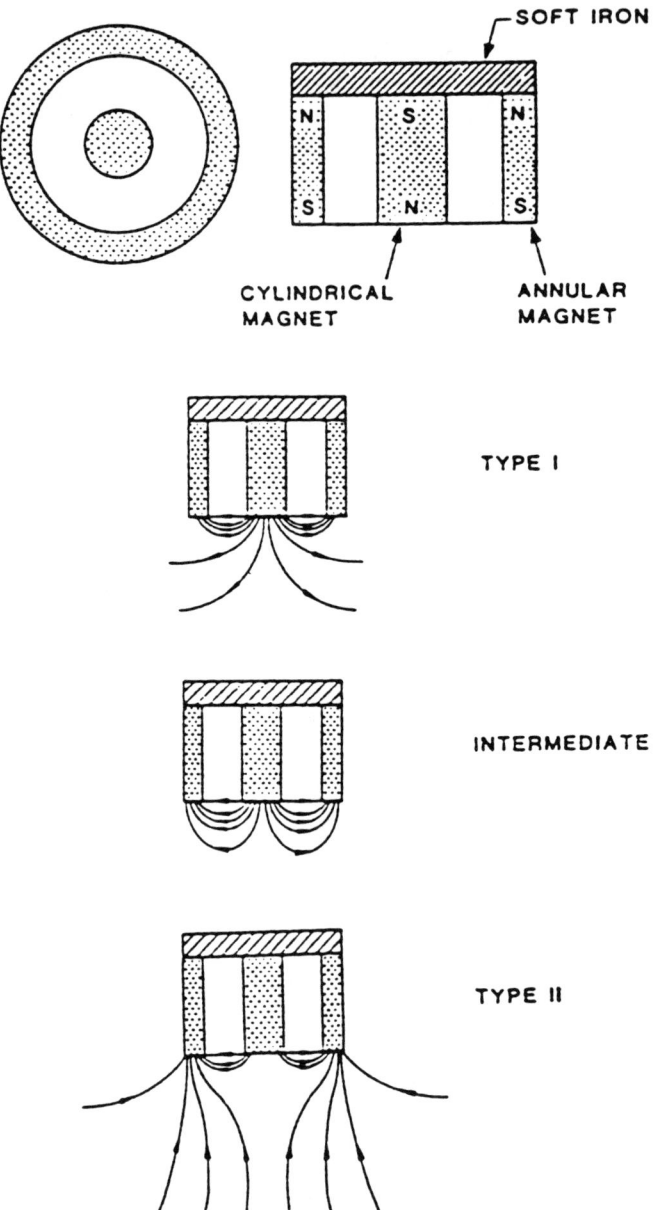

FIG. 12. Three types of magnetron configurations; Type I, Intermediate, and Type II. After Window and Savvides (17).

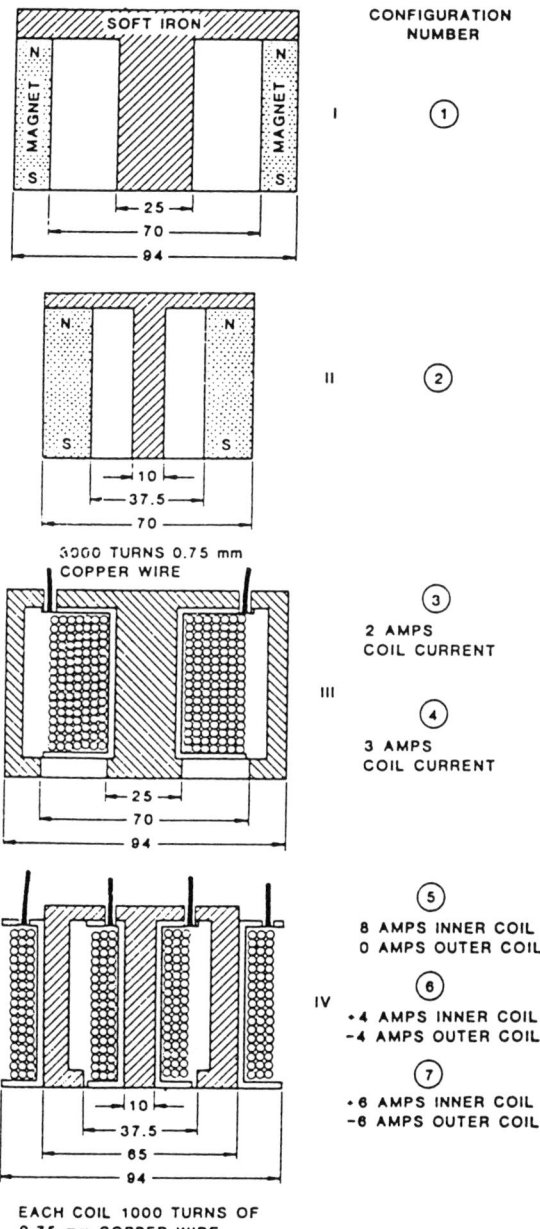

FIG. 13. The seven permanent and electromagnet unbalanced magnetron configurations studied by Window and Savvides (17).

TABLE I
PLASMA PROBE RESULTS FOR A 300 mA DISCHARGE CURRENT, OBTAINED USING A 100 mm DIAMETER PROBE[a]

Configuration	Axial Distance (mm)	Discharge Voltage (V)	Self-Bias Potential (V)	Grounded Probe Current (mA)	−100 V Probe Current (mA)
1	60	400	−25	264	−22
	100	400	−17	256	−20
2	60	595	−26	321	−53
	100	595	−16	318	−30
3	60	410	+2.7	−1.9	−4.3
	100	410	+1.9	−1.1	−1.8
4	60	400	+3.3	−1.9	−4.6
	100	400	+2.4	−1.1	−1.9
5	60	595	+0.4	−1.1	−3.7
	100	595	+0.4	−0.5	−2.0
6	60	540	−28	308	−66
	100	540	−20	311	−25
7	60	490	−33	307	−94
	100	490	−23	317	−42

[a]Configurations refer to those shown in Fig. 13. After Window and Savvides (17).

IV. Principles of Unbalanced Magnetron Sputtering

The primary objective in using an unbalanced magnetron is to increase the ionization density in the region near the substrates, and, in turn, the rate of removal during etching and ion bombardment during growth. What, however, constitutes a suitably "unbalanced" magnetron? In practical terms, the field away from the cathode in the substrate region must be strong enough to provide the desired level of ion bombardment without, however, sacrificing high deposition rates or creating undesirable film properties or microstructures.

As noted by Window and Savvides (17), most planar magnetrons are inherently unbalanced because it is difficult to perfectly match the strength of the inner and outer poles. However, except in extreme cases, this level of "unbalance" is not sufficient to produce the desired ion bombardment, and changes in the magnetic arrangement behind the cathode surface are necessary. The behavior of the Type II unbalanced magnetrons discussed by Window and Savvides (17), that is, configurations 1, 2, 6 and 7 in Fig. 13, deviates from that of conventional magnetrons in that a secondary discharge is formed near the substrates and is supported by electrons escaping from the primary magnetron discharge of the cathode. A glow discharge in the substrate region is readily observed for type II, but not for type I unbalanced magnetrons. It is this secondary discharge that is primarily responsible for the enhanced ion currents observed at the substrate.

One of the first questions that must be considered is, "How unbalanced should the cathode be?" The answer, of course, depends on the system geometry, the level of ion bombardment desired, and a myriad of other parameters. In their initial work, Windows and Savvides examined seven different configurations (17). Similarly, in their work to determine an appropriate level of unbalance for an experimental dual-cathode system, Rohde and co-workers tested more than a dozen different combinations of permanent magnets (27). Shown in Fig. 14 is the ion current density measured directly in front of the cathode, plotted versus the total sputtering pressure for six different magnetic unbalanced configurations (28). In general, the more "unbalanced" the field, the higher the ion current collected by the probe.

There is an additional parameter that must be considered in designing a multicathode system, and this is the magnetic interaction of the cathodes with respect to one another. This interaction can significantly alter the ion current observed at the substrate. In a dual-cathode system there are two possible ways to arrange the cathodes: They may mirror

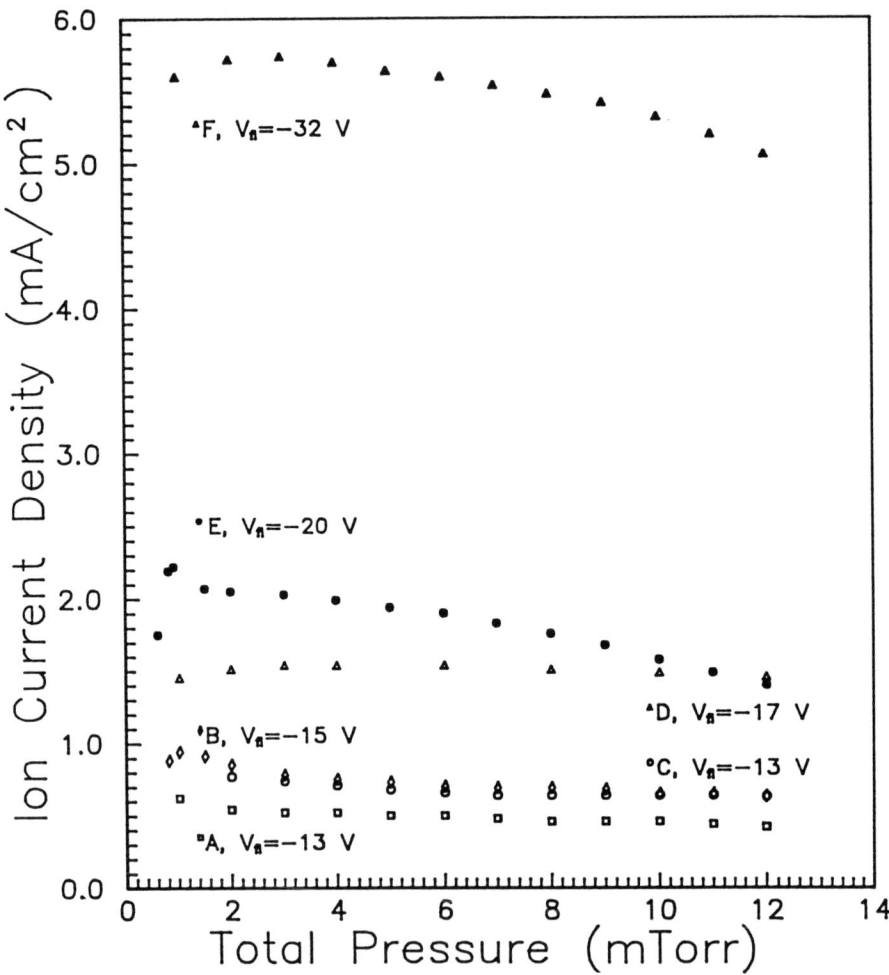

FIG. 14. Ion current density variations versus total pressure for six magnet configurations, presented in the form inner-magnet/outer-magnet material followed by the cathode arrangement: (A) AlNiCo/AlNiCo mirrored, (B) AlNiCo/AlNiCo opposed, (C) AlNiCo/AlNiCo + AlNiCo mirrored, (D) AlNiCo/AlNiCo + AlNiCo opposed, (E) Fe/AlNiCo + NdFeB mirrored, (F) Fe/AlNiCo + NdFeB opposed. The target power is 5 kW/target and the bias voltage is -100 V in each case. The floating potential, V_{fl}, noted in each case was measured at a pressure of 12 mtorr. After Rohde (27).

each other's magnetic polarity, as shown in Figs. 15a and 3; or they may be of opposite polarity, as shown in Fig. 15b. In each of the cases presented in Fig. 14, the "opposed" configuration produced a higher level of ion bombardment than the corresponding "mirrored" configuration.

While not particularly efficient, a workable configuration can be determined experimentally from correlation of the magnet geometry to the resulting film properties To what, however, does this correspond in terms of magnetic field strengths? While no measurements of magnetic field strength were reported by Windows and Savvides (17), research by Rohde (27) and colleagues indicates that a field strength of 50 mT (500 G) at the surface of the Ti target can produce deposition rates that are equal to those obtained using a balanced magnetron geometry and provide a suitable degree of ion bombardment on a substrate located 10 cm from the target. In this dual-cathode arrangement, only relatively low field strengths (<2 mT or 20 Gauss) are required in the center of the chamber (i.e., at the substrate) to produce a well-confined discharge. Of course, the magnetic field strength must vary continuously from one

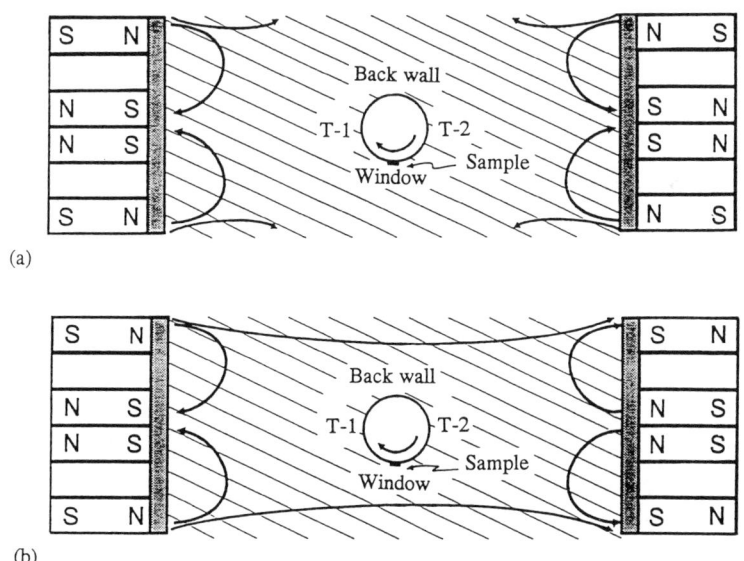

FIG. 15. Schematic illustration of the relative field lines for the two magnetic configurations possible in a dual-cathode system: (a) the mirrored and (b) the opposed configurations. Four standard plasma probe and sample conditions are also shown schematically in the center of the discharge. After Rohde et al. (35).

cathode to the other, and this variation is illustrated in Fig. 16. Shown in Fig. 16a is the parallel component of the magnetic field for the "mirrored" configuration, measured with respect to the vertical target face, Fig. 16b shows the perpendicular component. The parallel and perpendicular components for the opposed configuration are shown in Figs. 16c and 16d, respectively. In brief, the perpendicular component is most important in terms of controlling the density of the ion current at the substrate. In the "mirrored" case, this component changes sign at the midpoint between the two cathodes, providing a path for the electrons from the cathode region to escape to the grounded chamber walls. In the "opposed" case, the field is continuous in sign, and thus it is more difficult for the electrons to escape the center region and reach the chamber walls. Hence, since the electrons are better confined, more ionization occurs in the substrate region.

While these measurements provided a starting point for Rohde et al. (27), more detailed assessments of the fields could be made using a tabletop model consisting only of the magnets and an iron backing plate.

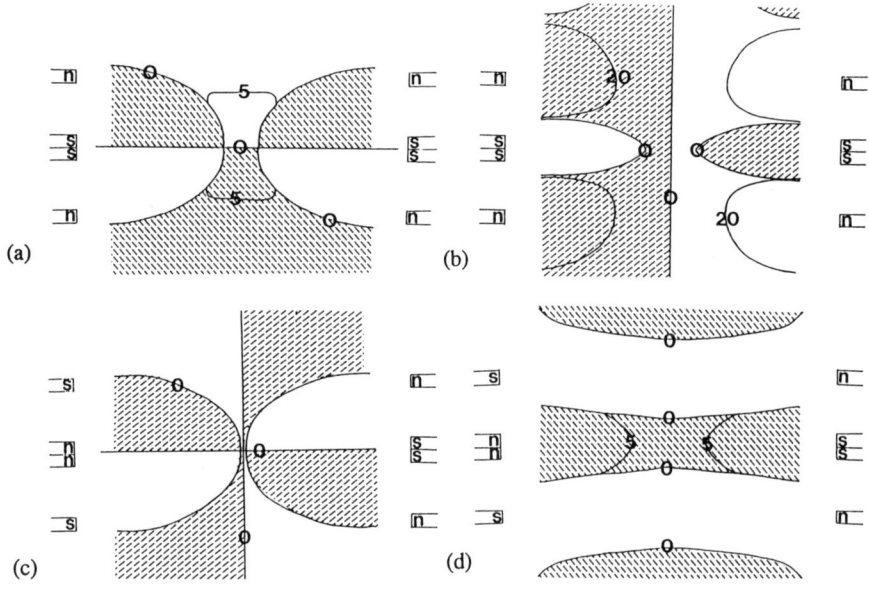

FIG. 16. Magnetic field for the two unbalanced magnetron configurations, measured in Gauss with respect to the target faces: (a) mirrored configuration—parallel component; (b) mirrored configuration—perpendicular component; (c) opposed configuration—parallel component; and (d) opposed configuration—perpendicular component. After Rohde et al. (35).

A simple experiment using this tabletop model and iron filings illustrates qualitatively the difference between the "opposed" and "mirrored" configurations. Using this model, interactions between the two cathodes could be observed at cathode separations as large as one meter. Although the model was useful for evaluating potential magnetic arrangements, this system was superseded by finite element analysis. Finite element analysis permitted the researchers to make detailed evaluations of the magnetic fields, both at the cathode and near the substrates, and in addition provided a means of accessing the combined effects of permanent and electromagnets. The finite element analysis yielded results that agreed well with the experimental results (29).

Finite element analysis of the magnetic fields has proven invaluable in the design and development of unbalanced magnetron cathodes, but for discrete locations magnetic field strengths can be estimated numerically using the method of Murphy and co-workers (30).

Many of the first unbalanced magnetron experiments (17, 28) were performed using larger, stronger (i.e., NdFeB or SmCo), or additional permanent magnets arranged in a variety of configurations inside and outside the discharge chamber. All of these UBM systems yielded increases in the current at the substrate, as well as increases in the magnetic induction at the substrate. However, once permanent magnets are installed, the degree of "unbalance" is fixed, although in many applications an additional degree of flexibility would be advantageous. And as demonstrated by Window and Savvides (17), electromagnets can just as easily be used to create an unbalanced magnetron geometry.

Thus, researchers began to experiment with the use of electromagnetic coils to create an unbalanced field geometry that could be varied according to the application. Experiments by Musil et al. (31) showed that electromagnetic coils could be used to produce suitably high current densities at the substrate. In this work, Musil et al. (31) also commented on perhaps one of the most interesting aspects of the unbalanced magnetron:

> The independent control of the deposition rate a_D and the modification of created films makes it possible to investigate basic relations between the film microstructure and process parameters. This is of fundamental importance for both the basic research and industrial production of new materials with prescribed properties.

Kadlec et al. (32) used an unbalanced magnetron to deposit TiN films on substrates at target-to-substrate distances of 200 mm and were able to obtain ion current densities as high as 6 mA/cm^2 for V_s from -5 to

−100 V. This was one of the first papers to address the problem of decreasing ion density with increasing target-to-substrate distance, which represents asignificant problem for many commercial applications. The study utilized an unbalanced planar magnetron equipped with two electromagnets and a circular Ti target. The geometry of this unbalanced magnetron is essentially that of a conventional magnetron with a magnetic coil (C_1) of 1200 turns and an additional outer coil (C_2) of 400 turns. The three modes of magnetron operation defined by Kadlec et al., illustrated in Fig. 17, are conventional magnetron (CM), unbalanced magnetron (UM), and double site-sustained discharge (DSSD) (32).

The relationship between the current in the inner (I_1) and the outer (I_2) coils determines the operational mode of the cathode. In the CM mode, the outer coil is not energized; thus, $I_2 = 0$ and the substrate current (I_s) is low. The UM mode is characterized by a current in the outer coil (I_2) of less than 3 A; in this mode, I_s increases and I_1 decreases as I_2 is increased. In DSSD mode, the coupling between the two electromagnets changes, and I_1 increases as I_2 is increased. The behavior of I_s varies with the applied bias (V_s); for $-V_s < 200$ V, I_s is nearly independent of I_2, and for $-V_s > 200$, I_s increases with increasing I_2. In the DSSD mode, for a discharge current (I_d) of 1 A and a discharge voltage of 500 V, it is possible to achieve $I_s > 5 I_d$ for $-V_s \approx 300$ V. The discharge at the substrates can become so strong that the negative values of I_1 are required to keep $I_d = 1$ A and $V_d = 500$ V constant; the onset of this occurs at about $I_2 = 10$ A and $-V_s = 265$ V. Another interesting feature of the DSSD mode is that I_s is nearly constant over large substrate-to-target distances, from approximately 25 to 175 mm. It is of

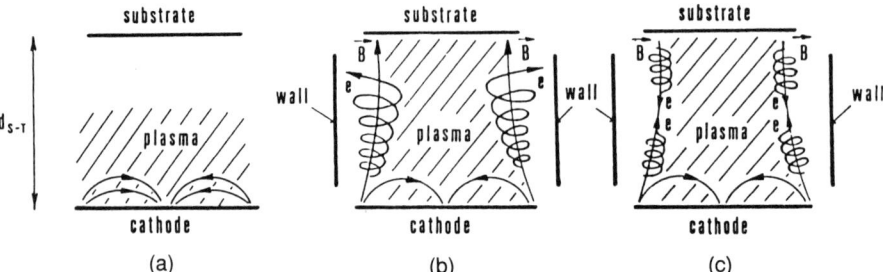

FIG. 17. Schematic representation of the plasma localization and confinement for three modes of magnetron operation: (a) conventional magnetron, (b) unbalanced magnetron, and (c) dual site sustained discharge. After Kadlec et al. (32).

note that all of these experiments were carried out at high Ar pressures, roughly 3 to 5 Pa (22 to 38 mTorr), and this condition acts to increase I_s.

The ratio of ions per deposited atom in these experiments ranged from 0.5:1 using CM to 2:1 for UM, and as high as 10:1 for DSSD (a level that is comparable to those obtained in ion plating). The properties of the films grown using the DSSD technique varied strongly with V_s. For $-V_s < 30$ V, the films were porous, soft, and of mixed (200), (111), and (220) texture. For 40 V $< -V_s < 100$ V, the films were dense, smooth, and brilliant gold in color. The substrate current density of these films ranged from 3.5 to 6.0 mA/cm^2 with increasing bias. The orientation shifted toward (111) texture and the lattice parameters increased with increasing negative bias. Measured hardness values range with increasing V_s from 2,270 to 2,470 kg/mm^2, which is good, but not exceptional. The critical load decreased from 64 N to 25 N with increasing V_s, which is often indicative of increasing film stresses. From this summary of physical properties, it would appear that the optimal bias is about -40 V, but analysis of the microstructure would be required to define this more fully. No mention is made of film properties for $-V_s > 100$ V, but it is presumable that increases in film stress resulted in rather brittle films with poor critical loads. For a more detailed discussion of the DSSD mode, the reader is referred to a related paper by Musil and Kadlec (33) that reiterates the advantages of DSSD magnetron sputtering in TiN and includes a detailed comparison among DSSD, UM, and CM modes, and ion plating.

Alternatively, electromagnetic coils could be added to an existing permanent-magnet conventional magnetron to create an unbalanced field. Several practical limitations need to be considered, however, when combining permanent magnets and magnetic coils. Conventional balanced magnetrons often use AlNiCo magnets in their cathodes, but the coercive strength of these magnets may be insufficient to resist demagnetization in the presence of a strong electromagnet. Thus, conversion to an unbalanced geometry via the addition of an external coil may not be as straightforward as it appears. Frequently, more costly rare earth magnets such as NdFeB or SmCo are required in addition to the electromagnetic coil. Cooling of the magnetic coils also may be necessary is some applications, and some researchers have observed that poor target utilization is a more noticeable problem with electromagnets than with permanent magnets (23).

While determining an appropriate field strength is important, perhaps more central to UBM technology is the assessment of the density of the ion current, or more appropriately, the ion-to-deposited-atom ratio during deposition. As discussed previously, it is important to evaluate both the flux and the energy of the ions in order to fully characterize the effects of ion bombardment. In their original paper, Window and Savvides reported ion-to-deposited-atom ratios for Cu films ranging from 0.00025:1 for configuration 5 (see Fig. 13) to 2:1 for configuration 2 (see Fig. 13) (*17*). In comparison, the ion-to-deposited-Ti-atom ratios for TiN films ranged from about 1.7:1 to 5.3:1 in a study by Kadlec *et al.* (*34*), and from 1.4:1 to 6.3:1 in a study by Rohde *et al.* (*35*). Thus, the range of flux densities varies widely from one material to another, and from one set of discharge conditions to another as well. In general, the ion-to-deposited-atom ratio can be estimated as follows:

$$J_i = \frac{J_s}{e}, \tag{8}$$

and

$$J_v = \frac{A_d \rho}{M_v m_u}, \tag{9}$$

$$\frac{J_i}{J_v} = \frac{M_v m_u J_s}{e a_d}, \tag{10}$$

where J_i = ion flux, J_v = flux of condensing atoms from the vapor, M_v = average mass of the condensing vapor-produced ion-to-deposited-Ti-atom species in amu, m_u = unit mass ($m_u = 1.66 \times 10^{-24}$ g/amu), ρ = average film density in g/cm^2, e = unit charge ($e = 1.6 \times 10^{-19}$ C), a_d = deposition rate of the condensing species in nm/s, and j_s = ion current density incident on the substrate in mA/cm^2.

For negative applied substrate bias potentials (V_s) that are large enough that the electron contribution to the measured current at the substrate can be considered negligible, the ion current density incident on the sample (j_s) can be estimated directly from the total ion current at the substrate (I_s) divided by the substrate area (A_s). However, for small substrate biases the electron contribution to the total substrate current must be accounted for, and in this case it is usually sufficient to estimate i_s from a linear extrapolation of $I_s/A_s = f(V_s)$. For the case of TiN films the preceding analysis yields $J_i/J_v \cong 71.3 \, (j_s/a_d)$.

Another figure of merit that is useful in comparing the ion bombardment from one set of deposition conditions to another, and perhaps more meaningful than J_i/J_v, is a combined term that accounts for the effects of both the ion energy and the ion-to-deposited-atom ratio. For this combined term an estimation of the ion energy eU_i is needed and can be calculated as follows:

$$eU_i = e(V_{pl} - V_s), \quad (11)$$

where V_{pl} is the plasma potential and V_s is the applied substrate bias potential.

In most cases in unbalanced magnetron sputtering it is reasonable to assume that the ions are singly charged, and thus, a combined energy and flux ratio term E_p, representing the average energy of the arriving ions per condensing atom, can be defined as

$$E_p = eU_i \frac{J_i}{J_v}. \quad (12)$$

Estimated values of these parameters from Kalec et al., for unbalanced magnetron deposition of TiN films under a variety of applied substrate bias potential, are presented in Table II (34). However, in this analysis Kadlec et al., like Window and Savvides (17), assumed a value of the plasma potential based on their experience with conventional magnetrons. Kadlec et al. assumed a plasma potential of $+10$ V, while Window and Savvides estimated the potential to range between $+5$ and $+20$ V, depending on the self-bias or floating potential.

While the measurement of the self-bias potential of a probe immersed in the plasma provides a good estimate of the direction, positive or negative, that the plasma potential shifts under varying discharge conditions, it does not directly indicate the plasma potential and hence does not aid in estimating the ion energy. The simplest way to access the plasma potential, and hence the effective io energy, is to use a small Langmuir-type probe. Detailed information regarding operating this type of plasma probe can be found in numerous books and articles (11, 36-38). In general, a set of $I-V$ characteristics can be plotted for the probe, and from this a number of plasma parameters can be estimated. The self-bias or floating potential (V_{fl}) is defined as the point where the net current collected by the probe is zero (i.e., the ion flux is equal to the electron flux), the ion saturation current (I_i^*) is that point where further increases in the applied negative potential no longer cause an increase in the current collected by the probe, and the electron saturation current (I_e^*) is that point at which the electron current remains constant with increasing positive applied potential. Neither the plasma potential (V_{pl}) nor the electron temperature (T_e) can be determined directly from the $I-V$ characteristics. The electron temperature can be calculated from the slope of the $I-V$ curve and by using the "tangent method" (37); plasma

TABLE II

DEPOSITION AND PLASMA PARAMETERS FOR TiN FILMS PREPARED UNDER A VARIETY OF SUBSTRATE BIAS POTENTIALS V_s IN A MULTIPOLAR UNBALANCED MAGNETRON SYSTEM[a]

Negative Ion Potential (U_i)	Ion Current (mA)	Film Thickness (μm)	Deposition Rate (nm/s)	Ion Current Density (mA/cm^2)	Ion-to-Atom-Ratio (V_i/V_v)	Energy per Atom (eV/atom)
30	0	5.1	0.77	1.8	1.7	70
35	80	3.6	0.67	1.9	2.0	90
40	180	3.9	0.79	2.0	2.0	100
50	280	3.4	0.63	2.2	2.5	150
60	380	3.7	0.66	2.4	2.5	180
80	530	3.7	0.69	2.7	2.8	250
100	610	3.2	0.59	3.0	3.7	400
120	680	3.0	0.56	3.4	4.3	560
150	790	3.0	0.53	3.9	5.3	850

[a] The ion potential U_i is given as $U_i = V_{pl} - V_s$, where the plasma potential V_{pl} is assumed to be $+10$ V. The cathode magnetic coil currents were $I_1 \simeq 1$ A and $I_2 = 4$ A for the inner and outer magnetron coils, respectively. In each case the substrate-to-target distance was 200 mm, total pressure was 0.09 Pa, substrate temperature was 500°C, target power was 3 kW, and deposition time was about 90 min. After Kadlec et al. (34).

potential can be estimated by locating the knee of the curve on a log–log plot of the I–V behavior.

The plasma potentials measured for conventional magnetrons tend to range from a few volts positive to a few volts negative of the ground potential. However, for unbalanced magnetrons, experiments by Rohde et al. have determined that under certain conditions negative plasma potentials in excess of $-20\,\text{V}$ can be estimated using this technique (27, 39), as shown in Table III.

In a dual-cathode unbalanced magnetron system, the plasma potential as well as the number of ions per deposited atom can be strongly affected by the magnetic relationship between the cathodes. For the mirrored case, the plasma potentials determined were either equal to the anode potential or slightly positive while for the opposed case, the plasma potentials measured ranged from -13 to $-20\,\text{V}$. There are two possible explanations for the observed negative plasma potentials: (1) either the magnetic trap setup between the cathodes provides a high enough level of confinement that the electrons cannot readily leave the near-substrate region, thereby shifting the plasma potential to a more negative value than would ordinarily be expected; or (2) the magnetic field is sufficiently strong in the near substrate region to invalidate the zero field assumptions made as a part of the probe analysis. However, dual-probe measurements designed to test the latter case also indicated negative plasma potentials. Assessment of the plasma potential in unbalanced magnetron sputtering appears to be fairly important, if the energy of the bombarding ions is to be correctly estimated, since at low bias potentials ($<150\,\text{V}$) an error of $\pm 15\,\text{V}$ in estimation of the ion energy will be significant.

For the dual unbalanced magnetron system of Sproul and co-workers, this combined energy term was estimated by Rohde et al., who compared E_p to the film density determined by cross-section transmission electron microscopy (XTEM) and preferred orientation, as presented in Table III (39). Only those films produced with an $E_p > 500\,\text{eV/atom}$ ($J_i/J_{Ti} = 4.4$) appeared to be fully dense. For $E_p < 500\,\text{eV/atom}$, the films had varying degrees of voiding along the grain boundaries. In comparison, a study by Hultman et al. examined several TiN films produced by Musil and Kadlec using XTEM to evaluate the relative film density (40). Films with a $J_i/J_{Ti} \geqslant 4$ were found to be fully dense, while films with a $J_i/J_{Ti} \leqslant 2$ were found to be highly underdense. While the V_{pl} in this case was not measured, if one assumes that the negative bias potential $V_s = 100\,\text{V}$ is an accurate estimate of the actual ion energy, the E_p values would be 400 and $200\,\text{eV/atom}$, respectively. While these figures are in good agreement

TABLE III
SUMMARIZED DEPOSITION PARAMETERS, PLASMA CHARACTERISTICS, AND PREFERRED ORIENTATIONS FOR TiN FILMS ON M-2 STEEL DEPOSITED IN A DUAL CATHODE UNBALANCED MAGNETRON SYSTEM

Magnet Configuration	Preferred Orientation	Bias Potential (V)	Plasma Potential (V)	Effective Potential U_i(V)	Deposition Rate (μm/h)	Ion Current Density (mA/cm^2)	Ion-to-Ti-Atom Ratio J_i/J_{Ti}	Energy per Ti Atom (eV/atom)
Mirrored	(311)	−75	+3	−78	4.6	2.3	2.1:1	164
Mirrored	(111)	−135	0	−135	4.0	2.0	2.1:1	284
Opposed	(311) & (111)	−75	−20	−55	5.0	3.5	3.2:1	176
Opposed	(220) & (111)	−135	−20	−115	3.8	3.9	4.4:1	506

[a]After Rohde et al. (39).

and provide a reasonable first approximation of the J_i/J_{Ti} and E_p values required to produce dense polycrystalline films of TiN, there is a significant difference in the way the two sets of films were produced. In the case of the Musil and Kadlec films, the substrate was placed directly in front of the cathode so that the majority of the flux arrived perpendicular to the surface. In comparison, the films from Sproul *et al.* were produced with the sample in rotation; thus the angle of the incident flux varied continuously (*35,39*). This geometry makes analysis of the exact film deposition conditions more difficult, but provides a situation more akin to most industrial processes.

Comparisons of the energy input into the film should help to quantify the effects of ion bombardment, to provide a more unified picture of the deposition environment and the benefits and limitations of unbalanced magnetron sputtering in particular applications. However, to date this picture is incomplete. One area in which knowledge is clearly lacking is our understanding of the role of neutral atom bombardment. Like the ions and electrons, undoubtedly the flux and energy of these neutral particles can play an important role in formation of film microstructures as well. However, the energy of the neutral vapor flux incident on the substrate is much harder to determine than that of a charged flux. While the thermalization distance, the average distance a particle ejected from sputtering cathode must travel in order to have an energy equal to the thermal energy of the plasma, can be estimated fairly easily (*41*), it is quite difficult to measure the energy of these energetic neutrals directly, even in a laboratory-scale system. Recently, Meyers and co-workers have developed an energy analyzer for neutral particles and have applied this in the area of magnetron sputtering (*42*). Future developments in this area should add to the fundamental understanding of unbalanced magnetrons.

Once the principles of unbalanced magnetrons are at least partially understood, the next logical step is to look toward the applications of this new technology. In a first attempt to assess the potential industrial applications of ion-assisting magnetron sources early in the development of unbalanced magnetron technology, Window and Harding predicted that perhaps unbalanced magnetron sputtering would have the greatest impact in the hard coatings industry, where the relatively high levels of residual stress would not be a detriment, and where achieving film densification at low substrate temperatures is a constant concern (*43*). As detailed in the following section, this early prediction appears to have been correct.

V. Applications of UBM Sputtering

Since the first articles describing unbalanced magnetrons appeared, a number of research groups worldwide have applied the concept of the UBM in a variety of applications. Initially, each unbalanced magnetron system was custom designed and fabricated, often by the authors themselves, so that a great many of the early articles describe the geometry and unique characteristics of the particular UBM configuration. Commercial UBM cathodes became available only recently, and currently there is no industry standard for the magnet material or the magnetic arrangement of the cathode. In general, the research applications of UBM technology can be grouped into four areas: (1) elemental thin films, (2) films for electronic and optical applications, (3) films for corrosion protection, and (4) films for wear and abrasion resistance. The majority of the literature is concentrated in the last of these four areas.

A. Elemental Thin Films

In general, the choice of elemental thin film materials is based not so much on their industrial applications as on their inherent simplicity. For instance, Window and Savvides deposited thin films of Si as a means of comparing the deposition parameters, film microstructures, and film properties achieved using unbalanced magnetron sputtering with those of Si films produced by other deposition methods. Window and Savvides found that, in general, the pressure and discharge current relationships for Cu and Si targets were quite similar, but the properties of the Cu and Si films were not reported (17,44).

Several of the early studies by Window and co-workers examined films of a number of elemental metals, including Fe, Mo, Nb, W, Ta, and Pt (45,46). In a later paper by Window and Harding, the effects of UBM-induced ion bombardment on the structure of Mo films on Si wafers were investigated, and it was determined that film stress is, as expected, strongly dependent on growth temperature (43). Films produced at high temperatures (400°C), but still well below temperatures where bulk diffusion would be sizable, showed the lowest levels of film stress; hence, it was predicted that the low-energy ion bombardment of the UBM technique in these cases must produce the stress relief observed. Similar decreases in film stresses were *not* observed for a low-growth-temperature film (30°C) subjected to a 400°C postdeposition anneal. The films deposited at the lowest temperatures and subject to the highest-energy ions showed the greatest degree of (110) preferred orientation and the highest stress levels, which agrees well with conventional magnetron studies. In addition, this paper detailed several generalized relationships for unbalanced magnetrons: (1) film stresses are linearly dependent on the ion energy; (2) compressive stress increases with increasing ion bombardment until plastic flow begins, after which further increases in ion bombardment act to reduce film stress; (3) film stress increases with increasing ion mass; (4) the preferred orientation of Zr films changes from that of the close-packed planes to that of the normal wire texture with increasing ion bombardment; and (5) film stress increases with increasing neutral atom bombardment in the same manner as observed for ion bombardment. However, it is important to note that in the foregoing generalizations, little distinction is made between the effects that are due to varying ion energy and those that are due to varying ion flux density or J_i/J_v (43).

In the area of elemental materials, only a few applications of UBM sputtering have been investigated; these are principally the deposition of thick oxygen-free high-conductivity copper (OFHC) on low carbon steel and of Al on Si wafers by Monaghan and Arnell (47, 48). The aim of the OFHC Cu research was to produce relatively thick films (>1 mm) with a very fine grain structure ($<1\,\mu$m), finer than can be produced using conventional metallurgical processing. In this work, a variety of deposition conditions were investigated, and dense, high-conductivity Cu films were produced under both applied negative biases and self-bias potential (V_{fl}) conditions. Noncolumnar microstructures were observed for substrate power densities in excess of $0.1\,\text{W}/\text{cm}^2$; however, for higher melting point materials, it is expected that higher power densities would be required to produce similar microstructures (48). The only caveat was that for thick films ($>50\,\mu$m), substrate heating over time gave rise to increased grain growth, but it is expected that this could be eliminated with the addition of more direct substrate cooling (47).

Monaghan and Arnell, sparked by reports of heteroepitaxial growth of Al on (111) Si wafers via ion cluster beam deposition, also investigated briefly the potential of using UBM sputtering to deposit Al films on (100) and (111) Si wafers. Highly conductive heteroepitaxially deposited Al thin films on Si could potentially reduce or even eliminate the problem of Si migration through the Al film, since the primary mechanism for this is grain boundary diffusion. While Monaghan and Arnell succeeded in producing highly (111)-oriented Al films, none of the films were single-crystalline. Analyses of film conductivity and diffusional resistance to Si were not provided, but it was speculated that because the films had a high dislocation density, the film conductivity was likely to be less than optimum. It is worth noting that both of these UBM studies were performed using a commercially available system, which is the exception rather than the rule. Both of these applications could be considered as electronic applications as well.

B. Films for Electronic and Optical Applications

Only a few other investigators have examined UBM deposition of electronic materials. In one of their earliest papers, Window and Savvides examined the effects of ion bombardment on the microstructure of Ni/Cr alloy films, with the goal of developing suitable materials for thin-films resistors (44). The primary requirements were, thus, a low temperature coefficient of resistance and long-term stability. In this preliminary investigation, two films were deposited onto glass substrates, one at 0 V and the other at −100 V bias. Improvements in film microstructure, room-temperature resistivity, and acid etch rates were observed for the high-bias specimen, but clearly this investigation was intended to be only preliminary in nature.

In a 1988 paper, Savvides and Window discussed the effects of UBM-enhanced ion bombardment on the microstructure, optical, and electronic properties of TiN films. In addition to applications as a wear-resistant thin film material, TiN also exhibits a high stability and low resistivity, making it a very promising diffusion barrier material for microelectronics applications. Detailed are measurements of film properties versus intensity of ion bombardment, including the temperature coefficient of resistance, electrical resistivity, optical reflectance, film microstructure, growth morphology, and superconducting transition temperature. In general, very high-quality TiN films were produced through careful optimization of high-density, low-energy ion bombardment during film growth. In the optimized case ($T_s = 500°C$, $V_s = 20$ to 50 eV), TiN films with room-temperature resistivities of 26 $\mu\Omega$-cm were manufactured, which is as least as low as the best reported values for polycrystalline TiN, and on the order of the best values reported for single-crystalline TiN (44, 49).

In the area of optical thin films, UBM deposition of InO and TiO_2 has been explored to some extent. The principal application of InO is as an optically transparent conductive thin film, and TiO_2 can be used as an optical dielectric. As in most previous cases, the majority of the published data in this area comes from custom research-scale UBM systems. In their study of InO and TiO_2, Spencer et al. found that in addition to producing changes in the refractive index of the TiO_2 films, the enhanced ion bombardment of the UBM also aided in increasing the reactivity of the oxygen in the plasma, resulting in increased reactive gas utilization (50). Using a closed-loop feedback gas control system, with the control signal coming from the spectral peak height of the metal species in the plasma, the flow of the reactive gas (O_2) could be effectively controlled to permit deposition in what is commonly termed the metallic cathode mode. In this mode, the arrival rate of the metal species at the substrate determines the growth rate, rather than the sputtering rate from a compound coated cathode; the latter is generally at least an order of magnitude lower than the sputtering rate from a metallic cathode. Thus, Spencer et al. were able to use this high-rate sputtering technique to cost-effectively deposit both InO and TiO_2. The results for InO were inconclusive, since films produced with and without ion bombardment showed no significant differences in either their resistivities or their refractive indices. The results for the TiO_2 were a bit more promising in that the refractive index increased with increasing ion bombardment; however, these films were deposited using a much smaller, compound coated target. A later paper by Howson et al. discusses some of these same experiments and the success of the gas control system (51). Resistivities as low as 4,000 $\mu\Omega$-cm were reported for the InO (51). The refractive indices of the TiO_2 ranged from 2.42 to 2.50 (at 633 nm), depending on the ion bombardment (50). These studies were unique in several regards: First, magnets were used behind the substrates for the TiO_2 films in order to further confine the plasma as it escaped the near cathode region; and second, the substrates were isolated in the discharge and thus were biased at the self-bias or floating potential (V_{fl}) of the plasma. Both papers mentioned that there was considerable heating of the substrates during deposition, which could be attributed in part to the large electron flux, in addition to the ion flux, incident on the substrates under self-bias conditions.

C. Films for Corrosion Protection

Again there are only a limited number of studies that have examined the applications of UBM sputtering in corrosion protection. These come primarily from Arnell and Bates, who looked at the potential of using unbalanced magnetrons to deposit metastable Al–Mg alloys using first, composite Al–Mg targets (52); second, co-deposition of Al and Mg from a pair of side-by-side UBMs (53); and finally, a large commercial multicathode UBM system with four Al cathodes magnetically linked in a closed-field configuration (54). The initial investigations used composite targets that were formed by drilling out sections from a solid Al or Mg cathode and replacing these with the appropriate amounts of Al or Mg to permit deposition of almost the full range of alloy compositions. The aim of this research was to produce noncolumnar Al–Mg alloy thin films, which have been shown to be effective in protecting steel substrates. The structure of the alloy films containing less than 20% Mg was that of a supersaturated solution of Mg in a (fcc) Al matrix; films containing between 20 and 70% Mg contained glassy regions; and the structure of films containing more than 70% Mg was that of a supersaturated solid solution of Al in a Mg matrix. Between the range of 45 to 50% Mg the films were entirely glassy, and above or below these limits there was a mixture of crystalline and glassy regions. All of the films were noncolumnar and exhibited higher hardness values than conventionally prepared alloys. Hardness values as high as 400 MPa were observed in this study.

One of the principal limitations of the composite cathode study was that a large number of cathodes was required to investigate the entire range of alloy compositions. In a follow-up research study, Arnell and Bates used two separate UBM cathodes in a side-by-side configuration to generate the required alloy variations (53). Again all of the films were noncolumnar, and even higher hardness values were reported, ranging from near 2,000 MPa for the pure metallic films to as high as 8,000 MPa for one particular film. Similar limits of solid solubility were observed for Al-rich films, and the onset of the glassy constituent was again at 20% Mg. At 45% Mg only the glassy phase was present. At about 50% Mg an $Al_{12}Mg_{17}$ intermetallic phase was detected and increased with increasing Mg content until at 60% Mg, the glassy phase was eliminated entirely. Above 90% Mg, the structure was that of a single-phase supersaturated solid solution of Al in a Mg matrix, which differs somewhat from the limit observed using composite targets. The potential applications of these alloys seem high, particularly in the aerospace industry, but to date information regarding their corrosional properties has not been reported.

Most recently, research in this area has moved closer to commercialization as test coatings of pure Al were deposited in a large commercial multicathode UBM sputtering system (54). Historically, Cd coatings have been used on many components in aerospace applications as a means of providing corrosion protection and improved tribological performance. However, with increasing concern about the environmental impact of Cd plating operations, alternative materials for corrosion protection have been examined. Some of the most promising alternatives appear to be Al and Al-alloy coatings, as these materials are cost-effective, provide adequate corrosion protection, and are more environmentally friendly than Cd. The principal limitation lies in the reduced tribological performance of Al as compared to Cd. Thus, it was the aim of this research to investigate the potential of using UBM-enhanced ion bombardment to produce dense Al films with improved tribological properties. As in the previous UBM studies, dense noncolumnar microstructures and enhanced hardness values were measured for the pure-Al films deposited in the work. This appears to be an important first step toward commercialization, and it is probable that this will be an area of increasing interest in the future.

D. Films for Wear and Abrasion Resistance

By far the largest volume of information regarding UBM-sputtered films is in the area of coatings for wear and abrasion resistance, including diamondlike carbon, TiN, and other transition-metal nitride thin films. In the research of Savvides (55) and co-workers (45) that ultimately led to the development of the unbalanced magnetron, a modified penning source was used that deposited diamondlike amorphous carbon (a-C) films. The a-C films with the highest degree of tetrahedral bonding were produced at the lowest sputtering power levels (i.e., the highest ion-to-deposited-atom ratios). These films also exhibited noncolumnar microstructures and unusually smooth surfaces, indicating that the potential for UBM deposition of a-C films should be high (44). Recently, Monaghan and co-workers have also investigated UBM growth of diamondlike carbon films, in this case using an industrial-scale four-cathode UBM system to deposit hydrogenated amorphous carbon films that also contained a small concentration of metal atoms (Me:C–H) (56). In this study, reactive sputtering with a plasma emission monitor was used to control the partial pressure of the acetylene gas added to the system, and thus the composition of the resulting films. An rf bias was applied to the substrates during growth to promote ion bombardment, and graded layer films were produced to enhance film adhesion to the M-2 tool steel substrates. The results were quite promising, with adhesion values as high as 115–125 N, wear rates about

$\frac{1}{5}$ those of similarly deposited TiN coatings, and coefficients of friction less than the 0.25 reported for layered films of Ti:C–H (56).

A wealth of materials are available detailing the deposition parameters and resulting film properties for TiN coatings, in part because TiN has proven to be a very successful material in the hard-coating industry and in part because TiN is very well characterized, and thus is an appropriate material in which to examine the strengths and weaknesses of UBM technology. In some of their earliest studies, Window and Savvides deposited TiN films to show the potential for using UBMs in electron, optical, and hard-coating applications (44,49). Since then a number of research groups have actively pursued UBM deposition of TiN, including Sproul and co-workers (27–29, 35, 39); Kadlec, Musil and co-workers (31–34, 40, 57, 58); Efeoglu, Arnell and co-workers (59); and others. The majority of studies examined films about 3 μm thick, which seem to perform well in most applications, with the most notable exception being films for bearing applications, which seem to be most effective at thicknesses near 1 μm. In general, the highest film hardness values are found in films deposited at low substrate temperatures and high ion-to-deposited-atom ratios, since under these conditions a high density of dislocations is created during deposition, resulting in high compressive stresses in the films. As is true for TiN_x deposited by other means, hardness typically increases with increasing N-content from well under $1,000 H_v$ for α-Ti to as high as $3000 H_v$ for stoichiometric TiN. Typical hardness values for near-stoichiometric TiN films range from near $1000 H_v$ for low ion-bombardment conditions, to an average value of about $2,000 H_v$, to more than $3,000 H_v$ for high ion-bombardment conditions. The ε-Ti_2N phase was seldom present in any of the UBM studies, and the films containing this phase usually had inferior hardness compared to TiN films. In addition, several researchers have found that the use of very strongly unbalanced magnetrons can be detrimental to film properties, so some compromise in deposition conditions is desirable (2759). Of course, hardness values are strongly dependent on the substrate material and microstructure as well as the film thickness and indentor load, and comparisons from one researcher to another are often not very meaningful. Likewise, film adhesion is frequently evaluated indirectly by comparing the critical load at which the film/substrate couple fails either cohesively or adhesively, and these measurements are often far more qualitative than quantitative. Typical critical load values reported for TiN films vary widely depending again on the substrate material and film thickness, but are usually on the order of 80 to 100 N for successful TiN films on tool steel substrates. The only reported data regarding the coefficient of friction provides an initial value of 0.2; however, in this case, the value increased to a value of 0.6 to 0.9 within 30 min (59).

Clearly, TiN films with comparable, if not superior, properties to those deposited using conventional magnetrons can be readily produced using a variety of unbalanced magnetron geometries. The primary advantages of UBM deposition, as illustrated through these TiN film studies, are the large ion fluxes available at the substrate, the low bias potentials required to produce dense films, and the ability to deposit adherent TiN films at large substrate-to-target distance. Target-to-substrate separations for conventional magnetrons usually range from about 3 to 6 cm, since these do not generate sufficient ion fluxes far enough from the cathode to produce dense, adherent hard coatings. This is not the case for UBM sputtering; for instance, dense TiN films can be deposited on M-2 tool steel substrates in rotation between two UBM cathodes using a bias voltage of less than 150 V, and an ion-to-deposited-Ti-atom ratio of more than 4.4:1 at an average target of 12 cm (35). For a stationary substrate, dense, golden TiN films were produced using only a -100 V substrate bias at target-to-substrate separations as large as 20 cm (57). The applications of UBM-deposited TiN films are as wide-ranging as TiN deposited by other methods. Unbalanced magnetron-deposited TiN films are already commercially available, and although the most common use of TiN is as a hard coating, UBM-deposited TiN is expected to have applications in the fields of optical coatings and electronics as well.

As has been shown for magnetron-sputtered coatings, TiN is not always the best choice for all wear- and abrasion-resistant applications. For instance, $(Ti_xAl_{1-x})N$ films have been shown to provide high-temperature oxidation resistance superior to that of TiN films and thus perform better in high-temperature drilling applications (60,61). Thus, it is only natural that researchers begin to explore the potential benefits of using UBM techniques to deposit these alternative transition-metal nitride, carbide and oxide coatings. Currently, only a few UBM studies of non-TiN hard coatings have been published; however, many research groups are actively pursuing such studies, and new results are no doubt in press. UBM-deposited $(Ti_xAl_{1-x})N$ coatings with a 5.5:1 ratio of Ti:Al have been examined by Efeoglu, Arnell and co-workers and have been shown to have critical load values about twice those of similar TiN films (62). Vickers hardness values for the $(Ti_xAl_{1-x})N$ films were slightly less than those of the previously deposited TiN films, being 3,020 and 3,150 kgf/mm^2, respectively. The tribological properties were also similar to the TiN films, starting out initially at 0.2 to 0.4 depending on the substrate surface roughness and rising quickly to a steady-state value of 0.6 to 0.8. The wear coefficients measured in this study showed that the TiN coatings showed about one-half the wear of the $(Ti_xAl_{1-x})N$ films. In a paper by Monaghan et al., a coating methodology is presented in which targets of several elemental metals can be combined in a multicathode arrangement to produce films with varying chemical compositions (63). However, only very brief descriptions of the resulting coating properties are provided. For instance, coating hardnesses of 3,000 to 4,000 H_v for TiAlN and 3,500 to 4,500 H_v for CrZrN, as compared to 2,200 to 2,500 H_v for TiN, are reported. However, representative critical load values for these films are not given. Perhaps more interesting, however, are the results reported for graded alloy nitride coatings. The initial layer is pure Ti, followed by TiN, and then by a gradual increase in the Zr composition of the film to form a thick TiZrN final layer. Similar procedures can be performed for CrZrN graded films. The goal is to achieve high film hardnesses without sacrificing film adhesion. Performance testing of these materials was not included, but these preliminary results look quite promising.

Similarly, discretely layered rather than graded layer coatings can also be used to enhance film hardness without sacrificing film adhesion, and a number of investigators have examined this alternative for films deposited using conventional magnetrons, typically with bilayer periods of 0.1 μm or larger. In single-crystal systems, coherent superlattices have been studied for many years in metallic and metal-nitride materials, and these superlattices have been shown to provide greatly enhanced film hardness and strength (64–66). Recently, this "superlattice" concept has been applied to polycrystalline films of TiN/NbN deposited by multi-cathode unbalanced magnetron sputtering (67). The bilayer periods investigated in this work ranged from 55 to 4 nm and resulted in film hardnesses as high as H_v 5,200 kgf/mm^2. Again, the preliminary results are very promising, and no doubt new results will soon be available in print.

In summary, the research applications of unbalanced magnetrons are just beginning to develop, and new applications of these UBM techniques are evolving each year. Perhaps the greatest benefit of UBM deposition appears to be the ability to combine the high deposition rates of magnetron sputtering with high-flux, low-energy ion bombardment. In addition, much greater target-to-substrate separations can be used with unbalanced magnetrons, increasing the range of substrate geometries that can be coated, as compared to those of conventional magnetrons. This is particularly notable for the multicathode systems, which utilize a large chamber volume that would be unthinkable for conventional magnetrons. However, each technique has limitations, and for UBM sputtering these are probably the higher minimum substrate temperatures ($\approx 250°C$), and the large density of defects created by the additional ion bombardment. The other serious limitation of UBM techniques is that, because of the inherent flexibility of the deposition process, the number of potential control parameters is so large that true optimization of film properties for a specific application can often be overwhelming, if not impossible, given the cost of time.

VI. Commercial Applications of UBM Sputtering

While the majority of the results presented previously were attained using custom-designed unbalanced cathodes, a number of research-scale unbalanced cathodes are now available from commercial magnetron suppliers. These sources range from about 2.5 cm to about 30 cm in diameter. Each cathode has advantages, and the differences between the chamber geometry and substrate-to-target distances will also significantly affect the performance of these cathodes in a given application. In particular, the location of grounded or electrically isolated components, such as "dark space" shields, can affect the "focusing" of the unbalanced magnetron as well as the overall ion density observed in the near-substrate region. This is because grounded surfaces near the cathode tend to collect stray electrons, which in turn will reduce the number of electrons available to cause ionization in the near substrate region.

One of the most promising commercial applications of UBM sputtering is in the deposition of hard, wear-resistant coatings such as TiN or TiC. These transition metal–based compounds are typically formed *in situ* using a pure metal target (e.g., Ti) and a reactive gas (e.g., N_2). In order to obtain deposition rates that allow unbalanced magnetron sputtering to be cost-competitive with other deposition techniques, most, if not all, commercial UBM systems utilize some type of interactive–reactive gas control. This class of technique is commonly referred to as "high-rate reactive sputtering" and has been used extensively in commercial magnetron systems (*68*).

Large-scale unbalanced cathodes have been developed by several manufacturers for industrial use. One such cathode is a cylindrical magnetron source that can be reconfigured to produce a cylindrical "unbalanced" magnetron. This is very well suited to coating large planar substrates. More common are large round or rectangular cathodes with unbalanced magnetic configurations that can be mounted internally, externally, or on a standard flange.

At present, the greatest variety of industrial-scale systems is in the area of large multicathode batch coating systems. Teer Inc. produced the first commercially available multicathode unbalanced systems in a number of different three-, four-, and five-cathode configurations. Because it has been established that an even number of cathodes is required to complete the magnetic confinement, Teer Inc. currently offers unbalanced systems with one, four, six, or eight unbalanced magnetrons. For two cathodes, an electron trap can be created between the opposed cathodes (35), but with four or more cathodes, the attractive fields must be set up between neighboring cathodes.

Hauzer b.v. currently produces two- and four-cathode unbalanced magnetron-sputtering systems. The four-cathode system is incorporated into a one-meter diameter chamber that utilizes the flexibility of electromagnets to produce stronger outer poles. This system is unique in that it combines unbalanced magnetron sputtering with arc evaporation in an integrated cathode design. The objective is to take advantage of the high film adhesion available from arc-deposited coatings—due in part to the density of highly ionized particles produced during arc evaporation—and combine this with the macroparticle-free microstructure of unbalanced magnetron-sputtered coatings (69, 70).

A variation on the multicathode unbalanced magnetron system is the Plasma Booster system developed by Leybold AG in Germany (71). In this system, there are two opposed unbalanced magnetron cathodes surrounded by electromagnetic coils similar to those found on the Hauzer system. Located equidistant between this pair of cathodes are anodes with magnets attached behind them. The magnets are used to confine the plasma in much the same way as in the four-cathode systems produced by Hauzer or Teer, Inc. However, since no potential is placed in these two anodes, they do not contribute to the sputtered flux. Leybold claims to have achieved high current densities in this system, up to $6\,mA/cm^2$, which is also typical of the current densities found in the Hauzer and Teer systems.

The commercial applications of unbalanced magnetron technology are growing rapidly. Unbalanced magnetron-sputtered TiN and other metal nitrides may provide a cost-effective, low-temperature alternative to CVD hard coatings on cutting tools and machine components. Since the substrate heating of UBM-coated components is much lower than for similar CVD-coated components, wear-resistant coatings can potentially be applied to a much wider range of substrate materials. Large multicathode systems also provide a cost-effective means of coating large batches of components, even those with complex three-dimensional geometries, since the shadowing observed in single cathode systems can be eliminated with carefully designed substrate rotation schemes.

One of the most promising industrial applications appears to be in the deposition of CrN and other thin-film materials to replace electrolytic hard chrome (72-74). While it is inexpensive, the electrolytic process has several serious drawbacks: (1) The electrolytic process can cause hydrogen embrittlement of some steels, most notably high-resistance steels; (2) the chemicals used in the plating process are generally toxic; and (3) the large volume of waste generated in the plating process presents a very significant environmental problem.

Another area of industrial interest is in multilayer thin-film materials. Multilayer thin films have been an area of intense academic research over the past decade (75, 76). With the prospect of gaining novel coating properties that were previously unattainable, several industrial coating firms have moved into the multilayer coating market.

A recent variation on the multilayer film theme is the deposition of nanolayer polycrystalline superlattice films. Chu *et al.* have shown that the hardness of very thin alternating layers of TiN and NbN can be increased well beyond the value of either of two normally very hard materials (TiN 2,200 Vickers and NbN 1,400 Vickers) to values exceeding 5,000 Vickers (65). The hardness of these films depends on the superlattice spacing, as do single-crystal superlattice films, but Chu *et al.* have shown that the hardness of the polycrystalline superlattice TiN/NbN films also depends strongly on the substrate bias current and voltage, and the composition of each layer. These polycrystalline superlattice films were deposited in a two-cathode closed unbalanced magnetron-sputtering system, and the hardness of the films could not have been obtained without the high ion bombardment present in that system, typically on the order of $5\,mA/cm^2$ during the deposition of the films (35). When the substrate ion current density was decreased, i.e., when the ion bombardment during the growth of the film was reduced, the hardness of the films decreased.

VII. Potential Future of UBM Technology

As the unbalanced magnetron moves from a laboratory curiosity to an industrial reality, greater understanding of the entire deposition process is needed in order to develop the robust, flexible, reproducible deposition processes demanded by industry. One of the most critical problems in thin films is how to quantify the effects of ions and energetic neutral particles. These concepts must be thoroughly understood so that the promise of "tailorable" film properties can be realized in industrial processing. Researchers such as Hultman et al. have begun to lay the framework by comparing the actual process conditions at the substrate for a number of differing deposition systems (77). This work looks beyond the characterization of individual coating processes to develop some degree of universality in ion-assisted processing. While simple Langmuir probes can produce a fairly accurate assessment of ion bombardment, the same is not true for neutral particle bombardment. A major stumbling block in calculating the total energy input into the growing film has been the difficulty of quantifying neutral particle bombardment. A quadrupole-based detector system that was recently developed by Meyers et al. (42) may provide some insight into the effects of neutral particles in coating processes.

To fully quantify the energy inputs into the growing film, the combined effects of temperature, ion bombardment, and neutral particle bombardment need to be considered. The early work in magnetic field modeling for optimization of single- and multicathode geometries is an area that will likely grow to include modeling of the effects of the various magnetic field arrangements on the local carrier densities within the plasma. Currently, mathematical modeling of deposition plasma in the absence of magnetic fields is possible, and combining this with finite element mapping of magnetic fields is not far off (78). This combination should provide a more direct path for evaluation of localized ion densities, plasma potentials, and other salient plasma parameters than the current method of inferring this information from modeling of the magnetic fields alone.

Along these same lines, work in the area of process modeling is starting to produce results that can be applied more directly to industrial coatings problems. Researchers have been active in developing models for sputtering processes, and as this field matures a natural link to more sophisticated process control should develop (79–82). Continued development of commercial unbalanced magnetron systems with more and more on-line process information should provide enhanced coating flexibility and reliability as well.

For research purposes, the range of available unbalanced magnetron geometries should improve. But, as the rapid acceptance of unbalanced magnetron technology has proven, it is a process that is well suited to industrial applications. As discussed previously, a number of multi-cathode, large-volume coating systems are currently available, including those systems that fall more accurately into the combined-technique category. The range of commercially available combined systems is likely to grow in the 1990s since the ability to combine several different coating methods in a single unit is attractive for a number of reasons; initial cost being the major limitation. The other area where there has been only moderate commercialization is in continuous-process unbalanced magnetron systems. Most of the large commercial systems currently available are batch-coating units, but as this field matures, the development of specialized continuous-coating systems is likely.

In summary, tremendous development and commercialization has occurred in the field of unbalanced magnetron sputtering since its inception in 1986. As this field progresses into its second decade, understanding of the relative strengths and weaknesses of unbalanced magnetron sputtering should allow users to provide a better match between promise and performance. Finally, as demand for environmentally friendly processing increases, greater utilization of physical vapor deposition technology can be anticipated. Despite high initial equipment investments, UBM sputtering offers high coating quality without the toxic chemicals required in chemical vapor deposition and the large amount of chemical waste of electroplating processes.

Acknowledgments

The author wishes to thank Ms. Wilma Ennenga and Dr. William Sproul for their assistance in preparation of this manuscript. And the support of this and other projects by NSF grant No. MSS-9358108, UNLs Center for Materials Research and Analysis, and the College of Engineering and Technology at UNL is very gratefully acknowledged.

References

1. For a more detailed discussion of glow discharges and sputtering, see for instance: B. Chapman, "Glow Discharge Processes," Wiley, New York (1980); "Thin Film Processes" (J. L. Vossen and W. Kern, Eds.), Academic Press, New York, 1978; or "thin Film Processes II" (J. L. Vossen and W. Kern, Eds.), Academic Press, Boston, 1991.

2. J. S. Chapin, *Research and Development* **25**(1), 37–40 (1974); U.S. Pat. Appl. Ser. No. 438,482 (1974).
3. S. L. Rohde and W.-D. Münz, in "Advanced Surface Coatings" (D. S. Ricerby and A. Matthews, Eds.), pp. 92–126. Blackie and Son Ltd., Glasgow, 1991.
4. S. Swann, *J. Vac. Sci. Technol. A* **5**(4), 1750–1754 (1987).
5. A. G. Spencer, C. A. Bishop, and R. P. Howson, *Vacuum* **37**(3/4), 363–366 (1987).
6. Y. Yong-Kang, *Surf. Coat. Technol.* **37**(3), 315–319 (1989).
7. W. D. Münz and J. Göbel, *Surface Engineering* **3**(1), 47–51 (1987).
8. K.-H. Müller, *Phys. Rev. B* **35**(15), 7906–7913 (1987).
9. K.-H. Müller, *J. Vac. Sci. Technol. A* **5**(4), 2161–2162 (1987).
10. F. Parmigiani, E. Kay, T. C. Huang, J. Perrin, M. Jurich, and J. D. Swalen, *Phys. Rev. B* **33**(2), 879–888 (1986).
11. I. Petrov, V. Orlinov, I. Ivanov, and J. Kourtev, *Contrib. Plasma Phys.* **28**(2), 157–167 (1988).
12. I. Petrov, V. Orlinov, I. Ivanov, J. Kourtev, and J. Jelev, *Thin Solid Films* **168**, 239–248 (1989).
13. P. J. Martin, *J. Vac. Sci. Technol. A* **5**(4), 2158–2161 (1987).
14. J. E. Greene, *J. Vac. Sci. Technol. A* **5**(4), 1947–1948 (1987).
15. M. E. Harper, J. J. Cuomo, and R. J. Gambino, in "Ion Bombardment Modification of Surface: Fundamentals and Applications" (O. Auciello and R. Kelly, Eds.), pp. 127–162. Elsevier, Amsterdam, 1984.
16. J. E. Greene, T. Motooka, J.-E. Sundgren, A. Rockett, S. Gorbatkin, D. Lubben, and S. A. Barnett, *J. Cryst. Growth* **79**, 19–32 (1986).
17. B. Window and N. Savvides, *J. Vac. Sci. Technol. A* **4**(2), 196–202 (1986).
18. J. A. Thornton, *J. Vac. Sci. Technol. A*, **4**(6), 3059–3065 (1986).
19. R. Messier, A. P. Giri, and R. A. Roy, *J. Vac. Sci. Technol. A* **2**(2), 500–503 (1984).
20. I. Petrov, L. Hultman, U. Helmersson, J.-E. Sundgren, and J. E. Greene, *Thin Solid Films* **169**, 299–314 (1989).
21. T. Hata, R. Noda, O. Morimoto, and T. Hada, *Appl. Phys. Lett.* **37**(7), 633–635 (1980).
22. M. Naoe, S. Yamanaka, and Y. Hoshi, *IEEE Trans. Magn.* **MAG-16**(5), 646–649 (1980).
23. S. Kadokura, T. Tomie, and M. Naoe, *IEEE Trans. Magn.* **MAG-17**(6), 3175–3177 (1981).
24. S. Kadokura and M. Naoe, *IEEE Trans. Magn.* **MAG-18**(6), 1113–1115 (1982).
25. Y. Hoshi, M. Kojima, M. Naoe, and S. Yamanaka, *IEEE Trans. Magn.* **MAG-18**(6), 1433–1435 (1982).
26. B. Window, F. Sharples, and N. Savvides, *J. Vac. Sci. Technol. A* **3**(6), 2368–2372 (1985).
27. S. L. Rohde, Ph. D. Dissertation, Northwestern University, 1991.
28. W. D. Sproul, P. J. Rudnik, M. E. Graham, and S. L. Rohde, *Surf. Coat. Technol.* **43/44**, 270–278 (1990).
29. M. S. Wong, W. D. Sproul, and S. L. Rohde, *Surf. Coat. Technol.* **49**, 121–126 (1991).
30. M. J. Murphy, D. C. Cameron, M. Z. Karim, and M. S. J. Hashim, *Surf. Coat. Technol.* **57**(1), 1–5 (1993).
31. J. Musil, S. Kadlec, J. Vyskoĉil, and V. Poulek, "Reactive deposition of hard coatings," presented at the 16th International Conference on Metallurgical Coatings, San Diego, April 17–21, 1989.
32. S. Kadlec, J. Musil, W.-D. Münz, G. Håkanson, and J.-E. Sundgren, *Surf. Coat. Technol.* **39/40**, 487–549 (1989).

33. J. Musil and S. Kadlec, "Reactive sputtering of TiN films at large substrate to target distances," presented at the 7th International Conference of Ion and Plasma Assisted Techniques, Geneva, Switzerland, May 31–June 2, 1989.
34. S. Kadlec, J. Musil, V. Valvoda, W.-D. Münz, H. Petersein, and J. Schroeder, *Vacuum* **41**(7–9), 2233–2238 (1990).
35. S. L. Rohde, W. D. Sproul, and J. R. Rohde, *J. Vac. Sci. Technol. A* **9**(3), 1178–1183 (1991).
36. N. Hershkowitz, in "Plasma Diagnostics," pp. 113–183. Academic Press, New York, 1989.
37. J. A. Thornton, *J. Vac. Sci. Technol. A* **15**(2), 188–191 (1978).
38. H. V. Boenig, "Plasma Science and Technology." Cornell University Press, Ithaca, New York, 1982.
39. S. L. Rohde, Y. K. Kim, and R. J. De Angelis, *J. Electronic Materials* **22**(11), 1327–1330 (1993).
40. L. Hultman, W.-D. Münz, J. Musil, S. Kadlec, I. Petrov, and J. E. Greene, *J. Vac. Sci. Technol. A* **9**(3), 434–438 (1991).
41. W. D. Westwood, *J. Vac. Sci. Technol.* **15**(1), 1–9 (1978).
42. A. M. Meyers, D. N. Ruzic, R. C. Powell, N. Maley, D. W. Pratt, J. E. Greene, and J. R. Abelson, *J. Vac. Sci. Technol.* **8**(3), 1668–1672 (1990).
43. B. Window and G. L. Harding, *J. Vac. Sci. Technol. A* **8**(3), 1277–1282 (1990).
44. N. Savvides and B. Window, *J. Vac. Sci. Technol. A* **4**(3), 504–508 (1986).
45. B. Windows, F. Sharples, and N. Savvides, *J. Vac. Sci. Technol. A* **6**(4), 2333–2340 (1988).
46. B. Window, *J. Vac. Sci. Technol. A* **7**(5), 3036–3042 (1989).
47. D. Monaghan and R. D. Arnell, *Surf. Coat. Technol.* **49**, 298–302 (1991).
48. D. Monaghan and R. D. Arnell, *Vacuum* **43**(1–2), 77–81 (1992).
49. N. Savvides and B. Window, *J. Appl. Phys.* **63**(1), 225–234 (1988).
50. A. G. Spencer, K. Oka, and R. P. Howson, *Vacuum* **38**(8–10), 857–859 (1988).
51. R. P. Howson, A. G. Spencer, K. Oka, and R. W. Lewin, *J. Vac. Sci. Technol. A* **7**(3), 1230–1234 (1989).
52. R. D. Arnell and R. I. Bates, *Vacuum* **43**(1–2), 105–109 (1992).
53. R. D. Arnell and R. I. Bates, *Le Vide, les Couches Minces—Supplément*, all No. 261—Mars–Avril, 309–311, (1992).
54. D. P. Monaghan, D. G. Teer, P. A. Logan, K. C. Laing, R. I. Bates, and R. D. Arnell, "An improved method for deposition of corrosion resistant aluminum coatings for aerospace applications," to be published in *Surf. Coat. Technol.*
55. N. Savvides, *J. Appl. Phys.* **58**(1), 518–521 (1985).
56. D. P. Monaghan, D. G. Teer, P. A. Logan, I. Efeoglu, and R. D. Arnell, "Deposition of wear resistant coatings based on diamond-like carbon by unbalanced magnetron sputtering," to be published in *Surf. Coat. Technol.*
57. S. Kadlec, J. Musil, W. D. Münz, and V. Valvoda, "TiN films deposited by unbalanced magnetron," presented at the 7th International Conference on Ion and Plasma Assisted Techniques, Geneva, Switzerland, May 31–June 2, 1989.
58. J. Musil, V. Poulek, S. Kadlec, J. Vyskocil, V. Valvoda, R. Cerný, and R. Kuzel, Jr., *Nucl. Instrum. Methods* **B37/38,** 879–901 (1989).
59. I. Efeoglu, R. D. Arnell, S. F. Tiston, and D. G. Teer, *Surf. Coat. Technol.* **57**(2–3), 117–121 (1993).
60. O. Knotek, W.-D. Münz, and T. Leyendecker, *J. Vac. Sci. Technol. A* **5**(4), 2173–2179 (1987).

61. D. McIntyre, J. E. Greene, G. Håkansson, J.-E. Sundgren, and W. D. Münz, *J. Appl. Phys.* **67**(3), 1542–1553 (1990).
62. I. Efeoglu, R. D. Arnell, S. F. Tinston, and D. G. Teer, *Surf. Coat. Technol.* **57**, 117–121 (1993).
63. D. P. Monaghan, D. G. Teer, K. C. Laing, I. Efeoglu, and R. D. Arnell, *Surf. Coat. Technol.* **59**, 21–25 (1993).
64. U. Helmersson, S. Todorova, S. A. Barnett, J.E. Sundgren, L. C. Markert, and J. E. Greene, *J. Appl. Phys.* **62**(2), 481–484 (1987).
65. P. B. Mirkarimi, L. Hultman, and S. A. Barnett, *Appl. Phys. Lett.* **57**(25), 2654–2656 (1990).
66. M. Shinn, L. Hultman, and S. A. Barnett, *J. Mater. Res.* **7**(4), 901–911 (1992).
67. X. Chu, M. S. Wong, W. D. Sproul, S. L. Rohde, and S. A. Barnett, *J. Vac. Sci. Technol. A* **10**(4), 1604–1609 (1992).
68. W. D. Sproul, *Surf. Coat. Technol.* **33**, 73–81 (1987).
69. W.-D. Münz, F. J. M. Hauzer, D. Schulze, and B. Buil, *Surf. Coat. Technol.* **49**, 161–167 (1991).
70. W.-D. Münz, D. Schulze, and F. J. M. Hauzer, *Surf. Coat. Technol.* **50**, 169–178 (1992).
71. D. Hofmann, P. Balhause, A. Feuerstein, and J. Snyder, "Sputter ion plating with plasma boosters: a breakthrough in hard coating technology," Society of Vacuum Coaters, 35th Annual Technical Conference Proceedings, 218–226 (1992).
72. A. Aubert, R. Gillet, and J. P. Terrat, *Thin Solid Films* **108**, 165–172 (1983).
73. H. Benien, J. Maushart, M. Meyer, and R. Suchentrunk, *Materials Science and Engineering* **A139**, 126–131 (1991).
74. H. Schulz and E. Bergmann, *Surf. Coat. Technol.* **50**, 53–56 (1991).
75. H. Holleck and H. Schulz, *Surf. Coat. Technol.* **36**, 707–714 (1988).
76. O. Knotek, F. Löffler, and G. Krämer, *Surf. Coat. Technol.* **54/55**, 241–248 (1992).
77. L. Hultman, W. D. Münz, J. Musil, S. Kadlec, I. Petrov, and J. E. Greene, *J. Vac. Sci. Technol.* **9**(3), 434–438 (1991).
78. D. Orlicki, V. Hlavacek, and H. J. Viljoen, *J. Mater. Res.* **7**(8), 2160–2181 (1992).
79. S. Berg, H.-O. Blom, T. Larsson, and C. Nender, *J. Vac. Sci. Technol. A* **5**(2), 202–207 (1987).
80. S. Berg, H.-O. Blom, M. Moradi, and C. Nender, *J. Vac. Sci. Technol. A* **7**(3), 1225–1229 (1989).
81. C. D. Tsiogas and J. N. Avaritsiotis, *Vacuum* **43**(2), 203–211 (1992).
82. S. K. Dew, T. Smy, R. N. Tait, and M. J. Brett, *J. Vac. Sci. Technol. A* **9**(3), 519–523 (1991).

The Formation of Particles in Thin-Film Processing Plasmas

CHRISTOPH STEINBRÜCHEL

Department of Materials Engineering and Center for Integrated Electronics, Rensselaer Polytechnic Institute, Troy, New York

I.	Introduction	289
II.	General Phenomena	291
III.	Particles in Deposition Plasmas	294
IV.	Particles in Sputtering Plasmas	301
V.	Particles in Reactive Ion Etching Plasmas	309
VI.	Modeling of Particles in Plasmas	312
VII.	Particle Contamination and Equipment Design	314
VIII.	Conclusions	314
	Acknowledgment	317
	References	317

I. Introduction

One of the main issues in present-day integrated circuit manufacturing is the yield obtained with a certain technology or sequence of processes. Given different technologies that produce circuits of comparable performance, the yields achieved will determine to a large extent whether the circuits can be sold at a reasonable profit. Thus, maximixing the yield of every process step is an overriding concern for any manufacturing technology.

For a particular processing step, such as deposition or patterning of a certain layer, a reduction in the yield may occur in one of two ways. Either control of the processing conditions themselves may have been insufficient, so that at the end of the step the wafer is out of specification, or in the course of the handling and processing of the wafer, an unacceptable level of contamination has been introduced.

Moreover, sources of contamination in wafer processing are manifold. First, the wafer may acquire chemical impurities from impure process chemicals or a contaminated process reactor. Furthermore, particles may be deposited on the wafer from the process ambient in the reactor or from the environment (and the people) that the wafer is exposed to before and after the process. The magnitude of the problem can be appreciated by realizing that for a process operating at 94% yield in DRAM manufacturing, a reduction in yield of 1% may lead to a loss in profits of as much as 2% (1).

With modern cleanroom operations, defects induced by the people and the environment can be controlled fairly well now, even though further improvements in that area continue to be pursued. On the other hand, the realization has been growing that the control of process-induced defects may become just as important. Such defects, especially in the form of particles deposited onto the wafer, may not only be due to particle release from the reactor walls during the process or to bringing the wafer into and out of the reactor, but they may in fact be a more or less unavoidable by-product of the process itself even in a clean reactor.

This paper will present a review of recent work related to the process-inherent formation of particle contamination in wafer processing, specifically as it arises in plasma-based processes such as plasma-enhanced chemical vapor deposition (PECVD) (2–14), sputtering (15–26), and reactive ion etching (RIE) (15,16,20–22,27–30), and including related theoretical modeling work (31–38). "Process-inherent" here means that particle formation occurs in the plasma itself, for example because of some undesired chemical reactions, and is not primarily a function of the cleanliness of the reactor. The phenomenon has also been referred to as dust or powder in a plasma. For the most part, we will be interested in work involving the observation of particles *in-situ* in the discharge, or collected on the substrate, and we will refer to work detecting particles downstream, in the exhaust, only to the extent that it is related to *in-situ* observations. (*In-situ* observations have been made largely on laboratory tools up to now, whereas for downstream detection of particles, commercial instruments are available (44)). Moreover, the paper will not deal with such issues as particle deposition from chamber walls or fixtures during venting of the reactor chamber (44,45), particle release directly from a sputtering target (46), or particle formation by condensation during a rapid pump-down (47,48), even though these issues are very important for the overall control of particle contamination on wafers.

It should also be noted that particles in plasmas play an important role in other areas of science such as in the study of interplanetary plasmas or in contamination control in plasma fusion devices (49). Furthermore, discharges have been used specifically for the purpose of producing particles in macroscopic quantities, which then served as a starting point for making powder-based materials *(50,51)*.

The organization of the present paper is as follows: We will begin by describing some general phenomena observed in connection with particle formation in thin-film processing plasmas. We will then discuss in turn particles in PECVD plasmas *(2–14)*, in sputtering plasmas *(15–26)*, and in RIE plasmas *(15,16,20–22,27–30)*. This will be followed by an overview of the theoretical modeling of various aspects of particles in processing plasmas *(31–38)*. We will then examine some issues of equipment design affecting particle formation *(39–43)*, and we will conclude with some remarks about the implications of this work for the control of process-induced particle contamination.

II. General Phenomena

In thin-film processing, plasma-generated particles were first observed *in-situ* in a silane–argon deposition plasma *(3)*, even though analogous observations had been made earlier in thermal CVD silicon deposition *(52)* (where reference was made to "smoke" in the gas phase). Since these early experiments, particles have also been seen in several other silane-based plasmas for the deposition of silicon *(4–13)* and silicon nitride *(14)*. In addition, particles were observed in the thermal CVD of aluminum *(53)*.

Perhaps one should not consider it too surprising that particles can be produced in a deposition system whose purpose is, after all, to form solid material from gaseous species, although preferably on the substrate rather than in the gas phase. However, it has become apparent that particles are also produced in etching plasmas, specifically in sputtering *(15–26)* and in RIE *(15,16,20–22,27–30)* plasmas, from a variety of substrates. In these latter systems, the species that end up forming particles are introduced into the gas phase via removal from the substrate.

The experimental technique most widely used to detect particles in these plasmas has been laser light scattering (LLS). LLS takes advantage of the fact that the cross-section for light scattering by a particle is a strong function of the particle radius *(54)*. The laser beam traverses the discharge parallel to the electrodes at a distance that is often adjustable. For typical laboratory lasers (e.g., He–Ne) and typical particle concentrations arising, it turns out that, as a rule of thumb, light scattering becomes readily visible when the particle diameter is of the order of one-fourth to one-half of the laser light wavelength. Of course, the minimum size detectable also depends on the laser power, on the method of light detection, and on the experimental detection geometry. In principle, the sensitivity in LLS is highest for scattering in the forward direction (54), but in practice the direction of light detection usually does not make a significant difference. One should also keep in mind that the detection limit in LLS from particles is a function of both particle size and concentration (54), and it is generally less than straightforward to separate these two effects from each other.

In many cases, particles are observed primarily in a quite localized region of the discharge, near the sheath edge at the powered electrode (or cathode) of a parallel-plate-type reactor. Two examples of this phenomenon are given in Figs. 1 and 2. (The surfaces of the two electrodes are usually oriented in the horizontal direction. We will assume this to be the case unless indicated otherwise. But note that the data for Fig. 1 represent an exception. They were taken in a reactor with vertical electrodes (4).) This formation of a cloud of particles immediately indicates that the particles are negatively charged and held suspended in

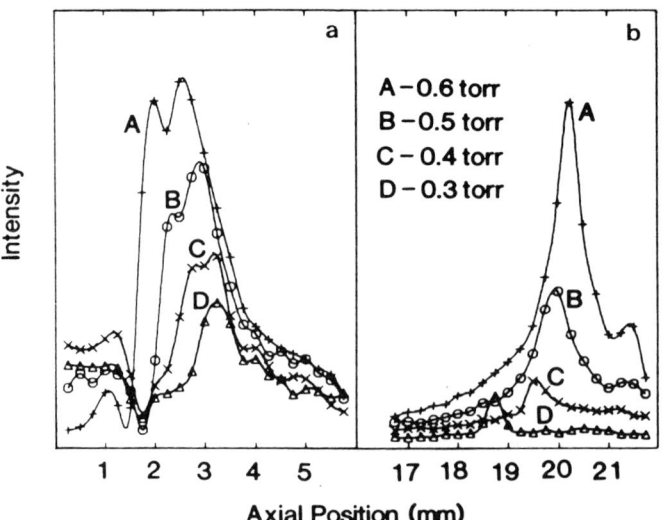

FIG. 1. LLS of particles in a silane/argon plasma. Parallel plate reactor with grounded electrode at 0 mm and rf electrode at 22 mm. The data in (a) and (b) were obtained in separate experiments and have different intensity scales. The peak signal at the cathode (b) is about 3 × the signal at the anode (a). (From Roth et al., Ref. 3.)

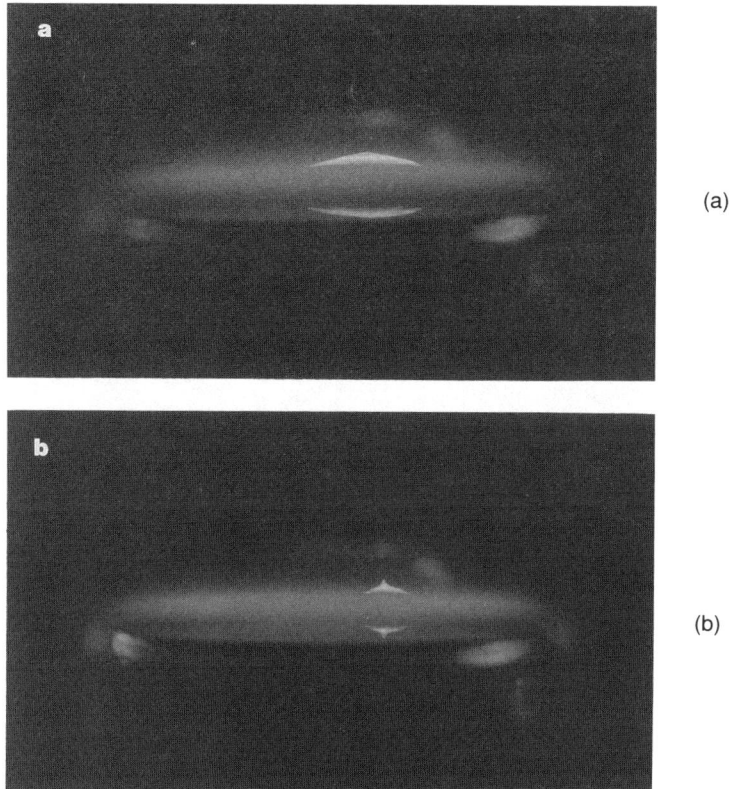

FIG. 2. Photographs of LLS of particle clouds in Ar sputtering of Si. Rf power = 200 W; flow rate = 200 sccm. (a) $p = 120$ mTorr; (b) $p = 183$ mTorr. Note that the cloud and its reflection from the wafer are visible. (From Yoo and Steinbrüchel, Ref. 29.)

the direction perpendicular to the electrodes by a balance of forces, including electrical repulsion from the negatively biased electrode, the gravitational force, and other forces due to collisions of plasma species (ions, neutrals) with the particles (see below). When the discharge is turned off, the particles fall down onto the substrate, or else they may be swept out into the exhaust by the gas flow.

The mechanism giving rise to the negative charge on the particles is basically the same as the one charging an electrically isolated surface in contact with the plasma to the floating potential. The steady state charge on a particle is determined by the condition that the fluxes of electrons and ions to the particle be equal. However, the potential actually acquired by a particle depends on its size and is only equal to the floating potential for a very large particle (*31–36*).

It is not clear yet at present what are the causes for confinement of a particle cloud in the horizontal direction (i.e., parallel to the electrode surface). The horizontal position of a particle cloud depends on such parameters as the reactor geometry, the gas pressure, the flow rate, and the flow pattern across the electrode (*22*). Also, it has been shown that the plasma potential may have a maximum near the sheath edge in the horizontal and vertical direction (*19*). In addition, geometrically induced discontinuities in the electrical potential of the discharge electrode may assist in particle confinement (*17*).

The volume of a particle cloud is often observed to increase with time (*6,22,27*). On the other hand, it should be noted that sharply localized confinement of particles into a cloud does not always occur (*cf.* Figs. 1 and 2). In some deposition systems, particles may occupy most of the discharge volume at a fairly uniform concentration (*10,13,14*).

Whereas particle charging and confinement have been described experimentally and modeled theoretically in some detail for a number of situations, particle growth has been less well characterized to date (*11,12*), and very little is known about the initial step of particle nucleation and the species responsible for it. In deposition plasmas, particle nucleation has been attributed either to certain radicals produced in the plasma from the parent gas (*6–8*) or to nuclei desorbed from the substrate (*14*). In sputtering and RIE plasmas, it is only clear that nucleation must be due to some yet-to-be-identified etch products removed from the substrate (*22,27–29*).

III. Particles in Deposition Plasmas

The earliest observations of particles in processing plasmas were made in deposition chemistries, specifically in silane-based plasmas. The problem with particles had manifested itself, on the one hand, by excessive light scattering in experiments investigating laser-induced fluorescence (LIF) of plasma species (*2*) and, on the other hand, by poor film quality and contamination problems in the deposition of amorphous silicon films (*6*).

The first detailed studies on deposition plasmas came from the group of Spears et al. (2–5). They used Ar plasmas containing silane in the percent range, at pressures of a few hundred mTorr. Their reactor was of the parallel-plate type (38 mm electrode diameter, 22 mm electrode separation), rf-excited at 12 MHz, electrically asymmetric, and with vertical electrodes. The gas flow was from top to bottom, i.e., parallel to the electrodes, at a rate of 20–50 sccm. The entire chamber could be translated horizontally so that the spatial dependence of LLS could be examined.

Spears et al. found that LLS vs. position in the axial direction had maxima near the sheath edges, with the one at the cathode being about three times as large as the one at the grounded electrode (or anode) (Fig. 2). LLS increased with decreasing flow rate and, somewhat surprisingly,

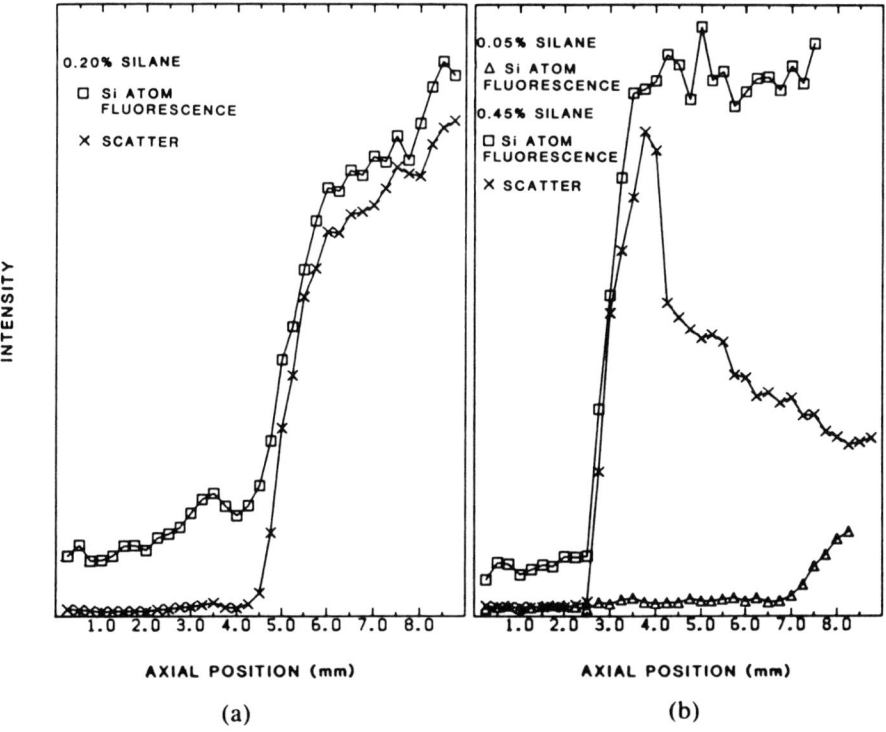

FIG. 3. LLS of particles and LIF of Si atoms for 251.4 nm excitation in a silane/argon plasma. Grounded electrode at 0 mm; $p = 0.5$ Torr; flow rate = 50 sccm. LLS for 0.05% silane was negligible and is not shown in (b). (From Spears et al., Ref. 4, ©1986 IEEE.)

also with decreasing SiH_4 content (above 0.2%) (*3,4*). In addition, LIF of Si atoms showed essentially the same spatial and gas composition dependence (Fig. 3). Optical emission peaked near the cathode, too, but was spatially less well resolved (*2*). At the lowest SiH_4 concentration, i.e., at 0.05%, LIF of Si atoms could still be detected, but there was no LLS (*4*).

These results indicated that there must be a close connection between the presence of Si atoms in the plasma and the formation of particles. This connection was emphasized by further experiments involving additional laser excitation (*4*). Laser excitation at 266 nm produced extra LIF signal (i.e., extra Si atoms), but LLS was unaffected; excitation at 532 nm had no effect on either LIF or LLS; excitation at 354 nm affected LLS but not LIF. Thus, depending on the photon energy, laser excitation of particles can either change their size distribution and/or concentration, or it can release Si atoms from particles already formed.

Spears *et al.* also attempted to gain information on the particle concentration and size distribution from the statistics of the scattered laser light pulses (*5*). They argued, on the basis of a theoretical model, that size distributions are quite narrow, that the average particle size depends on the position in the discharge, and that at a concentration of 10^8cm^{-3} the average particle size should be at least 16 nm. However, it should be noted that even for 363 particles in the scattering volume (which was considered a large number), the observed signal was only 31 photons. Thus, these photon statistics may be somewhat uncertain, particularly for lower particle concentrations.

In the work of Spears *et al.*, no mention is made of the time dependence of the phenomena observed. Presumably, their observations refer to some sort of steady-state condition. Various aspects of the time dependence have been considered by a number of other authors. Watanabe *et al.* (*6-8*) studied the effects of modulating the discharge on particle formation. Their reactor was also of the parallel-plate type but with horizontal electrodes (10 cm diameter, 2.7 cm separation). The upper (*6,8*) or lower electrode (*7*) was powered, and gas flow was either parallel (*6,8*) or perpendicular to the electrodes (*7*). The gas used was a SiH_4/He mixture at a pressure of around 500 mTorr (70 Pa). The rf excitation was at 6 MHz, modulated at 40 Hz to 1 KHz with a variable duty cycle.

Watanabe *et al.* found that in a 0.5% SiH_4/He plasma, LLS from particles was reduced by about a factor of 10 if the duty cycle was reduced to 80% or less (*6*). Furthermore, the electron density increased (having a weak maximum at a 60% duty cycle), the deposition rate normalized relative to the on-time increased, and the film quality improved. (Similar observations were made later by Verdeyen et al. (9).) In a subsequent study, Watanabe et al. showed that with a modulated discharge one can go to much higher rf powers, and thus obtain much higher deposition rates, without forming particles (7). They attributed this to a significant difference of the average lifetimes of different plasma species. SiH_3, as a precursor to film growth, was said to have a lifetime of ~ 1 ms and be lost mostly by diffusion, whereas SiH and SiH_2, as precursors to particle formation, were said to have lifetimes of ~ 0.1 ms and be lost by reaction. The effect of modulating the discharge was therefore to allow SiH and SiH_2 to react during the off-times, thus reducing greatly their concentrations, whereas SiH_3 remained largely unaffected.

Quite recently, Watanabe et al. reported experiments in which they followed LLS of horizontally and perpendicularly polarized light at 90 degrees with respect to the laser beam (with the two detectors being opposite each other) (8). A particle cloud was seen to develop first at the cathode after ~ 1 s and then expand towards the anode. After 2.5 s there was appreciable LLS near the anode as well, and particles began to move downstream, parallel to the electrodes, roughly at the flow velocity of ~ 13 cm/s (Fig. 4). Watanabe et al. concluded that the peak particle concentration near the cathode, after 1 s, was about 10^9 cm^{-3} and the average particle diameter was 60 nm. At later times, the particles grew larger and the particle cloud became less dense.

Additional time-dependent phenomena were explored by Bouchoule et al. (10). Their reactor was also of the parallel-plate type with horizontal electrodes, the upper electrode being powered. However, their electrodes were in the form of grids, and gas entered through the top electrode and flowed vertically downward. The entire electrode assembly was enclosed in an oven inside the vacuum chamber. Bouchoule et al. used a gas mixture of a few percent SiH_4 in Ar at a pressure around 100 mTorr. In addition, they collected particles on transmission electron microscopy (TEM) grids below the anode for further analysis.

Bouchoule et al. pointed out that particles appeared in LLS after a short time of much less than a minute (10). As time increased, the intensity of LLS showed periodic oscillations. The scattered light intensity was quite evenly distributed over almost the entire discharge volume, i.e., from about 5 mm below the cathode to about 5 mm above the anode. If the applied rf power was pulsed (~ 1 s on, 1 to 10 ms off), then for sufficiently long off times, either particles would not form, or if there were particles initially, they would disappear, being carried down-

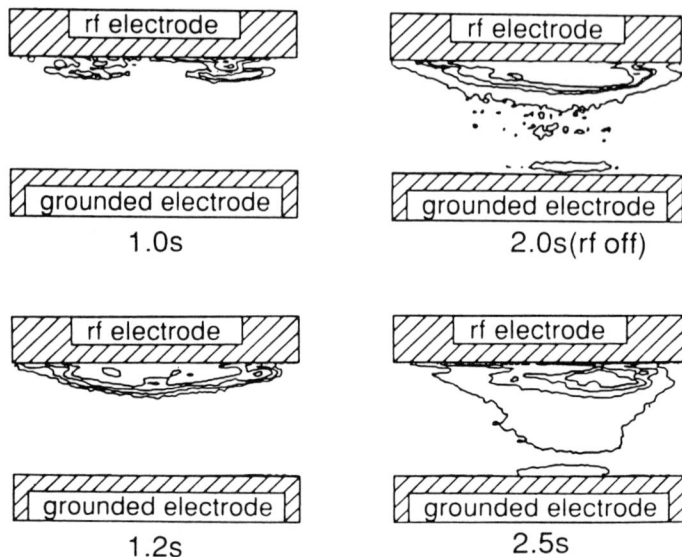

FIG. 4. LLS of development of particle cloud vs. time. 5% silane in He. $p = 80$ Pa (0.56 Torr); flow rate = 30 sccm. Intensity scale for contours is different in each figure. (From Watanabe et al., Ref. 8.)

ward with the gas at approximately the gas flow velocity. (These latter two observations are in agreement with those of Watanabe et al. (7,8).)

In the same work, effects of temperature and flow rate on particle formation were also noted (10). LLS from particles appeared sooner at higher flow rates. (By contrast, the results of Spears et al. (3,4) showed that LLS (in steady state) was higher at a lower flow rate.) Moreover, in the temperature range between 27°C and 150°C, a limiting flow rate was observed below which no particles appeared. This limiting flow rate increased with increasing temperature.

In a subsequent paper, Bouchoule et al. described the particle growth kinetics in more detail (11). Particles were collected and analyzed by TEM, after pulsing the discharge with on-times between 0.5 and 300 s. The gas was 4% SiH_4 in Ar at 117 mTorr and 30°C. During the first 5 s, the particle diameter increased linearly with time at a rate of ~ 10 nm/s (Fig. 5), and the size distribution was quite monodisperse. At $t \sim 10$ s, the growth rate decreased to ~ 2 nm/s. After ~ 30 s, the size distribution became much wider. These results were interpreted as evidence for very rapid initial nucleation, a short period of rapid growth accompanied by inhibited nucleation, and after ~ 30 s resumption of nucleation.

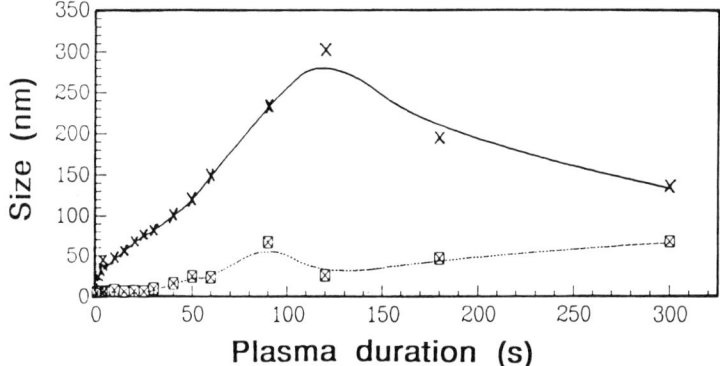

FIG. 5. Evolution of particle size distribution with time. 4% silane in Ar. $p = 117$ mTorr; total flow rate = 31.2 sccm. ($\times$) represents the mean value of the particle diameter and ($\square$) its standard deviation. (From Boufendi et al., Ref. 11.)

On the basis of the known size distribution (i.e., assuming the distribution of the collected particles was the same as the one for particles in the discharge), Bouchoule et al. estimated a fairly constant particle concentration of $\sim 10^7$ cm^{-3} up to ~ 50 s, after which time the concentration decreased rapidly (11). Bouchoule et al. did not present any electron microscope pictures of particles, but described them as multifaceted up to ~ 10 nm and roughly spherical above ~ 50 nm.

Further aspects of particle formation in a silane plasma were revealed in the work of Dorier et al. (12,13), especially as regards rf power, excitation frequency, and temperature effects. Their reactor was also a parallel-plate type, with the lower electrode rf-excited and the upper electrode at ground and heatable to 250°C (electrode diameter 130 mm, separation 20 mm). The reactor walls were kept at 100°C. The gas was pure SiH_4 at a flow rate of 30 sccm and a pressure of 0.3 mbar (230 mTorr), with the gas inlet being on the side of the reactor. The discharge volume was illuminated globally with white light, and light scattering was observed with a CCD camera.

Dorier et al. observed a threshold rf power, at 30 MHz, below which no particles appeared in light scattering. Above the threshold power, particles became visible near the cathode (the lower electrode). Even before the visual onset of powder in the plasma, a change in the electrical characteristics of the discharge was noticed in that the rf voltage at the cathode started to drop. Once a particle cloud was established, the discharge had reached a new steady state at a conductance that was higher than in the beginning. (For similar observations, see Ref. 10.) At this time, there was a continuous stream of particles radially outward from the cloud to the chamber walls. At a second, higher threshold rf power, particles also appeared near the anode. The threshold powers were higher either at higher anode temperature or at higher excitation frequency. If the rf power was raised further, there came a point where the amount of powder in the discharge was large enough so as to extinguish the plasma.

The distribution of scattered light intensity depended on the temperature difference between the two electrodes. An increase in the anode temperature led to a shift of the particle distribution towards the cathode (which was always kept at 15°C). Above an anode temperature of $\sim 120°C$, no particles were observed near the anode. (This effect of thermophoresis on particles has also been observed in a sputtering plasma (26).)

Finally we wish to mention the work of Anderson *et al.* on particle formation in a SiH_4/NH_3 rf discharge (14). The interest in this work here stems from suggestions made by the authors regarding the mechanisms of particle nucleation and growth, even though the emphasis of the work was more on making powder than on preventing particle contamination in thin-film deposition. The reactor geometry was different from that in all the other work discussed, in that it involved a cylindrical mesh cathode with the gas being fed in radially through pinholes in the axial anode. The chamber exhaust to the pump was below the electrode assembly. The gas pressure was 0.2–0.8 Torr, the total gas flow was 45–100 sccm, and the flow rate ratio of SiH_4/NH_3 was 0.06 to 0.5. Particles were collected in a filter installed in the chamber exhaust, usually over a period of ~ 2 hours. The collected powder was then dispersed in acetone and deposited onto grids for TEM analysis.

Anderson *et al.* found that particle size and composition depended on the input gas. For $SiH_4/NH_3 < 0.15$, particles were around 0.1–0.2 um in diameter and had a Si/N ratio of 0.5, corresponding to a $[Si(NH_2)_2]_n$ composition. For $SiH_4/NH_3 > 0.15$, particles were much smaller, ~ 30 nm, and more Si-rich. LLS showed particles everywhere in the discharge except in the cathode sheath.

The authors also noted that a fine, irregular film was deposited on the cathode during the course of an experiment. Furthermore, particle generation was somewhat larger if the cathode was dc coupled rather than capacitively coupled. Since in the dc coupled case the electron current to the cathode is larger, Anderson *et al.* concluded that nucleation is heterogeneous, with cluster nuclei being desorbed by electron excitation from the cathode. Subsequent growth was said to involve sequential reaction of positively charged species $[Si(NH_2)_2]_n$ with SiH_4 and NH_3 to form $[Si(NH_2)_2]_{n+1}$.

IV. Particles in Sputtering Plasmas

Particle formation turns out not be restricted only to systems exhibiting a rather complex chemistry such as deposition plasmas. It also occurs quite readily in systems chemically as simple as the sputtering of an elemental target in a noble gas discharge. Particles have been observed in Ar plasmas sputtering Si (15–17,19–22), SiO_2 (15,18,22), graphite (15,23), Lexan and Teflon (15), as well as Al and Cu (24–26). Additionally, particles appeared in N_2 and O_2 plasmas sputtering Si (15). Such particles consisted primarily of the target material (16,17), so that it is clear that they must have nucleated and grown in the gas phase from target atoms removed by the sputtering process.

This aspect of material removal from the substrate presents a new complication in sputtering and RIE. Since all surfaces subject to ion bombardment may release atoms into the plasma, one has to make a trade-off between two conflicting requirements. If the substrate is the same size as the rf-driven electrode, then only substrate material will be etched away but the etching may be rather nonuniform across the substrate (21,22). If, on the other hand, the substrate is placed on a larger electrode of a different material, then the etch rate may be more uniform but different types of atoms will be released into the plasma and may contribute to particle nucleation and growth (17,19).

The first observations of particles in an etch plasma were reported by Selwyn et al. for the RIE of Si (27). Subsequent work has focused more on sputtering plasmas. In the author's own laboratory, it was established that particle formation was not unique to a particular plasma or a specific substrate material. For example, Durham and Petrucci observed particles in the sputtering of Si in Ar, O_2, and N_2, SiO_2 in Ar, graphite in Ar, and Lexan and Teflon in Ar (15,16). This work was done in a parallel plate rf reactor (5-in. electrode diameter, 4-in. separation), with the substrate wafer having the same size as the electrodes. A guard ring around the cathode served to confine the discharge. Gas flow was mostly parallel to the electrodes.

LLS usually showed a well-localized particle cloud at the sheath edge near the center of the wafer, except in the case of graphite + Ar, where the cloud was more extended and appeared to consist of especially fine particles (15). A cloud typically became visible after 1–2 minutes at 200 mTorr and 200 W (after ~ 15 minutes for Si with O_2 or N_2). The appearance of a cloud was faster at higher rf power and at higher pressure. The lateral position of the cloud was a function of the pressure. The cloud could be made to move towards the exhaust and even off the wafer by reducing the pressure to the 10 mTorr range. When the rf power was turned off, the particles in the cloud settled vertically down onto the wafer. From a particle count on the wafer and an estimate of the cloud volume, it was concluded that the particle concentration in the cloud was 10^6-10^7cm^{-3}.

Electrical characteristics of the discharge also had a marked effect on the appearance of a particle cloud (16). With all other plasma parameters kept constant, a lower rf excitation frequency (2.4 vs. 13.56 MHz) resulted in less efficient particle generation and a much smaller cloud volume. In addition, detuning the discharge changed the shape of the cloud, even to the point where particles were expelled from the cloud onto the wafer or into the exhaust. Furthermore, although particles had to be formed from sputtered material, no one-to-one correlation could be found between the etch rate and the appearance time of a particle cloud.

These initial results were followed by more detailed studies for the sputtering of Si and SiO_2 in Ar (20–22). Yoo observed threshold behavior vs. both rf power and pressure for the appearance of a particle cloud, in that no particle cloud formed even after a very long time if the rf power or the pressure were below certain limiting values (Fig. 6) (21). Above the threshold, cloud appearance was generally faster for Si than for SiO_2 (22). The threshold conditions roughly coincided with a significant reduction of the etch rate and, in the case of Si, the onset of redeposition in the center of the wafer where the cloud appeared. (But note that the etch rate along the wafer edge was always substantially positive.) The cloud also impeded etching right underneath it. The size and location of the cloud depended on pressure and flow rate (22). At a certain low pressure or low flow rate, the cloud would move off the wafer towards the exhaust. (This implies that formation of a stable particle cloud over the wafer is not just a function of the residence time.)

This latter effect could be used to move particles directly onto a TEM grid placed downstream from the wafer. (Particles deposited on the wafer after the discharge was shut off could also be analyzed by scanning electron microscopy, SEM.) At the onset of cloud appearance, particles were typically ~ 200 nm in diameter and quite monodisperse. After

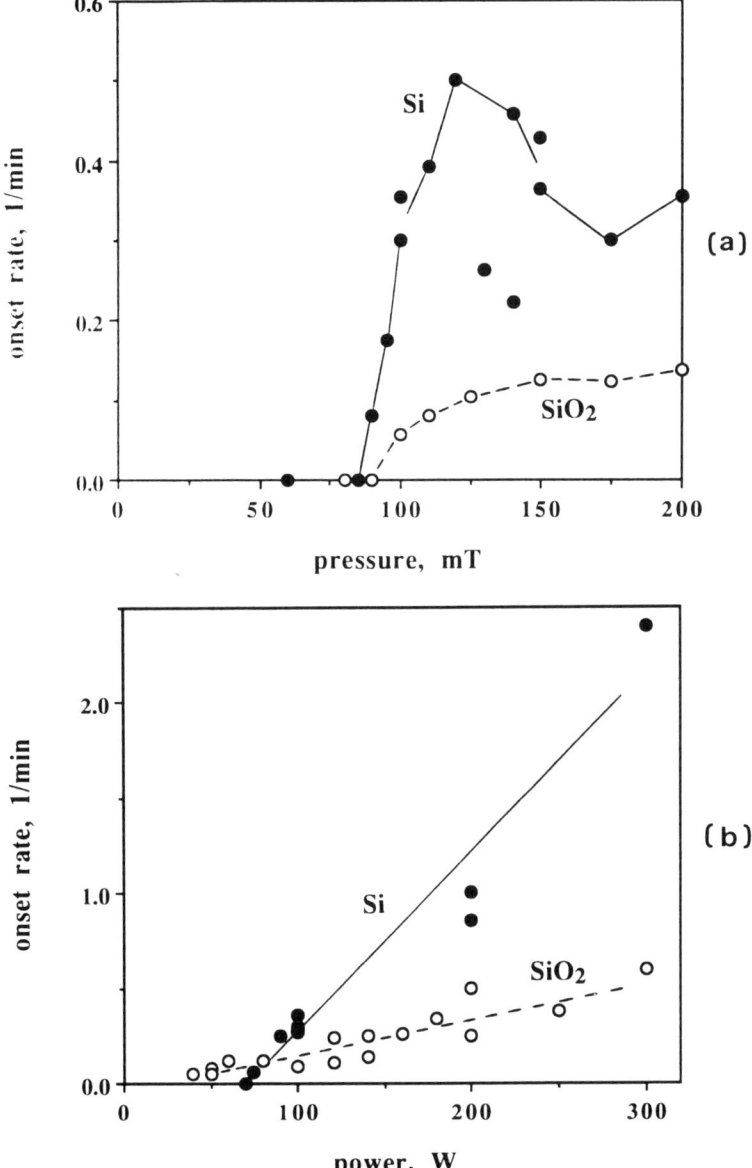

FIG. 6. (a) Onset rate (i.e., inverse of onset time in LLS) vs. pressure for sputtering of Si and SiO_2 in Ar at rf power of 200 W and flow rate of 200 sccm. (b) Onset rate vs. rf power for sputtering of Si and SiO_2 in Ar at pressure of 100 mTorr and flow rate of 200 sccm. (From Yoo and Steinbrüchel, Refs. 20–22.)

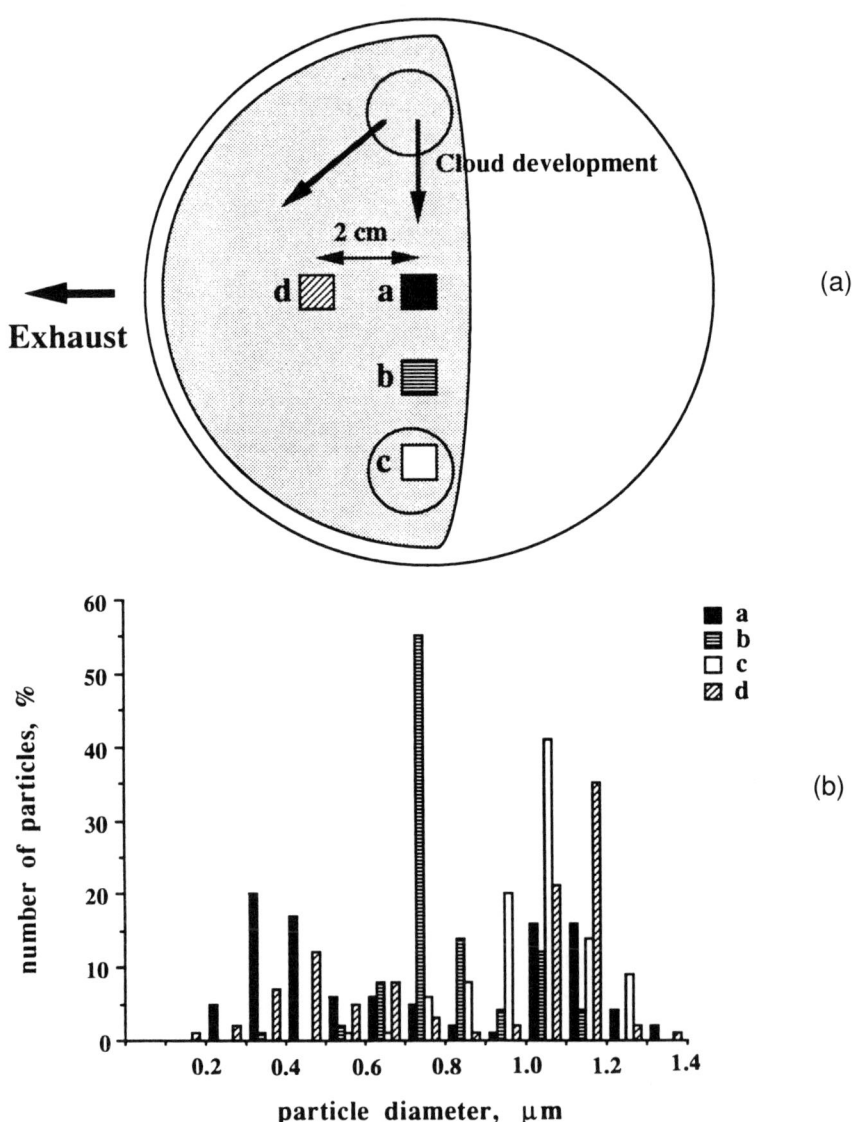

FIG. 7. (a) Top view of cloud development with time. Ar sputtering of Si: $p = 120$ mTorr, flow rate = 200 sccm, rf power = 200 W. The circles at location c indicate the cloud size at onset (after 2 min). The cloud then increases in size until it covers about half the wafer after 20 min (shaded area). Locations a–d are sampling spots for SEM analysis of particle size distributions. (b) Particle size distributions from the different sampling spots after 20 min. Note that c is the oldest part of the cloud, and a the youngest part. (From Yoo and Steinbrüchel, Ref. 22.)

further development of the cloud, which generally included a certain increase in volume, the average particle size had increased and the size distribution had become wider and dependent on the exact location within the cloud (Fig. 7) (22). The "oldest" part of the cloud showed the largest particles.

Fairly large particles deposited on the wafer had an "orange-peel" type surface and sometimes had broken up into cone-shaped pieces (Fig. 8) (22). The cause for this texture was revealed by TEM pictures of much smaller particles collected downstream. These had a porous, spherulitic structure (Fig. 9) reminiscent of columnar growth in thin films. TEM diffraction also showed them to be amorphous. Most importantly, such small particles could be collected downstream either before the onset time of LLS from a cloud or below the threshold conditions for cloud appearance (22).

Selwyn, who pioneered particle studies in etching plasmas, also

FIG. 8. SEM of particles collected on wafer after sputtering of Si in Ar. Note orange-peel-type surface texture of particles and cone-shaped columnar structure of particle fragments. (From Yoo and Steinbrüchel, Ref. 29.)

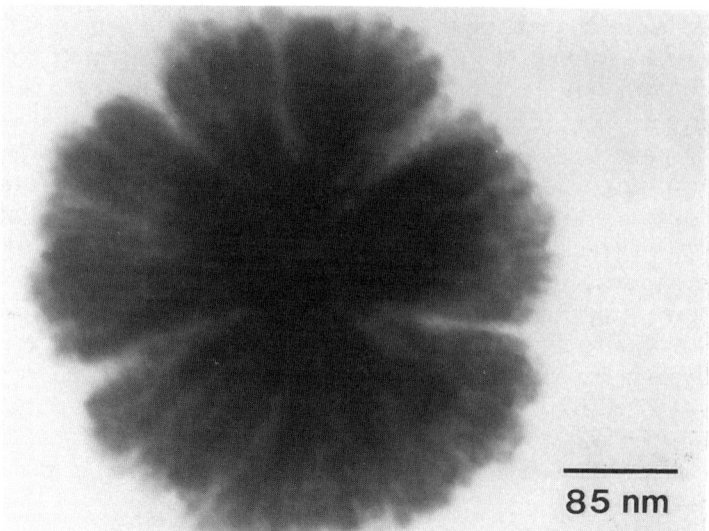

FIG. 9. TEM of particle collected downstream in Ar sputtering of Si before onset of LLS ($p = 110$ mTorr; rf power = 200 W; flow rate = 200 sccm; $t = 1.25$ min; TEM grid placed 2.5 cm from guard ring). (From Yoo and Steinbrüchel, Ref. 22.)

investigated particles in the sputtering of Si (*17*) and SiO_2 (*18*) in Ar. The reactors in these studies had rather large electrodes (diameter of 56 cm or 100 cm) and were pumped either through the center of the powered bottom electrode (*17*) or on the side (*18*). In the sputtering of Si, a few wafers were placed on the graphite-covered cathode (*17*), whereas in the sputtering of SiO_2, the entire cathode was covered with SiO_2 (*18*).

Using LLS, Selwyn showed that particles were not only trapped in a cloud at the sheath edge over the center of a wafer, but also in a ring over the edge of the wafer (Fig. 10) (*17*). Reducing the rf power reduced the "strength" of these particle traps, and at pressures below ~ 50 mTorr no central cloud could be observed. Particles from the sputtering of Si wafers on the graphite-covered cathode indeed contained Si as well as substantial amounts of C. Selwyn also pointed out that downstream monitoring of particles in the exhaust line to the vacuum pump may correlate poorly with the presence of particles in the plasma as long as the plasma particle traps are not full.

Graves *et al.* examined particle formation in the dc and rf sputtering of metals, in particular Al and Cu (*24–26*). Their reactor was a small

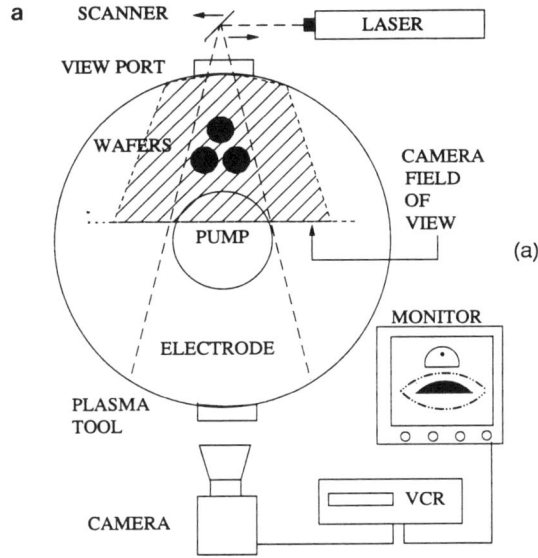

FIG. 10. (a) Top view of sputtering reactor of Selwyn et al. (Ref. 17). Dashed lines represent the typical amplitude of the laser scanner, and the shaded region designates the approximate field of view of the camera. The electrode is covered by a graphite plate. (b) Photograph of rastered LLS during sputtering in 200 mTorr of Ar. A particle dome above the center and a particle ring above the edge can be seen for each wafer. The laser scan plane cuts through the bottom of the dome.

parallel-plate type (electrode diameter 8.9 cm, separation 2.5 cm) with the top electrode powered. In the dc sputtering of Al, LIF of Al atoms exhibited a maximum at ~ 4 mm from the cathode and decreased linearly towards the anode. The peak concentration of Al atoms was as high as 4×10^{11} cm^{-3}. Evidence for particles was deduced from enhanced LIF near the anode, due to laser ablation of Al atoms from particles. LLS near the anode corresponded closely to the LIF profile (24). Furthermore, using vertically polarized laser light with two detection angles in LLS, Graves et al. determined that with particle sizes in the range of 100–200 nm, particles tended to be larger but had a lower concentration at the anode sheath (25). Under some conditions, non-spherical particles and rodlike agglomerates could be observed. In addition, optical emission and ion density were reduced in the presence of particles.

In subsequent experiments on the rf sputtering of Al and Cu, Graves et al. investigated the effect of varying electrode temperatures on particle formation (26). They demonstrated a thermophoresis effect on particles, in that particles tended to move towards the colder electrode, and they suggested that this effect might be useful for the control of particle contamination on a wafer. They also noted from optical emission spectroscopy (OES) and electrical measurements that when particles were present, the discharge had characteristics similar to the case of an electron-attaching gas.

Carlile et al. combined LLS and diagnostics with a tuned Langmuir probe on particle-forming rf sputtering plasmas (19). Their reactor was also a parallel-plate type (electrode diameter 10 cm, separation 8.5 cm), with a graphite-covered cathode on which a 5 cm Si wafer was placed. The plasma was further confined by permanent magnets along the sidewall on the outside of the reactor. Typical plasma conditions were a pressure of 15 mTorr of Ar, a gas flow of 20 sccm, and an rf power of 250 W. Particles readily appeared in a cloud at the sheath edge over the center of the wafer as well as in a ring over the edge of the wafer. These regions of particle trapping coincided with localized maxima of the plasma potential V_p of ~ 5 V above its surrounding value, as indicated by the Langmuir probe data. After a period of about an hour, the central maximum in V_p had become more extended, whereas the maximum along the wafer edge had disappeared. With a bare Al cathode, neither particles nor a maximum in V_p were detected.

In a very recent paper, the structure of particles obtained from low-frequency rf sputtering of graphite in Ar was described (23). These carbon grains were amorphous and had a spherulitic texture similar to the one noted above for Si particles. The size distribution of the carbon particles was said to be narrowly peaked.

V. Particles in Reactive Ion Etching Plasmas

The first observations of particles in an RIE plasma were reported by Selwyn et al. (27). These experiments were conducted in a parallel-plate reactor with two 80 mm Al electrodes separated by 52.5 mm. A 78 mm Si wafer was placed on the bottom cathode. Particles were detected by LLS and Cl atoms by two-photon LIF. Cl atoms could originate either from dissociation of the etch gas in the plasma or via photodetachment from Cl-containing negative ions. LLS and LIF could be done in a spatially resolved fashion by translating the entire reactor relative to the optical system.

Examining various gases at a pressure typically around 200 mTorr, Selwyn et al. found that a particle cloud appeared at the cathode sheath edge in a matter of minutes in CCl_2F_2/Ar (Fig. 11), $O_2/CCl_2F_2/Ar$, or

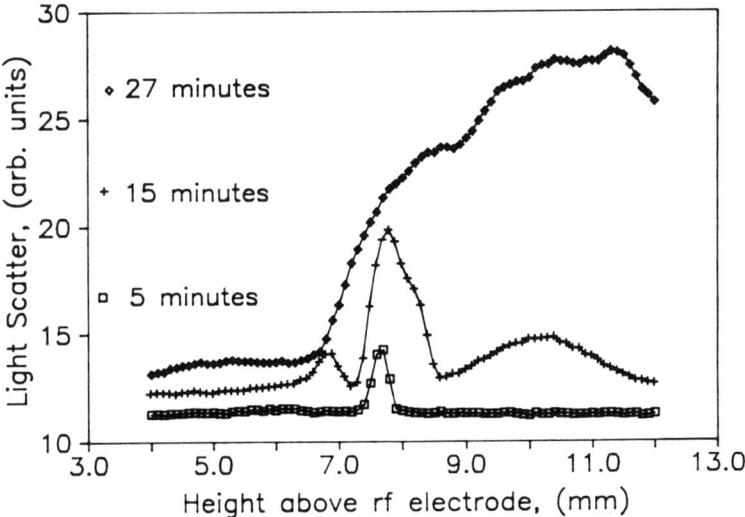

FIG. 11. LLS from particles observed in the cathode sheath region of a 10% CCl_2F_2/Ar plasma in the RIE of Si at 200 mTorr. The Si wafer is located on the cathode at 0 mm. Note the rapid increase of the LLS signal at the sheath edge, 7 mm above the wafer. The plots have been displaced vertically for clarity. (From Selwyn et al., Ref. 27.)

$SF_6/Cl_2/Ar$. On the other hand, no particle cloud appeared even after 7 hours in Cl_2/Ar, CF_4, CF_4/Ar, or CCl_2F_2/Ne. Particle production was reduced from SiO_2 as compared to Si, or in pure CCl_2F_2 as compared to 10% CCl_2F_2/Ar. Particle production was favored at higher pressure, lower flow rate, and in a dirty reactor.

Selwyn et al. also noted the similar spatial behavior of LLS and LIF of Cl particularly as regards the onset near the cathode (Fig. 12) (27). The localization of the LLS signal at the sheath edge indicated that the particles were negatively charged, and thus the similar localization of the LIF signal suggested that the Cl atoms were due mostly to photodetachment from Cl-containing negative ions. This led Selwyn et al. to the conclusion that Si- halide etch products, with their propensity to form negative ions, may be involved in the formation or nucleation of particles.

SEM pictures of $\sim 20\,\mu m$ particles and particle fragments, obtained after ~ 6 hours in a 10% CCl_2F_2/Ar plasma at 200 mTorr, illustrated the orange-peel surface texture and the spherulitic, radially fibrous structure of the particles (27).

In the same reactor, O'Neill et al. performed LLS and FTIR measurements on particle-forming plasmas of CCl_xF_y-type gases (28). In a CCl_2F_2 plasma, all species from CF_4 to CCl_4 were detected by FTIR. In a plasma of 10% halocarbon/Ar at 200 mTorr and 130 W, particles appeared soonest with CCl_4 and last with CF_4, although the etch rate was highest with CCl_2F_2. LLS was reduced in an O_2-containing plasma. XPS showed substantial amounts of Al and Cl in all collected particles.

In the present author's laboratory, Durham (15) and Petrucci (16) observed particle clouds in the RIE of Si in CF_4, CCl_2F_2, and mixtures of CCl_2F_2 and Ar, as well as in the RIE of graphite, photoresist and Teflon in O_2. (The reactor was the same as described earlier, and plasma conditions were typically 200 mTorr pressure and 200 W rf power.) With Si, particles formed more readily at 13.56 MHz than at 2.6 MHz, whereas with the carbonaceous materials there was no significant difference between different excitation frequencies.

Yoo investigated the particle cloud formation from Si and SiO_2 in 10% CCl_2F_2/Ar RIE plasmas in more detail (21,22,29). He found that the threshold power and threshold pressure for particle cloud appearance in LLS were higher in the RIE of Si and SiO_2 than in the sputtering of Si and SiO_2 in Ar. Also, particles appeared sooner in sputtering than in RIE, although for given plasma conditions the etch rate was much higher in RIE. Moreover, in the RIE of Si, particles appeared only at pressures considerably higher than where the etch rate peaked, i.e., only

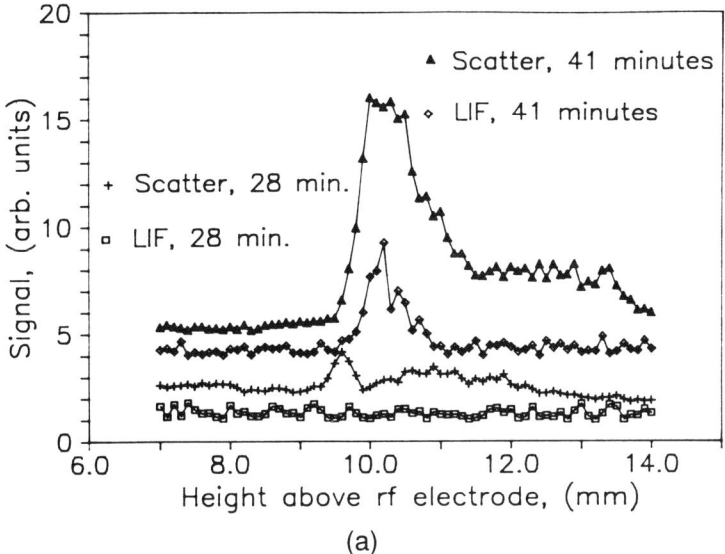

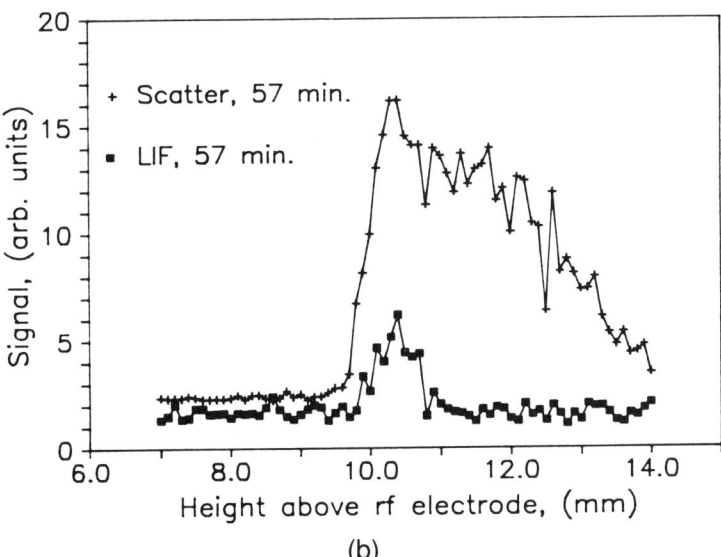

FIG. 12. Simultaneous LLS and two-photon LIF measurements in the cathode sheath boundary region of a 10% CCl_2F_2/Ar plasma in the RIE of Si ($p = 90$ mTorr). Data are shown at three different times. The plots have been displaced vertically for clarity. (From Selwyn et al., Ref. 27.)

in a plasma with a noticeable tendency for polymerization. In fact, a surface film deposited on the Si wafer could be detected by SEM long before particles appeared in the gas phase. At pressures below the etch rate maximum, the wafer surface remained clean and no particles were observed in the plasma.

These observations led Yoo and Steinbrüchel to conclude that perhaps Si atoms were primarily responsible for particle nucleation and growth rather than halogenated, volatile etch products in both sputtering and RIE (21,22). This suggestion is supported by very recent results on OES of sputtering and RIE plasmas. These indicate that substantial amounts of Si atoms are present in Ar plasmas sputtering Si or SiO_2, but not in corresponding 10% CCl_2F_2/Ar RIE plasmas (29).

The group of Carlile has reported statistically designed experiments of particle counts on a Si wafer in a SF_6/Ar RIE plasma (30). The reactor in this work was a parallel-plate type (electrode diameter 8 in., wafer diameter 10 cm). The two plates were powered separately at 13.56 MHz and 100 KHz. These authors found that the particle count went up with rf power and etch time. Also, particle production had a maximum vs. flow rate and was independent of pressure between 20 and 140 mTorr. The particles contained mostly Si, F, and S.

VI. Modeling of Particles in Plasmas

Theoretical modeling of particles in plasmas has focused on three aspects, which will be discussed in turn. These three aspects are particle charging, electrical effects of particles on a plasma, and forces on particles.

The physical mechanism why particles in a plasma should be charged is quite easily appreciated. An initially uncharged particle subject to the greatly different fluxes of electrons and ions will become charged to such a negative potential that finally the two fluxes to it are equal. If the particle is large, specifically larger or at most about equal to the electron Debye length, which is of the order of some tens to 100 μm in processing plasmas, the potential the particle will reach is the floating potential V_f (31–35). The situation is more complicated for smaller particles with a diameter of, say, 1 μm. There one must consider explicitly the trajectories of the ions around this very small spherical probe, in order to calculate the trajectories and the resulting potential self-consistently. The form of this potential turns out to be almost coulombic very near a particle, being screened farther out over a distance similar to the ion Debye length $\lambda_{D,i}$ (which is much smaller than $\lambda_{D,e}$) (*33–35*). For typical plasma densities, the potential value assumed by a particle is of the order of $2kT_e$, and the total charge on the particle is a few thousand electron charges (*34,35*). Also, trajectories leading to trapping of positive ions near the particle need to be taken into account (*35,37*). Furthermore, the potential around the particle has a very significant effect on the cross-section for collisions between the particle and other plasma species (*35*). The prediction of efficient screening of the particle potential over a distance comparable to $\lambda_{D,i}$ is consistent with the remark that if the screening distance were $\lambda_{D,e}$, then the lateral forces between particles might be too large to allow for confinement in a cloud at the particle concentrations observed experimentally (*16*).

Collisions between a particle and plasma electrons may have a noticeable effect on the electron energy distribution (EED) (*31,32*). For 1 μm "dust" particles at concentrations N_D of $10^3–10^5$ cm^{-3}, the EED is reduced at high energies, and thus the ionization rate coefficients are reduced. If the particle distribution is nonuniform, the discharge current will be constricted and increased between island of contamination. These effects are predicted to become significant when $N_D P > 10^5$ Torr cm^{-3} (*32*), which will certainly be the case in typical situations where particle clouds have been observed, based upon experimental estimates of N_D. Indeed, this prediction is in accord with observations of the effect of particles on the plasma impedance (*10,13*) and on the etch rate (*21*).

There are four forces on a partially shielded, charged particle that can be identified: the gravitational force, the electrical force due to the electric field in the plasma, and viscous drag forces due to collisions between the particle and neutral or ionic plasma species (*36,38*). The modeling results suggest that without an ion drag force, particles would accumulate in the center of a symmetric parallel plate discharge (*36*). Segregation of small particles (~1 μm) at a sheath edge can be attributed to a balance between the electric field force and the ion drag force (*36,38*). A peak in the plasma potential in the direction perpendicular to the electrodes is not necessary for spatial confinement (*36*). For large particles (>10 μm), the gravitational force will be more important than the ion drag force (*38*). The neutral drag force is said to be small compared to the other forces (*38*), although this statement is clearly in need of qualification considering the experimental results pointed out earlier with regard to thermophoresis (*13,26*) and the dependence of particles clouds on gas flow patterns (*8,22*).

VII. Particle Contamination and Equipment Design

Several of the factors discussed earlier determining the formation of a particle cloud and particle contamination in a processing plasma have a direct bearing on equipment design. For example, as emphasized by Selwyn (39–41), any material or geometrical discontinuity on the wafer-holding electrode may give rise to a particle trap above it, in the plasma. Such discontinuities include the edge of a wafer on a flat electrode, a well for recessed placement of the wafer, a ceramic insert in the wafer well, or a screw hole in the electrode surface. All of these may cause a local maximum in the plasma potential (42). Careful consideration of these as well as other issues in equipment design can lead to a significant reduction of particle contamination on the wafer (41). A specific suggestion made by Selwyn is to have electrodes with grooves in them surrounding the wafers (40). These grooves act as beneficial particle traps and can be used to guide particles safely away from the wafers and into the exhaust.

Of course, an important related issue is particle release from the reactor walls during a process. This may occur, for example, because of mechanical stress on the walls (18) or because particles may become charged highly enough to overcome adhesive forces and enter the gas phase (43). Such effects may also be controllable by careful equipment and process design.

VIII. Conclusions

From the preceding discussions it should be evident that the formation of particles in general and localized particle clouds in particular is a ubiquitous phenomenon in processing plasmas. We now wish to put in perspective at least some of the large number of observations, while at the same time pointing to a few additional interesting issues that are still unresolved.

First, let us note that particle formation occurs in all sorts of different plasmas under the appropriate conditions. The plasma chemistry by itself or the plasma–substrate system appear to play a less important role than the specific plasma conditions themselves. Even in a system chemically as simple as the sputtering of an elemental target in a noble gas discharge, particle formation may be very rapid.

The spatial segregation of particles into a cloud at a sheath edge, as well as the physical mechanisms contributing to this segregation, is fairly well documented. Important aspect of this are that particles in the discharge must be negatively charged and that they experience a viscous ion drag force towards an electrode. It seems noteworthy, upon comparing many different situations, almost all of which involve a parallel plate reactor, that particle clouds are in general more sharply defined and more localized in the direction perpendicular to the discharge electrodes in sputtering and RIE plasmas rather than in deposition plasmas. It is only in some deposition plasmas where particles appear more or less uniformly through essentially the entire discharge volume. And, most significantly, in all these examples the gas flow is perpendicular to the plane of the electrodes (*7, 10, 11, 14*). In other, chemically very similar plasmas but with gas flow parallel to the electrodes, there is a more or less pronounced localization of particles at the sheath edges (*3–6, 8, 13*). Also, in sputtering and RIE plasmas the lateral position of the particle cloud depends on the pressure and on the flow rate (*22*).

This suggests that in the theoretical modeling of particles in discharges, the drag force due to neutral species has not been taken into account yet in a realistic fashion. In fact, the mechanisms giving rise to transverse confinement of a particle cloud over the central portion of a wafer or, for that matter, giving rise to a maximum in the plasma potential in the transverse direction at the sheath edge are unclear at this time.

The concentration of particles in a cloud is a quantity that is not known very well in most situations. LLS usually does not yield direct information on this point except when the size distribution is known or can be estimated reasonably well (*5, 11, 25*). Alternatively, one may infer the particle concentration in the cloud from the areal density of particles collected on the substrate and from an estimate of the cloud volume, given that the particles are observed to settle vertically on to the substrate when the discharge power is turned off (*15, 21*). Estimates of particle concentrations in a cloud have ranged from $\sim 10^5$ to $10^9 \, \text{cm}^{-3}$ (*5, 8, 11, 15, 21, 25*), which is certainly at a level where they will have a noticeable effect on the plasma properties (*32*).

Another aspect that has not yet been studied well experimentally or described theoretically is the very first stage of particle formation, especially nucleation and, to a somewhat lesser extent, early growth of particles. Observations of very small particles collected downstream clearly indicate that particle nucleation is more or less continuous (*10, 11, 14, 22*), even under below-threshold conditions where a particle cloud never becomes detectable by LLS (*22*). Moreover, particles are on average largest in the part of a cloud that has been present for the longest time (22).

The plasma species responsible for nucleation are uncertain at present. It has been argued that in SiH_4 deposition plasmas nucleation may be heterogeneous due to nuclei desorbed from the cathode (14) or homogeneous due to SiH and SiH_2 (6–8). However, the fact that in most cases (with the caveat regarding gas flow patterns noted above) particles in LLS appear at a quite uniform concentration throughout essentially the entire discharge volume seems to favor homogeneous nucleation. In the sputtering and RIE of Si and SiO_2, Si atoms may well be the nucleating species (21,22,29). Furthermore, the similarity in spatial profiles of Si atoms and particles in SiH_4 plasmas (2–4) is also consistent with the notion that Si atoms may be involved in the particle nucleation and growth in those plasmas, rather than being merely plasma by-products.

For particles in sputtering and RIE plasmas, it is clear that whatever the nucleating species are, they must originate from the substrate as products of the etching process. Hence the etch rate and the various types of etch products are important factors. But equally important will be the transport of etch products across the sheath and into the bulk plasma, the collisions and reactions they undergo in the process, and their eventual transport out of the discharge onto the reactor walls or into the exhaust. It is the interplay among all these different factors that determines whether nuclei can form and grow into large enough particles so as to be confined in a cloud rather than being swept out of the discharge. This is another issue requiring further experimental as well as theoretical elucidation.

In the same context it should also prove interesting to examine particle formation in magnetically enhanced discharges. The question there would be whether, for the purpose of particle nucleation, the reduced number of gas-phase collisions at the much lower pressures can be compensated for by the much higher etch rates.

It seems fair to conclude that the work described here has yielded considerable insight into the mechanisms responsible for particle nucleation and growth in processing plasmas. Of course, the ultimate application of this work should be to the control of particle contamination in microelectronics manufacturing. Hence, one may ask what these studies have contributed to the solution of this very important practical problem. First, it is clear that certain guidelines regarding choice of materials, geometrical layout, etc., in equipment design can be deduced that should help insure that if particles are generated in a process, they do not end up on the wafer (*39–41*). Second, in light of the threshold phenomena observed (*12,13,20–22*), it appears likely that for many processes relatively "safe" regions of plasma parameter space exist where particle formation is insignificant, or at least does not give rise to particles localized in a cloud over the wafer. Third, certain fairly simple process modifications such as modulating the discharge power (*6–9*) or choosing a soft shut-down of the plasma,—that is, gradually reducing the pressure rather than turning off the rf power (*21,22*)—could be effective in reducing particle contamination on the wafer.

Acknowledgment

The work on particles in processing plasmas in the author's laboratory has benefited from support by IBM East Fishkill. The author is grateful to his student W. J. Yoo for preparing several figures and to G. S. Selwyn for providing the photograph in Fig. 10.

References

1. J. F. O'Hanlon and H. G. Parks, *J. Vac. Sci. Technol. A* **10**, 1863 (1992).
2. R. M. Roth, K. G. Spears, and G. Wong, *Appl. Phys. Lett.* **45**, 28 (1984).
3. R. M. Roth, K. G. Spears, G. D. Stein, and G. Wong, *Appl. Phys. Lett.* **46**, 253 (1985).
4. K. G. Spears, T. J. Robinson, and R. M. Roth, *IEEE Trans. Plasma Sci.* **PS-14**, 179 (1986).
5. K. G. Spears, R. P. Kampf, and T. J. Robinson, *J. Phys. Chem.* **92**, 5297 (1988).
6. Y. Watanabe, M. Shiratani, and Y. Kubo, *Appl. Phys. Lett.* **53**, 1263 (1988).
7. Y. Watanabe, M. Shiratani, and H. Makino, *Appl. Phys. Lett.* **57**, 1416 (1990).
8. Y. Watanabe, M. Shiratani, and M. Yamashita, *Appl. Phys. Lett.* **61**, 1510 (1992).
9. J. T. Verdeyen, J. Beberman, and L. Overzet, *J. Vac. Sci. Technol. A* **8**, 1851 (1990).
10. A. Bouchoule, A. Plain, L. Boufendi, J. Ph. Blondeau, and C. Laure, *J. Appl. Phys.* **70**, 1991 (1991).
11. L. Boufendi, A. Plain, J. Ph. Blondeau, A. Bouchoule, C. Laure, and M. Toogood, *Appl. Phys. Lett.* **60**, 169 (1992).
12. A. A. Howling, Ch. Hollenstein, and P.-J. Paris, *Appl. Phys. Lett.* **59**, 1409 (1991).
13. J. L. Dorier, Ch. Hollenstein, A. A. Howling, and U. Kroll, *J. Vac. Sci. Technol. A* **10**, 1048 (1992).
14. H. M. Anderson, R. Jairath, and J. L. Mock, *J. Appl. Phys.* **67**, 3999 (1990).
15. J. A. Durham and Ch. Steinbrüchel, in "Proc. 8th Symp. Plasma Process." (G. S. Mathad and D. W. Hess, Eds.). The Electrochemical Society, Pennington, New Jersey, 1990, p. 207.
16. J. L. Petrucci, Jr., and Ch. Steinbrüchel, in "Proc. 8th Symp. Plasma Process." (G. S. Mathad and D. W. Hess, Eds.). The Electrochemical Society, Pennington, New Jersey, 1990, p. 219.

17. G. S. Selwyn, J. E. Heidenreich, and K. L. Haller, *J. Vac. Sci. Technol. A* **9**, 2817 (1991).
18. G. S. Selwyn, J. S. McKillop, and K. L. Haller, *J. Vac. Sci. Technol. A* **8**, 1726 (1990).
19. R. N. Carlile, S. Geha, J. F. O'Hanlon, and J. C. Stewart, *Appl. Phys. Lett.* **59**, 1167 (1991).
20. W. J. Yoo and Ch. Steinbrüchel, *Appl. Phys. Lett.* **60**, 1073 (1092).
21. W. J. Yoo and Ch. Steinbrüchel, *J. Vac. Sci. Technol. A* **10**, 1041 (1992).
22. W. J. Yoo and Ch. Steinbrüchel, *J. Vac. Sci. Technol. A* **11**, 1258 (1993).
23. B. Ganguly, A. Garscadden, J. Williams, and P. Haaland, *J. Vac. Sci. Technol. A* **11**, 1119 (1993).
24. G. M. Jellum and D. B. Graves, *J. Appl. Phys.* **67**, 6490 (1990).
25. G. M. Jellum and D. B. Graves, *Appl. Phys. Lett.* **57**, 2077 (1990).
26. G. M. Jellum, J. E. Daugherty, and D. B. Graves, *J. Appl. Phys.* **69**, 6923 (1991).
27. G. S. Selwyn, J. Singh, and R. S. Bennett, *J. Vac. Sci. Technol. A* **7**, 2758 (1989).
28. J. A. O'Neill, J. Singh, and G. G. Gifford, *J. Vac. Sci. Technol. A* **8**, 1716 (1990).
29. W. J. Yoo and Ch. Steinbrüchel *J. Vac. Sci. Technol. A* **12** (to be published, 1994).
30. M. M. Smadi, G. Y. Kong, R. N. Carlile, and S. E. Beck, *J. Vac. Sci. Technol. B* **10**, 30 (1992).
31. M. J. McCaughey and M. J. Kushner, *Appl. Phys. Lett.* **55**, 951 (1989).
32. M. J. McCaughey and M. J. Kushner, *J. Appl. Phys.* **69**, 6952 (1991).
33. R. N. Nowlin and R. N. Carlile, *J. Vac. Sci. Technol. A* **9**, 2825 (1991).
34. J. E. Daugherty, R. K. Porteous, M. D. Kilgore, and D. B. Graves, *J. Appl. Phys.* **72**, 3934 (1992).
35. S. J. Choi and M. J. Kushner, *Appl. Phys. Lett.* **62**, 2197 (1993).
36. T. J. Sommerer, M. S. Barnes, J. H. Keller, M. J. McCaughey, and M. J. Kushner, *Appl. Phys. Lett.* **59**, 638 (1991).
37. J. Goree, *Phys. Rev. Lett.* **69**, 277 (1992).
38. M. S. Barnes, J. H. Keller, J. C. Forster, J. A. O'Neill, and D. K. Coultas, *Phys. Rev. Lett.* **68**, 313 (1992).
39. G. S. Selwyn, *J. Vac. Sci. Technol. B* **9**, 3487 (1991).
40. G. S. Selwyn and E. F. Patterson, *J. Vac. Sci. Technol. A* **10**, 1053 (1992).
41. M. C. Chuang, F. Martinet, and G. S. Selwyn, presented at 39th Nat. Symp. of Am. Vacuum Society (Nov. 9–13, 1992, Chicago).
42. S. M. Collins, J. F. O'Hanlon, S. E. Beck, and L. Hong, presented at 39th Nat. Symp. of Am. Vacuum Society (Nov. 9–13, 1992, Chicago).
43. J. Goree and T. E. Sheridan, *J. Vac. Sci. Technol. A* **10**, 3540 (1992).
44. J. J. Wu and R. J. Miller, *J. Appl. Phys.* **67**, 1051 (1990).
45. G. Strasser, H. P. Bader, and M. E. Bader, *J. Vac. Sci. Technol. A* **8**, 4092 (1990).
46. C. E. Wickersham, Jr., J. E. Poole, and J. J. Mueller, *J. Vac. Sci. Technol. A* **10**, 1713 (1992).
47. J. J. Wu, D. W. Cooper, and R. J. Miller, *J. Vac. Sci. Technol. A* **8**, 1961 (1990).
48. J. Zhao, B. Y. H. Liu, and T. H. Kuehn, *Sol. State Technol.* **33**(9), 85 (1990).
49. C. K. Goertz, *Rev. Geophys.* **27**, 271 (1989).
50. P. Ho, R. J. Buss, and R. E. Loehman, *J. Mater. Res.* **4**, 873 (1989).
51. G. M. Chow, C. L. Chien, and A. S. Edelstein, *J. Mater. Res.* **6**, 8 (1991).
52. Z. M. Qiau, H. Michiel, A. van Ammel, J. Nijs, and R. Mertens, *J. Electrochem. Soc.* **135**, 2378 (1988).
53. M. G. Simmonds, W. L. Gladfelter, N. Rao, W. W. Szymanski, K. Ahn, and P. H. McMurry, *J. Vac. Sci. Technol. A* **9**, 2782 (1991).
54. H. C. van de Hulst, "Light Scattering by Particles." Dover, New York, 1957.

Author Index

A

Akhiezer, A. I., 147
Allen, J. E., Jr., 16
Allis, W. P., 137
Amemiya, H., 124
Anderson, H. M., 300–301
Andosca, R. G., 218
Arkhipenko, V. A., 137
Arnell, R. D., 273, 275–276, 278, 279
Asmussen, J., 124, 154

B

Bakai, A. S., 147
Barbour, J. C., 209
Bardos, L., 216
Barker, R. A., 70
Bates, R. I., 275–276
Ben Daniel, D. J., 133
Boswell, R. W., 40
Bouchoule, A., 297–299
Brown, S. C., 122–145
Buchanan, D. A., 124, 165
Budden, K. G., 36
Budnikov, V. N., 137
Bulat, E. S., 218

C

Calderon, M. A. G., 151
Carl, D. A., 124, 125, 159, 184, 185–186, 217
Carlile, R. N., 308, 312
Castagna, T. J., 105, 124
Caughman, J. B. O., II, 182
Chen, F. F., 40, 46, 50, 80, 150
Chiang, C., 222
Chorney, P., 65
Chu, T. K., 151

Chu, X., 283
Consoli, T., 122, 159
Cook, J. M., 60

D

Dandl, R. A., 122, 124, 151, 152, 155
Datlov, J., 143
den Hartog, E. A., 79
Dorier, J. L., 299
Dreicer, H., 151
Durham, J. A., 301, 310
Dusek, V., 125

E

Eckert, H. U., 52
Efeoglu, I., 278, 279
Eggarter, E., 16

F

Ferrari, L. A., 137, 151
Flamm, D. L., 60
Forster, J., 90, 169, 187
Fraser, D. B., 22

G

Gabriel, C. T., 98–99
Geisler, M., 155
Gibbons, E. F., 143
Ginzburg, V. L., 127, 137
Godyak, V. A., 18, 22, 80, 147, 148
Golant, V. E., 149
Gonda, S., 91–92, 96
Gorbatkin, S. M., 124, 125, 159, 184, 194, 203–205, 208

AUTHOR INDEX

Gottscho, R. A., 124, 159, 165, 194
Gould, R. W., 65
Graves, D. B., 306–308
Greene, J. E., 245

H

Hara, T., 96–97
Harding, G. L., 271, 272
Harper, M. E., 245
Hata, T., 249
Hayashi, M., 16
Henry, D., 40
Herak, T. V., 218
Hershkowitz, N., 80
Hirao, T., 209, 213, 215
Hittorf, W., 52
Holber, W. M., 90, 124, 165, 169, 182, 187, 194
Hopwood, J., 125, 194
Howson, R. P., 275
Hultman, L., 269, 284
Hussein, M. A., 194

I

Iizuka, S., 91–92, 96

J

Jaeger, F., 33, 147
Jurgensen, C. W., 71

K

Kadlec, S., 263–264, 265, 266, 267, 269, 271, 278
Kawamura, T., 146
Kennedy, T. N., 221
Kiuchi, M., 209
Komori, A., 45, 51
Koslover, R., 78n
Kushner, M. J., 124, 194

L

Lax, B., 122
Lee, Y. H., 203, 204
Lieberman, M. A., 62, 124, 147, 159
Liu, J., 76
Lowenhardt, P. K., 51

M

McVittie, J. P., 98–99
McWilliams, R., 78n
Machida, K., 221
Mantei, T. D., 163
Matsuo, S., 122, 124, 125, 152, 159, 167–168, 178, 209, 217–218
Matsuoka, M., 81, 82 and n6, 84, 124, 165, 194
Maximov, V. N., 18
Mayer, T. M., 70
Messiaen, A. M., 146
Messier, R., 246–248
Meyers, A. M., 271, 284
Miller, D. B., 122, 143, 152, 159, 169
Minomo, S., 165, 184, 185
Miyake, K., 216
Moisan, M., 65
Monaghan, D., 273, 276, 280
Müller, K.-H., 244, 245
Murphy, M. J., 263
Musil, J., 122, 124, 125, 136, 137, 151, 152, 165, 263, 265, 269, 271, 278

N

Namura, T., 102
Naoe, M., 250–253
Neumann, G., 155, 184
Nihei, H., 125

O

Oikawa, H., 221
O'Neill, J. A., 310
Ono, K., 81, 82 and n6, 84
Oomori, T., 194, 203–204, 207–208

P

Pai, C. S., 222–224
Pelletier, J., 124
Perez, J. M., 151
Perry, A. J., 40
Peterson, L. R., 16
Petrov, I., 248
Petrucci, J. L., Jr., 301, 310
Piejak, R. B., 59
Pongratz, S., 155
Popov, O. A., 124, 125, 147, 159, 160, 168, 184, 194, 204, 218
Porkolab, M., 150
Porteous, R. K., 83–84

R

Reinke, P., 91–92
Rohde, S. L., 259, 261, 262, 266, 269
Rossnagel, S. M., 125, 180, 182, 187, 194

S

Sadeghi, N., 79, 82, 85, 90 and n
Sakudo, N., 122, 166
Samukawa, S., 27n, 100–101, 102, 104, 125, 160–162, 163, 165, 168, 179, 186
Sato, N., 91–92, 96
Savvides, N., 253, 255, 259, 261, 263, 266, 267, 272, 274, 277, 278
Schwar, M. J. R., 80
Selwyn, G. S., 301, 305–306, 309–310, 314
Shaqfeh, E. S. G., 71
Shirai, K., 91–92, 96, 159
Smullin, L. D., 65
Spears, K. G., 295–296, 298
Spencer, A. G., 275
Sproul, W. D., 269, 271, 278
Steinberg, A. R., 60
Steinberg, G. N., 60
Steinbrüchel, C., 312
Sternberg, N., 22
Stevens, J. E., 38, 45, 124, 159, 160, 167, 173, 176, 184, 186
Suzuki, K., 122, 152, 163
Swift, J. D., 80

T

Thomson, J. J., 59
Thornton, J. A., 246
Torii, Y., 166
Trivelpiece, A. W., 65
Tsunokuni, K., 102
Tuma, D. R., 65

V

Vandenplas, P. E., 146

W

Waldron, H., 218
Wang, Y., 194
Watanabe, Y., 296–297, 298
Window, B., 253, 255, 259, 261, 263, 266, 267, 271, 272, 274, 278
Woods, R. C., 82

Y

Yapsir, A. A., 96, 97, 98
Yoo, W. J., 302, 310–312
Yunogami, T., 105

Z

Zakrzewski, Z., 65
Zheleznyakov, V. V., 136

Subject Index

A

Absorption
  of left hand polarized wave, 127, 129, 136–138, 167, 187–190
  level of, 178
  microwave power, 124, 126, 131, 150, 162, 172
    in electron cyclotron resonance, 126–128, 188–190
  non-linear, 130, 148, 150–151
  at "pure" ECR, 131–132
  of right hand polarized wave, 126, 129, 136–138, 167–168, 170, 188–190
  at second harmonic resonance (SHR), 189–190
  surface, 162, 194
  at upper hybrid resonance, 139–142
  of whistler wave, 129, 134–136, 151, 184, 188–190
  zone, microwave power of, 126, 150, 162–163, 176, 185, 194, 206
Acceleration
  electrons, 148
  ions, 159
Acoustic waves, 151
Alumina window, 166, 167
Amorphous boron films, 159
Amorphous silicon films, 159
Angular distribution, ion, 73
  See also Energy distribution
Anodization, plasma, 96
Antenna, 126, 153–154
  coaxial, 153–154
  horn, 153
  multiple, symmetrically distributed, 154
Area
  effective, 22
  loss, 17
Argon plasma, 162–163, 173–178, 181, 203, 205, 209–213

B

Biasing, 90
  dc, 91–92
  rf, 92–95
Bohm velocity, 8, 17
Boltzmann relation, 18
Boron nitride coatings, 122
Breakdown (ignition)
  of electron cyclotron resonance, 122, 132
  of microwave, 122, 126, 170–172, 178
  resonance curve, half-width, 128
  second harmonic resonance, 142

C

Cavity
  multimode, 154
  resonant, 130, 143, 154, 155, 167–168, 178–179
Charge transfer, collisional, 15, 72, 82
Child law, see Sheath, Child law
Circular polarizer, 169
Circulator, 184, 210
Collisions
  electron-ion, 149
  electron-neutral, 126, 128, 130–132, 146–151, 203
  inelastic, 129
Conservation
  of energy, 17, 67
  of particles, 19
Contamination
  particle, plasma, 289, 291, 314
  process-inherent, 290
Coupling aperture, 167–169, 178–179
Critical plasma density, 126–128, 131–141, 146–147, 150–151, 179, 185–186, 190–191

SUBJECT INDEX

Cutoff
  cavity/chamber, 154, 167–168
  below cutoff, cavity/chamber, 190–192, 197, 200–201
  frequency, 139
  plasma density
    left hand polarized wave, 137–138, 187, 190–191
    right hand polarized wave, 140

D

Damage, 10, 71, 96
  atomic displacement, 96
  charging, 91
    effect of plasma uniformity, 99
    effect of rf bias, 102
  contamination, 98
  radiation, 104
  x-rays, 105
Debye length, 20
Density
  ion current, 22
  plasma, determining, 22
Deposition, 8
Diagnostics, plasma
  Doppler-shift spectroscopy, 77, 79, 82
  electrostatic energy analysis, 76, 81, 90
  Faraday cup, 100
  Langmuir probes, 80
  laser-induced fluorescence (LIF), 78
  magnetic pickup coil, 38, 45
Diamond coatings, films, 122, 159
Diamond-like carbon, coatings, films, 122, 159
Diffusion, ambipolar, 8, 17–18
  See also Transport
Discharge
  capacitive, 5, 90
    limitations of, 9
    typical parameters, 9
  design examples, 23, 51, 57
  design principles of, 13
  heating, 23
    ohmic, 24, 54
    stochastic, 24, 54
    wave-particle, 25
  high density, see Discharge, high efficiency

  high efficiency, 10, 13
    remote, 12
    low pressure, 13
    magnetically enhanced, 10
  MERIE, 10
  rf diode, see Discharge, capacitive
  RIE, see Discharge, capacitive
  triode, 10
Doppler
  broadened resonance, 132–134, 146, 173–177
  shifted resonance, 132–134, 136, 155, 162, 169–178
Doppler effect
  for ECR resonance, 37
  for optical measurements, 77
Doughnut-shaped plasma, 174

E

$\vec{E} \times \vec{B}$ drift, 241
ECR, see Electron cyclotron resonance discharge
eedf, see Energy distribution, electron
Elastic scattering, collisional, 15, 72
Electric field
  ambipolar, 159, 200
  perpendicular to static magnetic field, 126–127, 139, 148–151
  threshold, 127, 149–151
Electron
  collision frequency, 126–128
  density, 125–126, 128, 130, 132–140, 145–147, 159, 167, 171–190, 194–195, 200–202, 205–208
  drift (oscillation) velocity, 127, 149
  effective collision frequency, 128, 133, 146–151
  energy distribution function, 124, 127–128, 132–133, 146–151
  energy distribution tail (fast/hot electrons), 148, 150–151, 190–193, 196–198, 201–205
  heating time, 148–149
  mean free path, 149, 237
  motion in magnetic field, 240
  motion in electro/magnetic field, 240
  temperature, 14, 126, 128, 132–133, 146–151, 159

SUBJECT INDEX

determining, 19
thermal velocity, 127, 131–133, 135, 146–151
Electron cyclotron resonance
 definition, 126–129
 "pure," 126–128, 131–132
Electron cyclotron resonance discharge, 12, 25
 cavity, 31, 40
 configurations, 26
 design example, 23
 distributed ECR, 31, 39
 heating, 31
 microwave systems, 26
Energy
 ion
  bombarding, 7, 9, 11, 19, 71, 81
  control of, 90
  threshold, 71
  transverse, 87
 loss per electron-ion pair created, 15–17
Energy distribution
 electron, 14
 ion, 71, 89
  control of, 90, 92
  measurements of, 76, 81, 90
  transverse, 87
  variation with frequency, 90
 Maxwellian, 14, 72
Etching
 anisotropic, 8, 73, 87
 isotropic, 8
 model for, 70
Evanescent zone, 136
Excitation, collisional, 14, 73

F

Floating potential, 186, 192–193, 198–199, 202–206
Flux, ion, 9, 11
Frequency
 cutoff, *see* Cutoff
 electron cyclotron, 126–130, 131–133, 135–137, 150–151
 electron gyration, 27
 electron plasma, 34
 ion cyclotron, 129–130
 lower hybrid, 41

microwave, 126, 128, 130–140, 143, 145–147, 151
plasma, 130–131, 133–136, 139, 147, 150, 238
resonant, in helical resonator, 63
upper hybrid, 40

G

Gas
 distributed ring, 211, 224
 flow rate, 212–215, 217, 219–226
Geometrical resonances, 129, 145–146
Growth rate of film
 silicon dioxide, 217–226
 silicon nitride, 209, 213–215, 217

H

Harmonics resonance
 higher, 129
 second (SHR), 142–145, 189, 190, 209–216
 upper (UHR), 129, 139–141
Heating, electron, *see* Discharge, heating
Helical resonator discharge, 12, 60
 modes in, 62
Helicon discharge, 12, 40
 antenna, 42, 44
 coupling, 46
 configurations, 41
 density measurements, 49
 design example, 51
 matching network, 42
 modes, 42
Helmholtz pair, 157–158, 162–163, 170–172, 211
Hitachi ECR source, 152–153, 155, 157–158, 163–165, 168–169, 180, 182–183, 185, 188
Hydrogen content, 209, 213, 215

I

ICP, *see* Inductive discharge
iedf, *see* Energy distribution, ion

Inductive discharge, 12, 52
  coil coupling, 55
  configurations, 52
  design example, 57
  efficiency, 58
  low density, 58
  operation of, 55
  power absorption, 54
  regimes, 54
Ion
  bombardment, 157, 198
  cyclotron resonance, 129
  probe current, 134, 138, 140–144, 161, 171–175, 181–182
Ion-assisted deposition, 243–249
  effect of ion energy, 244
  effect of ion-to-vapor-flux, 245
  microstructural changes, 246
Ion energy, 192, 198
  average, 267
  per deposited atom, 267
Ionization
  balance, 139, 180, 182, 202
  collisional, 14, 73, 86
    single step, 19
  efficiency, 177, 180, 187
  two-step, 190
  zone, 163, 194
Ion-to-deposited-atom ratio, 266

L

Landau damping, 50, 131, 135–137, 139, 146, 192, 197
Langmuir probe, 138, 161–163, 171–177
Left hand polarized waves
  absorption, propagation, reflection, 136–139, 176, 186–187, 190
  definition, 127
  transformation into plasma waves, 137–139, 186–187, 190
Line-averaged plasma density, 174, 176–177, 184–185, 187–189, 206

M

Magnetic coils, 126, 152–154, 157–165, 170–172, 210–211

Magnetic field
  absorption, 160–162, 171–173
  axial, 17
  breakdown, 170–172
  divergent, 157–165, 179
  "high-field-side," 134–136, 151, 159–163, 177–178, 186–187
  mirror, 142, 147, 155, 160, 180
  mirror-like, 158–160, 163
  multidipole, 12, 42, 52
  resonant, 126–129, 131–134, 136–140, 171–173, 194
  uniform, 133, 157–159, 165, 177, 180, 211
  vacuum window, 159–162, 171–172, 188, 203
Magnetron, 165
Magnets, permanent, 153–155, 163–165
Matching
  in ECR, 27
  in helical resonator, 64
  in helicon, 42
  in inductive discharge, 53, 58
  in surface wave discharge, 69
Mean free path, collisional, 14, 15
Microwave
  breakdown, 170–172
  modes, 129, 152–155, 164–176, 178–179, 186
  power, absorbed, forward, reflected, 126–130, 143–144, 147–151, 167–169, 172–174, 176–181, 184–193, 203, 207, 210
  window, 126, 137, 152, 154, 159–160, 165–169, 210–211
Mode
  helicon, 42
    absorption, 50
    damping of, 50
    measurements, 45
    standing, 49
  slow wave, 61, 65
    in helical resonator, 62
  surface, 65
  See also Wave

N

Narrow plasma mode, 160–161, 176–178, 188–190, 208

SUBJECT INDEX

Nitrogen, 138, 140–144, 177, 180, 185–186, 189–193, 195–202, 204, 206–207, 209–217
Non-linear microwave absorption processes, 148–152, 184–187
NTT/Matsuo source, 152–153, 155, 157–162, 167–168, 178–180, 183, 188–189, 194, 209, 217–218, 221–222

O

Oxygen, 173, 176–177, 187–188, 207, 216–226

P

Particle
  cloud, plasma, 292, 294, 307
  contamination, plasma, 289, 314
  plasma
    charging, of particle, 292, 312
    cloud, 292, 294, 307
    concentration, 296, 297, 313, 315
    deposition, 294
    detection, in-situ, 291, 295, 299, 301, 309
    etching, 309
    forces, on particle, 313, 315
    formation, 291
    laser light scattering, 291, 295, 301, 309
    modeling, 308
    nucleation, 294, 298, 300, 315
    size, 292, 299, 300, 304
    sputtering, 301
    structure, 305, 306, 310
    trapping, of particle, 306
Phase velocity, 126, 130–132, 134–136, 139, 146, 179
Plasma
  density, *see* Electron density
  frequency, *see* Frequency
  modes, 160–163, 184–189, 205–211
  potential, 159, 175, 180, 192–193, 195–196, 200–202, 204–206
  resonance, 131, 134–137, 147, 150–151, 185–187, 190
  thin-film processing, particle, 289, 290

Potential
  distributed, 16, 84
  floating, 16, 91
  plasma, 8, 86, 91
    measurements of, 81
    methods for measuring, 80
  surface, 101
Power
  absorption, 36, 50, 54, 67
  balance, 17, 67
  minimum, for discharge, 37, 68

R

Refractive index
  microwave, 126, 131, 135–136, 139
  SiO_2 film, 218–221, 224–225
  Si_3N_4 film, 209, 213, 215, 217
Resonant
  cavity, *see* Cavity
  frequency, *see* Frequency
rf bias, 182–183, 210–211, 218, 221–223
RIE, *see* Discharge, capacitive
Right-hand polarized wave, 126–129, 131, 134–140, 151, 159–165, 169, 172–178, 186–190

S

Scattering, elastic, 14
Second harmonic resonance, 142–144, 189, 209, 212–213, 216
Sheath, 6, 18
  Child law, 8, 22, 94
  distributed, 21
  magnetically enhanced, 10
  thickness, 20
  voltage, 20
Silane (gas), 209–217
Silicon nitride film
  deposition rate, *see* Growth rate of film
  thickness, 209, 212–213
  uniformity, 177, 212–213
Silicon oxide film
  deposition rate, *see* Growth rate of film
  thickness, 220–223
  uniformity, 177, 218, 220–223, 225

Skin depth
　collisional, 55
　collisionless, 55
Sources, *see* Discharges
Sputtering, 236
　dc-diode, 237
　magnetron, 239, 241–243
　　benefits of, 242
　　$\vec{E} \times \vec{B}$ drift, 241
　　shadowing in, 241
　rf, 238
　triode, 238
　unbalanced magnetron, 259
Stochastic heating, 146–148
Substrate (wafer) holder, 153, 163–164, 170, 182–184, 209–211, 221–224
Surface wave discharge, 65
　modes, 65
　power balance, 67

T

TCP, *see* Inductive discharge
Temperature
　electron, *see* Electron
　ion, 71, 89
　process, 209, 211, 215, 216, 219, 223–224
TEOS, 223–226
Thin-film processing, plasma, particle, 289, 290
Threshold
　electric field, 149–150
　microwave power, 176–177
　plasma density, 176–177
Transformer-coupled plasma, *see* Inductive discharge
Transport
　ion, 73
　plasma, 69
　See also Diffusion
TTMSB, 223–226
Tuner (three-stub, plunger) microwave, 152–154, 168–170, 177, 210

U

Unbalanced magnetron sputtering, 255–271
　applications of, 272–281
　　corrosion protection, 275–277
　　electronic and optical uses, 274–275
　　elemental thin films, 272–274
　　wear and abrasion resistance, 277–281
　commercial applications of, 281–283
　future trends, 283–285
　initial development, 255
　ion bombardment in, 265–271
　magnetic fields in, 259–265
　precursors, 249–255
　principles of, 259–261
　type I, 255
　type II, 255
Uniformity
　plasma, 99
　　magnetically enhanced, 10
　　relation to damage, 101
　process, 13
Upper hybrid resonance, 139–142, 151

V

Vacuum field, 127, 149–151

W

Wave
　absorption, 34
　extraordinary, 40
　evanescent, 36
　focusing and defocusing, 38–39
　left hand polarized, 31
　measurements, 38
　right hand polarized, 25, 28, 31
　spatial decay of, 54
　whistler, 39, 42
　WKB, 34
　See also Mode
Wavenumber, 126, 131–133, 135, 139, 146
Whistler wave, 134–136, 160–163, 176–177, 185–190, 196–198, 207–208, 209, 213
Window, dielectric, 11, 13

ISBN 0-12-533018-9